Library of
Davidson College

THE ARCHAEOMETALLURGY OF THE ASIAN OLD WORLD

Vincent C. Pigott

Editor

University Museum Monograph 89

UNIVERSITY MUSEUM SYMPOSIUM SERIES
VOLUME VII

MASCA Research Papers
       in Science and Archaeology Volume 16

THE ARCHAEOMETALLURGY OF THE ASIAN OLD WORLD

Vincent C. Pigott

Editor

Published by
The University Museum
University of Pennsylvania
1999

Editing
Museum Applied Science Center for Archaeology and
Publications Department
University of Pennsylvania Museum

Design and Layout
Garret Schenck

Printing
Science Press
Ephrata, Pennsylvania

Cover logo: Copper wand, perhaps depicting an ibex, from Hissar III, Iran. H. 12 cm. F. Schmidt, *Excavations at Tepe Hissar Damghan* (Univ. of Penn. Museum, Univ. of Penn. Press, 1937), pl. 48.

Library of Congress Cataloging-in-Publication Data

The archaeometallurgy of the Asian old world / Vincent C. Pigott Editor.

    p. cm. – (University Museum monograph ; 89) (University Museum symposium series ; v. 7) (MASCA research papers in science and archaeology ; v. 16)
Includes bibliographical references.
 ISBN 0-924171-34-0
 1. Metal-work, Prehistoric–Asia. 2. Metallurgy–Asia–History. 3. Asia–Antiquities. I. Pigott, Vincent C. II. Series. III. Series: University Museum symposium series ; v. 7 IV. Series.
 GN799.M4 A73 1999
 950'.1–dc21
                                  99-006963

ISSN 1048-5325

Copyright © 1999
UNIVERSITY OF PENNSYLVANIA MUSEUM
of Archaeology and Anthropology
Philadelphia
All rights reserved
Printed in the United States of America

Printed on acid-free paper

TABLE OF CONTENTS

THE ARCHAEOMETALLURGY OF THE ASIAN OLD WORLD:
INTRODUCTORY COMMENTS 1
    by Vincent C. Pigott

    CHAPTER 1.   Copper and Bronze in Cyprus and the Eastern Mediterranean 15
                            James D. Muhly

    CHAPTER 2.   The Coming of Iron in the Eastern Mediterranean:
                            Thirty Years of Archaeological and Technological Research......................27
                            Jane C. Waldbaum

    CHAPTER 3.   Aspects of Early Metallurgy in Mesopotamia and Anatolia 59
                            Tamara Stech

    CHAPTER 4.   The Development of Metal Production on the Iranian Plateau:
                            An Archaeometallurgical Perspective 73
                            Vincent C. Pigott

    CHAPTER 5.   Metal Technologies of the Indus Valley Tradition in Pakistan
                            and Western India 107
                            Jonathan M. Kenoyer and Heather M.-L. Miller

    CHAPTER 6.   The Early Iron Age in South Asia..153
                            Gregory L. Possehl and Praveena Gullapalli

    CHAPTER 7.   The Transition to Iron in Ancient China 177
                            Bennet Bronson

APPENDIX I..199
    translated by Henry N. Michael

CONTRIBUTORS TO THIS VOLUME 207

To those pioneering scholars, named and unnamed, who sought the origins of Old World metallurgy in the realms of the Eastern Mediterranean and across Asia:

Earle R. Caley, William Rostoker, Cyril Stanley Smith, R. F. Tylecote, Theodore A. Wertime

The Archaeometallurgy of the Asian Old World

Introductory Comments

Vincent C. Pigott

In 1988 I was invited to organize the American participants for the fourth in a succession of USA–USSR archaeological exchanges that were sponsored by the International Research and Exchange Board (IREX). It was to be focused on the topic of the origins and development of metallurgy. I accepted the invitation and was appointed the Project Coordinator for IREX of "The 4th USA–USSR Archaeological Exchange/Symposium: The Development of Ancient Metallurgy in the Old World," which was held September 27–October 8, 1988, in Tbilisi and Signakhi, Georgia.

An international gathering of scholars assembled from the United States and various Soviet countries. We started in Tbilisi and then moved east to Signakhi where we spent almost a week delivering papers and visiting archaeological and historical sites in the region. The papers delivered were primarily focused on archaeometallurgical developments in the Old World, but there was some discussion of New World developments as well.

Plans were made to pursue the publication of the entire symposium proceedings, but obstacles posed by faulty communication between the organizers and by the participants being spread, as they were, over the whole of what was then the Soviet Union, not to mention the barrier of translation, proved so unsurmountable that in the end no volume of the presented papers could be assembled. Instead, the papers contained in this volume are updated versions of those dealing with archaeometallurgy and presented by participants from the United States, with an additional, invited, contribution by Jonathan M. Kenoyer and Ph.D. candidate Heather M.-L. Miller (Dept. of Anthropology, University of Wisconsin at Madison), writing about the development of copper-base metallurgy in South Asia.

A short summary of the symposium was published by L. I. Avilova and N. N. Terekhova (1989) in Russian in *Sovetskaya arkheologiya*. This summary is valuable for its coverage of the papers by Soviet scholars which have not been published. An English translation of this summary article, produced by Henry N. Michael of the Museum Applied Science Center for Archaeology (MASCA), appears at the end of this volume as Appendix I.

Deciding on an appropriate title for this volume proved to be a difficult task, and the one chosen may not communicate precisely the geographical breadth of the work. The basic organizing principle of the volume is simple. The papers are very much in the spirit of what was presented to our colleagues in Signaki: rich in archaeological and analytical data but scant on analysis based on social, economic, political, or social context. While not free of interpretation or scholarly biases, these papers focus on the excavated metal artifacts and other evidence related to production (with some discussion of the processes involved) and on the analysis of those data categories. They present the data, whether convincing or only suggestive, upon which are based our current reconstructions of how metallurgy began and developed on Asia's western border and then across Asia itself over the eight millennia prior to the beginning of the common era.

The evidence for the archaeometallurgy of two major metallurgical traditions—copper and its alloys, and iron—is reviewed by scholars who are among the most knowledgeable in their fields. There are individual contributions covering—moving from west to east—the eastern Mediterranean, Cyprus, Anatolia and Mesopotamia, the Iranian Plateau, the Indus Valley, South Asia, and China (iron only). Only in my paper on the Iranian Plateau is the evidence for both copper-base and iron metallurgy treated in a single chapter.

Treatment of the metallurgical developments across Asia and for all periods is, however, incomplete. There is no coverage of copper-base metallurgy in Egypt (see Garenne-Marot 1984, 1985) and the Levant (see Hauptmann and Weisgerber 1990; Levy and Shalev 1989; Moorey 1994; Maddin 1988; Golden 1998), nor in Southeast Asia (see contributions by Higham; Pigott and Natapintu; Muhly; Murowchick;

Stech and Maddin; and White published in Maddin 1988; also Higham 1996; White and Pigott 1996; Pigott in press; Pigott et al. 1997), nor in China (Barnard and Sato 1975; Bagley 1987; Franklin 1983; Wagner 1996; Murowchick 1989). Moreover, iron metallurgy in Mesopotamia (see Moorey 1994) and eastern Anatolia (see McConchie 1998) has not been addressed nor has the coming of iron in Southeast Asia, a topic which has yet to be reviewed in detail by any scholar.

The present volume lays the groundwork for future scholarly review of a trio of issues of fundamental importance to an anthropological assessment of the role of metals and metallurgy in the culture areas of the Asian continent, namely: (a) the societal impact of the development of metallurgical technology, and how in turn metallurgy may have been affected by its social context; (b) the validity and applicability of the concept of the "metallurgical province," as defined by Chernykh (1980; see also White 1988; Pigott in press); (c) the volatile debate concerning the role of diffusion as the cultural dynamic that fostered technological innovation. On the last point, the papers here stir up the heated controversy that has raged for decades, ever since the study of archaeometallurgical developments in the Asian Old World commenced. That particular fire has been further fueled with the two-volume publication of Victor Mair's major international conference proceedings entitled *The Bronze Age and Early Iron Age Peoples of Eastern Central Asia* (1998; see also Mair 1995).

Each of the authors has provided an abstract so the gist of their contributions need not be reviewed here. However, a few comments relevant to each paper are merited to bring to the reader's attention some of the points made which I, as editor, find of particular interest and which provide some food for thought.

THE USE OF NATIVE COPPER

It is illuminating to find that it is only in Southwest Asia, where metallurgy was born, that the use of native copper prior to other metals is documented. Tamara Stech's comment (in this volume) on malachite and native copper's co-occurrence certainly states the Southwest Asian situation aptly: "Metal in the form of finished artifacts was more highly regarded than copper ore... Malachite and native copper may have come together from their sources, which are presently unknown, but their differences and the rarity of the metal seem to have been appreciated." In Southwest Asia native copper use is supported by finds at a substantial number of sites with Neolithic contexts. Its use, starting with the remarkable finds from Çayönü, does not necessarily signal the beginning of metallurgy (see summary in Stech, this volume), but I would argue it can at least be seen to be part of the continuum of metal use that characterizes Southwest Asia exclusively. In a region replete with weathered copper sulfide ore bodies there were substantial deposits of this natural metal, and quantities would have been noticeable and accessible at the surface.

Archaeology in Southwest Asia continues to shed light on possible very early smelted metal. What are we to make of the copper-base metal artifacts from Nevali Çori in Anatolia (Hauptmann et al. 1993:543–544 in Stech, this volume)? Do they represent smelted metal from the Aceramic Neolithic? The already available but ambiguous evidence from Çatal Hüyük raises that possibility. Recent additional study of material from the site by Gerhard Sperl (1997) at the Erich-Schmid Institut für Festkörperphysik in Leoben, Austria, contributes new evidence of early "smelting." Stech (this volume) makes a relevant observation: regardless of how the earliest metals came into being, the evidence "suggest[s] that metal was not integral to developing social complexity, an idea supported by the paucity of metallurgical remains attributable to the next three millennia" (following the evidence from Çayönü for native copper).

At this juncture, we must make brief mention of lead, which can be smelted from its ores at temperatures decidedly lower than for any other metal of antiquity. As Stech (in this volume) writes, "It is possible that lead showed the possiblities of smelting." The early incidence of lead in Neolithic contexts (i.e., the ore in the form of beads at Çatal Hüyük [Sperl 1990], and the metal in bracelet form at Yarim Tepe [Merpert et al. 1977:82]) makes it clear lead was among the repertoire of materials being "played with" by early artisans. The time at which smelting appeared on the scene and the issue of what importance metal had for the societies using it in its earliest form are two questions unanswered at present. Fortunately, Anatolian archaeology is alive and well and will continue to yield new and intriguing pieces to the puzzle of metallurgical origins.

Outside Southwest Asia, however, the picture is different. James Muhly (in this volume), for example, speculates on native metal's early use on Cyprus, but there are no documented finds. Metallurgy, when it emerges in South Asia (with the exception of what may be a native copper bead from Neolithic levels at Merhgarh in Pakistan), as well as in Southeast and East Asia, appears as a full-blown smelting technology. It is difficult to imagine that across the vast expanse of Asia, where epochs of tectonic and metallogenic activity resulted in the emplacement of sulfide ore bodies which have weathered over the millennia, native copper is not present on an appreciable scale. It is probable that geological research in these regions has rarely focused on the question of the presence or absence of native copper—which is of little importance to modern economic geologists. Moreover, the remoteness of much of the region, inadequate access due to political considerations, and the difficulties of language barriers have all hindered getting the full story on native copper's presence. Are these regions by geological chance devoid of native copper? or did cultures outside Southwest Asia not exploit native copper reserves in their localities? or has archaeology (and the necessary analytical follow-up) simply not told the tale? My sense is that the last scenario is the most accurate.

One could suggest that native copper was the first and perhaps the only native metal exploited where metallurgy was born, though the presence of gold in Neolithic context in the Levant cannot be ignored (Gopher et al. 1991). After that (if one were to take a diffusionist standpoint), it was smelting technology (of copper and lead) that moved east. Even if native copper exists in the vast regions east and north of Southwest Asia, it may therefore be the case that it was never used by peoples who were adopting a rapidly spreading smelting technology. Outside of Southwest Asia, acceptance of the innovation perhaps was driven more by the novelty and usefulness of smelted metal than by a history of experience with working native copper. The reader may contrast with this the strongly anti-diffusionist stance taken by Mark Kenoyer and Heather Miller writing about South Asia in this volume (see also discussion in Possehl and Gullapalli, this volume).

ARSENICAL COPPER METALLURGY

James D. Muhly and most scholars in the Old World have adopted the use of the term "arsenical copper" to refer to the alloy of the two metals. For the Old World, and particularly the ancient Near East, this term is preferable to the term "arsenical bronze," because of the continuing controversy over the sources of tin. In the Old World the term "bronze" has in more recent archaeometallurgical discussions been used to designate the tin-copper alloy. The use of the term "arsenical bronze" is apt to confuse the broader readership who might then assume that tin and arsenic were somehow both involved in the alloy under discussion.

Muhly refers to mid–fourth millennium arsenical copper metallurgy in Chalcolithic Palestine as the best documented and earliest example of this technology. This tradition is also in evidence at Susa and on the Iranian Plateau during this period (e.g., at Tepe Yahya, Tal-e Malyan), if not before (e.g., at Tepe Hissar). Citing, for example, Tepe Yahya is somewhat complicated by the question of whether smelting was used deliberately to produce an arsenical alloy or whether native coppers and copper arsenides were simply melted together. The Anatolian evidence, as Muhly indicates, is later fourth millennium in date, though we may find arsenical copper occurring at increasingly early dates throughout Anatolia as excavations proceed.

Clearly, arsenical copper metallurgy was widely distributed across Southwest Asia and into Central Asia and was known also at Indus Valley Tradition sites. The technology, however, based on current evidence, was not a part of metalworking traditions in ancient China or Southeast Asia. Why this is the case is a question worthy of investigation.

In Southwest Asia, arsenic-rich ores were sufficiently available that the technology to smelt them emerged as miners exploited the surface-deposited richer oxidic ores, then sought what lay just below—grey coppers in the zone of secondary enrichment. I see arsenical metallurgy as a technological phenomenon that varied widely based on local geological conditions but which underlay some of the earliest experiences with smelting. Metalworkers in Iran exploiting the Anarak-Talmessi deposits could have produced arsenical copper simply by melting accidental or purposeful mixtures of native copper and copper arsenides. Early smelting experiments across the Asian Old World may have involved one of several processes: (a) the low temperature reduction of copper arsenates; (b) the co-smelting of accidental (or purposeful?) mixtures of oxidic ores with sulfarsenide ores; (c) the reduction of roasted sulfarsenides. Any of these processes would have yielded arsenical copper—essentially a new metal with improved mechanical properties. Muhly (this volume) suggests that arsenical ores of copper were recognized for their properties. When smelted, they yielded a metal which (depending on final arsenic content) could be superior to unalloyed copper. On these grounds, therefore, he argues, arsenical copper can be considered a deliberate alloy because metalworkers were selecting arsenical ores purposefully.

Since arsenical copper metallurgy comes on the scene in the fourth millennium B.C., is widespread across Southwest Asia, and persists well into the third millennium B.C., it is clearly a long-lived and dominant metallurgical tradition, which suggests it is likewise one deliberately chosen. If and where inadvertent co-smelting was the order of the day, based on the scenario described above, then arsenical copper metallurgy may have evolved more as a consequence of geological stratigraphy than of purposeful selection of ores. We may well be engaging in a moot debate here, one that neither archaeology nor the laboratory is ever going to resolve.

A more valuable line of inquiry with regard to the topic of arsenical copper is that which attempts to understand the production processes in order to explain the composition and structure of the metal. Focused on Old World archaeometallurgy, William Rostoker's research was one such laudable initiative (Rostoker et al. 1989; Rostoker and Dvorak 1991) as was Paul Budd and colleagues' research into how arsenical copper was smelted (Budd et al. 1992; Budd 1993). In the New World, the research of Heather Lechtman (1996) and her work with Sabine Klein (Lechtman and Klein 1999) is of direct relevance to Old World problems. I do find it curious that after more than thirty years of modern archaeometallurgical research only now are scholars spending time on this theme. We have an inclination to pass over the fundamental research on the way to what we see as more sophisticated questions. Often asking and then resolving the most basic questions helps enormously in achieving a thorough understanding of issues like the nature and end-product of production processes. I have delved a bit into this discussion in my chapter.

THE DEVELOPMENT OF TIN BRONZE METALLURGY

In Old World archaeometallurgy no single issue has been more hotly debated than the occurrence of tin and tin bronze in ancient Southwest Asia. James A. Charles (1975) summed it up best in his article title—"Where Is the Tin?" Moreover, the debate will continue because, despite the identification within the last 10 to 15 years of extensive and massive tin deposits in Central Asia east of Mesopotamia (Ryzanov 1979; Rossovsky et al. 1987) and in Afghanistan (Chymriov et al. 1973; Shareq et al. 1977; Cleuziou and Berthoud 1982), we still know little more than we did prior to these discoveries regarding the use of these rich sources. The mere presence of tin bronze in fourth millennium levels at Mundigak does not tell us where the tin was mined, who mined it, whether it, in fact, moved west to Mesopotamia and Anatolia, or (if it did) how it moved and who moved it.

Of import also is the fact that lapis lazuli (hailing from Badakhshan Province in northeast Afghanistan, also from the Chagai Hills of Pakistan, not to mention the less likely sources in the Pamirs and the Urals) was distributed throughout Mesopotamia, Iran, and Central Asia by the close of the fourth millennium B.C. (Casanova 1992:49). Lapis and tin, both of which have been found distributed along river courses in Afghanistan, could have been traveling together along exchange networks (Stech and Pigott 1986:45–46). Iran, curiously, has very little indication of tin bronze use in the fourth and third millennia. Was Iran bypassed in this period? Laboratory analyses at MASCA have indicated that tin bronze appears to be the alloy of choice at Tal-e Malyan in Fars Province by the early second millennium. If tin was moving by caravan across Iran along the Great Khorasan Road, why wasn't it being used by Iranian Plateau peoples along the way?

The remarkable metallurgy of Tepe Hissar in Iran clearly indicates the technological sophistication of a site where lapis was also being shaped. The same can be said for Shahr-i Sokhta, which is not that far from the lapis source in the Chagai Hills in Pakistan (Casanova 1992:49). Is the lack of early bronze on the Iranian Plateau indirect evidence of a sea-based trade with South Asian middlemen (and later Harappans) at the helm? As Kenoyer and Miller (this volume) point out, Shortugai in northeastern Afghanistan was well situated in a region rich in gold, lapis, and tin, yet only lapis among this trio of materials appears in any appreciable quantity (Francfort 1989). In Mesopotamia, however, with the advent of the Bronze Age ca. 3000 B.C., the movement of gold, lapis, and also tin to Sumerian consumers is clearly documented in the riches of the Royal Tombs of Ur (Zettler and Horne 1998). The circumstantial evidence for this trio of precious, exotic, and highly desired materials traveling to Mesopotamia from points east, whether those were Afghanistan, Pakistan, and/or Central Asia, remains difficult to ignore (Muhly 1977:76).

Promising new research on pre-Islamic tin production in Central Asia has been carried out since 1997 by a German team from the German Archaeological Institute, the German Mining Museum, and the TU Bergakademie (Alimov et al. 1998). The focus has been on occurrences of cassiterite in quartzite veins around Karnab, Uzbekistan, and a very large deposit of stannite at Muschiston, Tadjikistan, at an altitude of about 3000 m. Many grooved stone hammers have been found on the surface at Karnab, where the earliest mining is at present documented by a calibrated radiocarbon date of around 800 B.C. Based on three radiocarbon dates and a ceramic sherd of the Andronovo culture found in a mine, ancient mining is also attested at Muschiston, from the first half of the second millennium B.C. These first results have confirmed extensive ancient mining at Karnab and Muschiston, though production figures cannot yet be estimated (E. Pernicka, pers. comm.).

Muhly (this volume; cf. Stech, this volume), challenging (for the first time by any scholar, I might add) the hypotheses proposed by Tamara Stech and myself (1986) that tin was somehow reaching Troy via Mesopotamia, proposes instead that "it seems much more reasonable to assume, following the work of Manfred Korfmann (1987), that Afghan tin came to Troy (and the north Aegean) across the Black Sea, from Kolchis to Troy." The same could hold true for Central Asian tin, though routes through the Caucasas would not have been the easiest. Could tin have moved along the Caspian littoral? This hypothesis would explain tin's absence from Iran while allowing that lapis, which is not Central Asian, could move from Afghanistan (or Pakistan) across Iran to Mesopotamian consumers. Relevant to this topic is Andreas

Schillinger's (1997, in press) recent master's thesis (University of Tübingen) on the earliest tin bronze in the Black Sea region (dated to the second and third quarters of the third millennium B.C.). In addition, the Black Sea Trade Project under the direction of Fred Hiebert of the University of Pennsylvania Museum will clearly contribute to this interesting debate (Hiebert et al. 1998; in press).

To complicate this question of tin sources even further, the fieldwork in the Bolkardağ Mining District of the Central Taurus Mountains directed by K. Aslihan Yener, now of the Oriental Institute, University of Chicago, revealed a region replete with ancient mining in south-central Turkey (e.g., Yener et al. 1989; Yener and Vandiver 1993a, b; Vandiver et al. 1993). Excavations of the Kestel mine, which the project geologists and mining engineers have cogently argued is a tin mine (Willies 1995; cf. Muhly 1993 and this volume), and of the settlement site of Göltepe (2 km distant), where the tin was being processed, have given us the first archaeological indication of the nature of activities that characterize a Bronze Age tin source—one clearly capable of having yielded considerable amounts of tin.

This is not the place to review the substantial literature on and the remarkable evidence from the mine and site. In my view, however, the arguments for a tin mine and processing facility—which are made by mining specialists and based on laboratory analysis; combined with the uncovering of a major metal production quarter at nearby Göltepe (a site of significant hectarage with a walled summit)—are persuasive. The tens of thousands of ground-stone tools for ore processing and numerous crucibles with tin-bearing residues combined with kilos of powdered tin ore excavated in the metalworker's quarter at Göltepe comprise very strong evidence for tin processing (Yener and Vandiver 1993a). The larger question remains of just how much tin the Kestel-Göltepe complex was capable of producing. Here again the same questions that I asked above can be asked about this Turkish source. Who were the people who did the mining? Where was the processed product going and how? How extensive an area was this source supplying? These questions may remain, for the present, unanswerable but the Kestel-Göltepe complex is significant beyond its association with tin because, prior to its revelation, few could have predicted its existence or its location in a major mining district. There are other such discoveries to be made.

Suffice it to say that there is major research yet to be done, in the field and in the lab, on Southwest Asia and its ever-vexing tin bronze question. For example, if the thirteenth century B.C. oxhide tin ingots of the Ulu Burun shipwreck off the coast of Turkey (Bass 1986) cannot as yet be sourced, the unanswered questions surrounding the sources of ancient Near Eastern tin will remain as such for some time to come.

The question of sources is not the only enigmatic aspect of the tin bronze question. Much is taken for granted metallurgically concerning just how the alloy is manufactured. In my chapter I have discussed a summary by Michael Wayman et al. (1988) of the research on the various processes by which tin bronze might have been manufactured. It is interesting to note how little work has actually been done and that archaeometallurgists, myself included, have been taking much for granted about this rather complex process. Discussions of analyses of tin bronze artifacts, unlike discussions of arsenical copper artifacts, rarely question how the alloy came into being. We might do well to spend a little more time investigating the fundamental production processes which yielded the metal we so often look at under the microscope. Unfortunately, there is a shortage of capable specialists (at least in the United States) willing to analyze and interpret the industrial residues of the smelting process. The few we do have simply cannot take on any more work, especially when it may be research which lies well outside their scholarly interests. By focusing only on metals-based analysis we are in effect only answering half the question of how an ancient metal artifact was produced.

I will put forward a final, but particularly important, point with regard to tin in cultural context. Tamara Stech and I made in 1986 what we felt were cogent arguments, in an article in the journal *Iraq* entitled "The Metals Trade in Southwest Asia in the Third Millennium B.C.," regarding, in part, the role of tin as a status metal. Generally, at least in the Southwest Asian context, the laboratory analysis of tin bronze artifacts has yet to reveal that there is a close correlation between artifact type and tin content in those classes of artifacts—weapons and other edged tools—that function by cutting or piercing, hitting or pounding. These are the artifacts that would function most effectively from being made of a copper alloy with a tin content close to 10% that could then be work-hardened by hammering to produce a metal with useful mechanical properties. However, it is these classes of artifacts which in fact do not evidence conscious, deliberate alloying and mechanical working to maximize performance. Perhaps we have missed something in our rather random studies of patterns of alloying. (Little has been done on a large-scale and systematic basis across Southwest Asia and through the relevant time periods, and the substantial diversity of analytical techniques used in elemental analysis may be clouding the waters somewhat also).

Thus, reasons other than metallurgical need to be sought to explain the acceptance and spread of tin bronze as a technological innovation. Tamara Stech extends this argument of tin bronze as status metal in this volume. For example, as she makes clear, "the introduction of tin in bronze in both Anatolia and Mesopotamia did not occasion its entry into general circulation." She states (this volume) that

in inland Southwest Asia, regular bronze consumption in the Bronze Age occurred only in a few places, particularly in graves. This would suggest that the arrival of tin bronze was neither rapid nor widespread and did not herald a "technological revolution" of some sort. Rather, indications suggest that it was restricted to special, often elite, contexts.

She cites the evidence from Kish Cemetery Y which suggests that tin bronze associates with special burials at a point in time when Kish's First Dynasty was "preeminent in Mesopotamia." The reasons for "bronze's relative rarity must have been cultural rather than technological," as must be true also for the purposes to which the metal was put. Tin "was not regarded strictly as a technological enhancement to the properties of copper, but a cultural support to the bases of power, both political and economic."

I would add to the discussion a paper of Carole Gillis (1999) in which she suggests that specially treated vases covered with tin foil, which when heated resembled gold, were placed in tombs as markers of wealth and prestige. I will leave it to the readers to assess Tamara Stech's provocative discussion of this issue.

ON THE COMING OF IRON

With treatments by Jane Waldbaum, Bennet Bronson, Gregory Possehl and Praveena Gullapalli, and myself, iron does not receive quite as extensive coverage in this volume as do copper and bronze. However, the questions surrounding the coming of iron are no less significant. Waldbaum's overview makes it clear that a major technological change swept across the eastern Mediterranean world and into Southwest Asia and beyond, beginning in the closing centuries of the second millennium B.C. She examines in detail five major untested assumptions concerning the origins, development, spread, and use of iron. I wish to touch directly only on two of these here.

To her discussion concerning the assumption that the "superiority" of iron over bronze was instrumental in iron's introduction, I would add complementary research which I conducted with Robert Maddin and Reed Knox on ninth century B.C. iron from the destroyed Iron Age citadel of Hasanlu in northwestern Iran (Pigott 1981, 1989). Metallography indicated that the iron in the samples analyzed was very heterogeneously carburized wrought iron. The carburization present was, most probably, a result of the smelting process and was diminished further by decarburization during forging which would proceed from the exterior surface inward. Depth would depend on the temperature and the amount of time spent in oxidizing circumstances. No indications of intentional treatment to enhance the properties of the iron were apparent. It is interesting to note that in any artifact that retained a carburized interior core and which was decarburized and softer at the edges, the act of sharpening the edges could, over time, lead to increasingly harder and sharper edges as the softer metal was abraded away and the harder metal exposed.

As with bronze, as discussed above, we need to consider factors in addition to iron's putative superiority (as steel) in seeking an explanation for the rapid and widespread acceptance of this technological innovation. We cannot totally rule out, however, the possibility that among those who first began to make iron on some scale in the late second millennium B.C. there were smiths whose techniques of smelting yielded blooms that were true steel. Based on such achievements the reputation of this remarkable new metal could have spread rapidly, promoting the widespread production of iron that was, perhaps more often than not, not the desired steel. Unless archaeology can offer up evidence supporting or disproving this possibility, I would tend not to give it much credibility, given the current poor state of our knowledge of the earliest iron technology in Southwest Asia.

But how did people first come to know the metal iron in smelted (not meteoritic) form? Waldbaum presents ample evidence for the presence of iron in Bronze Age contexts. One of the truly fascinating arguments which has been put forth for some time that could explain some of this precocious iron is the "iron out of copper or lead smelting" theory. Though this is not the place to launch into this discussion, Theodore Wertime (1964:1262; 1973) was among the early proponents of the idea that through an accidental combination of furnace materials (too much charcoal, iron-rich copper ores, and iron oxide flux) and conditions (hard driven, high temperature smelting) metallic iron would form accidentally during copper smelting. Various other scholars have discussed this possible route to iron's discovery (e.g., Gale et al. 1990;[1] van der Merwe and Avery 1982; Pigott 1982:21; Cooke and Aschenbrenner 1975; Tylecote 1970:290; Smith 1966). If, in truth, iron came on the scene via the copper smelting tradition then it can be seen to be a true innovation. As I have written elsewhere (Pigott 1982:21), "Iron as an innovation represented a new line of development, a recombination of previously existing knowledge which resulted in an entirely new technological configuration (Wallace 1972:469; Bee 1974:1740)."

Regarding the third assumption that Waldbaum discusses—namely, that of the "Hittite monopoly"—it is possible to suggest that it was Hittite bronzesmiths who were among the first to transform the accidental occurrence of iron during copper smelting into the process of deliberately smelting iron. Something has to explain the Hittites' remarkably early familiarity with iron, even if excavations of Hittite sites have yielded little in the way of actual iron artifacts.

We can take this speculation somewhat further. Given that the Hittites were great borrowers and assimilators of other cultures' traits, perhaps they actually borrowed iron smelting from those who had already discovered it. One such group could have been the Chalybean tribes which occupied the Pontic region of Anatolia and were never assimilated by the Hittites. The eastern end of the Pontic region borders closely on ancient Colchis, the only region in Southwest Asia to offer substantial evidence of early iron smelting. Though the dates remain controversial, iron smelting installations have been excavated in Colchis in modern Georgia that are tentatively dated to the mid to late second millennium B.C. (Khakhutaishvili 1976). Colchis and the Pontic region are rich in iron because a band of laterite soil (also rich in nickel) runs along the entire northern and western coast of Anatolia, across Colchis (see Fig. 4.12, this volume). Such laterites can be smelted (though not with ease). Nickel-rich laterites found in Greece were being exploited as early as the thirteenth century B.C. (Varoufakis 1982:317 in Waldbaum, this volume).

In addition, enormous marine placer deposits of iron-rich black sands are found off the coast where the Pontic region and Colchis meet (Tylecote 1981). There are indications in the Colchis iron smelting sites that black sands were being used. Thus the evidence for iron smelting found in the Colchis region combined with the presence of the iron-using Chalybes in the Pontic region offers a possible scenario for transmission of iron technology to the Hittites (Pigott 1996a:162–163). We can speculate that perhaps forthcoming archaeological work will lay to rest or offer strong support for this thesis.

A distinct step forward in our understanding of iron smelting in eastern Anatolia in the Iron Age and later has been made by Oktay Belli (1991) who has reported on no less than three rich iron ore mining districts, all of which contained iron slag heaps, countless tuyéres, and other indicators of iron processing. In Southwest Asia, other than the evidence from Colchis, which is less than well understood, these three regions of eastern Anatolia comprise the only large area characterized by the presence of substantial evidence for iron production. While many periods are represented, the earliest indications are of Urartian exploitation, signaled by what O. Belli says are finds of Urartian pottery at the industrial sites.

However, in a recent Ph.D. thesis at the University of Melbourne focused on ancient iron producing communities in northeastern Anatolia, M. McConchie (1998; see also Burney 1996) argues that the Urartian association with the substantial remains of iron production (e.g., slag and tuyéres) is not unequivocal. This thesis not only brings to light important new evidence of iron production and use in contexts ranging from the Early Iron Age to the Early Roman period but includes the most thorough and scientifically detailed analytical study of ancient Near Eastern iron accomplished to date.

Elsewhere in Southwest Asia our evidence is chiefly small finds at sites, some of which may not be iron smelting installations. Gradually our understanding of the emergence of the ancient technology of iron and its role in a variety of cultural contexts and periods is improving. However, it remains a major enigma of ancient Near Eastern archaeology that the region where iron was born has been so reluctant to offer up hard evidence of iron's smelting and blacksmithing.

CONCLUDING REMARK

I will close with mention of what must rank in terms of its scholarly importance among the most significant archaeological discoveries of this century, namely, the cemeteries containing the naturally mummified remains of more than 100 Caucasian individuals in the Tarim Basin of Xinjiang Province, western China (Mair 1995, 1998). Current scholarship suggests the dates that they were in use from as early as ca. 2000 B.C. down into the first millennium B.C. The implications for our understanding of metallurgy are staggering. I have discussed elsewhere that the presence of these Caucasian people so far to the East offers evidence, for the first time, in support of the notion that the transmission of knowledge of metallurgical crafts could have occurred with east-west contact (Pigott 1996b). If the present cemeteries can be dated as early as 2000 B.C., was there yet earlier contact? Does this have implications for the coming of copper/bronze to China? This is an event currently dated as early as the latter half of the third millennium B.C. in the Longshan culture on the middle and lower reaches of the Yellow River. An Zhimin (1993) feels that the Longshan is linked to the Qijia culture of northwestern China, and could have benefited from transmission of knowledge of metalworking along the Silk Road. The fact that Chinese archaeologists themselves are open to this possibility is a remarkable turnabout from their long-standing position on the indigenous development of metallurgy in China and bodes well for fruitful future discussions.

As for iron, the Tarim Basin mummies are even more significant, given that iron is known in Xinjiang in the tenth century B.C. (D. Kamberi, pers. comm.), the point in time when iron tends to be known in a number of cultural contexts across Southwest and South Asia. How can this be explained? In a word: diffusion. Iron appeared earliest in the eastern Mediterranean world but by the tenth–ninth centuries it is present in Southwest, Central, and South Asia. We know it in Southeast Asia ca. 700–500 B.C. (Pigott et al. 1997) and in the Chinese heartland by the sixth century B.C. (Wagner 1993). If one subscribes to diffusionist arguments, then there is the appearance of a horizon sloping from west to east. The movements of Caucasian populations eastward through Central Asia is certainly one potential source of knowledge of iron technology for which we now have to ponder the implications (cf. the anti-diffusionist perspective on the coming of iron in South Asia in Possehl and Gullapalli, this volume).

Regardless of how iron metallurgy first arrived or began in China, its development in the Chinese heartland and its fluorescence during the Han Dynasty is what Bennet Bronson terms a "remarkable transformation." Bronson offers an in-depth overview of the coming of iron in China in this volume. He first reviews the evidence chronologically and then treats the major technological innovations for which ancient China is renowned, e.g., the blast furnace and the casting of molten iron. The latter is a process which first appeared in Central Asia only in the twelfth to fourteenth centuries A.D., after which it finally appeared in the West. Complementary to Bronson's study is a recent volume by Donald Wagner (1993), entitled *Iron and Steel in Ancient China*, which is the most current and comprehensive overview of this subject. While Wagner (pers. comm.) has already indicated he is rethinking the discussion in that volume of the origins and spread of this tradition, the wealth of data he has compiled (social, historical, linguistic, and archaeological) makes it eminently clear that the ancient Chinese context is perhaps the best in which to study the nature of a true industry as well as the social impact of profound metallurgical/technological innovation and change (Pigott 1996b).

***

In closing, this introduction is not the place to debate the many complex issues for discussion and research that abound in archaeometallurgy today. Many of these critical questions have been addressed in this volume by the preeminent specialists in the field. In this commentary my only aim is to put these issues before the readers prior to their reading this volume. In addition, I hope the high quality of scholarship that has characterized our field will continue as new evidence allows these issues to be further investigated throughout the Asian Old World—where the world's metallurgical traditions were born.

ACKNOWLEDGMENTS

I would like to thank Robert McC. Adams, C. C. Lamberg-Karlovsky, and Robert H. Dyson for making possible my involvement in the 4th USA–USSR Archaeological Exchange and their efforts in fostering good relations and scholarly exchange between American scholars and their counterparts in the former Soviet countries. Sadly, Valery P. Alekseyev has died since the time of the IREX symposium in 1988. He had been instrumental in the success of the 4th Exchange/Symposium. Evgeni Chernykh and his team of talented researchers not only offered the benefit of their years of research and analysis during the symposium but did much to organize it and keep it running smoothly. Our Georgian hosts offered their remarkable and generous hospitality both in Tbilisi and at Signakhi. They gave us a truly memorable experience of their country (and its cuisine) and of its intriguing archaeology. Finally, thanks must go to IREX (and Wesley Fisher, then of IREX) for their generous funding of the symposium participants from the United States. This support made the unique gathering of scholars a reality.

NOTES

1. Recent research now brings into question just how many (if any) of the iron artifacts excavated at the Hathor Temple at Timna are to be considered as made from iron formed as a by-product of copper smelting (Merkel et al. 1998). Final conclusions will have to await full publication of the analyses.

[Final ms. received 12/98.]

REFERENCES CITED

An Zhimin
1993   Shilun Zhongguo de Zaoqi Tongqi (A tentative discussion of China's early copper/bronze implements). *Kaogu* 12:1110–1119.

Alimov, K.; Boroffka, N.; Bubnova, M.; Burjakov, J.; Cierny, J.; Jakubov, J.; Lutz, J.; Parzinger, H.; Pernicka, E.; Ruzanov, V.; Weisgerber, G.
1998   Vorislamische Zinngewinnung in Mittelasien. Vorbericht der Kampagne 1997. *Eurasia Antiqua* 4. In press.

Avilova, L. I., and Terekhova, N. N.
1989   Sovetsko-amerikanskiy simpozium "Drevneyshaya metallurgiya starogo sveta" (The Soviet-American Symposium "Early Metallurgy in the Old World") *Sovetskaya arkheologiya* 3:290–296.

Bagley, R. W.
1987   *Shang Ritual Bronzes in the Arthur M. Sackler Collections.* Washington, DC: The Arthur M. Sackler Foundation, and the Harvard University Press.

Barnard, N., and Sato Tamotsu
1975   *Metallurgical Remains of Ancient China.* Tokyo: Nichiosha.

Bass, G. F.
1986   A Bronze Age Shipwreck at Ulu Burun (Kaş): 1984 Campaign. *American Journal of Archaeology* 90:269–296.

Bee, R. L.
1974   *Patterns and Processes.* New York: The Free Press.

Belli, O.
1991   Ore Deposits and Mining in Eastern Anatolia in the Urartian Period: Silver, Copper and Iron. Pp. 16–41 in *Urartu: A Metalworking Center in the First Millennium B.C.E.*, ed. R. Merhav. Jerusalem: The Israel Museum.

Budd, P.
1993   Recasting the Bronze Age. *New Scientist* Oct. 23:33–37.

Budd, P.; Gale, D.; Pollard, A. M.; Thomas, R. G.; and Williams, P. A.
1992   The Early Development of Metallurgy in the British Isles. *Antiquity* 66:677–686.

Burney, C.A.
1996   The Highland Sheep are Sweeter. In *Cultural Interaction in the Ancient Near East: Papers Read at a Symposium Held at the University of Melbourne, Dept. of Classics and Archaeology (29-30 September 1994)*, ed. G. Brunnens. Louvain: Peeters Press.

Casanova, M.
1992   The Sources of Lapis-lazuli Found in Iran. Pp. 49–56 in *South Asian Archaeology 1989*, ed. C. Jarrige. Monographs in World Archaeology No. 14. Madison, WI: Prehistory Press.

Charles, J. A.
1975   Where Is the Tin? *Antiquity* 49:19–24.

Chernykh, E. N.
1980   Metallurgical Provinces of the 5th–2nd Millennia in Eastern Europe in Relation to the Process of Indo-Europeanization. *Journal of Indo-European Studies* 8(3/4):317–335.

Chymriov, V. M.; Stazhilo-Alekseev, K. F.; Mirzad, S. H.; Dronov, V. I.; Kazikhani, A. R.; Salah, A. S.; and Teleshev, G. I.
1973   Mineral Resources of Afghanistan. Pp. 44–85 in *Geology and Mines and Industries of the Republic of Afghanistan.* Kabul.

Cleuziou, S., and Berthoud, T.
1982   Early Tin in the Near East: A Reassessment in the Light of New Evidence from Afghanistan. *Expedition* 24(3):14–19.

Cooke, S. B., and Aschenbrenner, S. E.
1975   The Occurrence of Metallic Iron in Ancient Copper. *Journal of Field Archaeology* 2:251–266.

Francfort, H.-P.
1989   *Fouilles de Shortughaï. Recherches sur l'Asie Central Protohistorique.* 2 vols. Mèmoires de la Mission Archéologique

Française en Asie Centrale Tome II. Paris: Diffusion de Boccard.

Franklin, U. M.
1983 On Bronze and Other Metals in Early China. Pp. 279–296 in *The Origins of Chinese Civilization,* ed. D. N. Keightley. Berkeley: University of California Press.

Gale, N.; Bachmann, H. G.; Rothenberg, B.; Stos-Gale, A.; and Tylecote, R. F.
1990 The Adventitious Production of Iron in the Smelting of Copper. Pp. 182–191 in *The Ancient Metallurgy of Copper,* ed. B. Rothenberg. London: Institute for Archaeometallurgical Studies.

Garenne-Marot, L.
1984 Le cuivre en Egypte pharonique: Sources et métallurgie. *Paléorient* 10(1):97–126.
1985 Le travail du cuivre dans l'Egypte pharonique d'apres les peintures et les bas-reliefs. *Paléorient* 11(1):85–100.

Gillis, C.
1999 The Role of Tin and the Smith in the Aegean Late Bronze Age. In *Metals in Antiquity,* Proceedings of the International Symposium, Harvard University, Sept. 10–13, 1997, eds. S. M. M. Young, A. M. Pollard, P. Budd, and R. A. Ixer. Oxford: Archaeopress of British Archaeological Reports.

Golden, J.
1998 *The Dawn of the Metal Age: Social Complexity and the Rise of Metallurgy in the Southern Levant circa 4500–3500 B.C.* Ph.D. dissertation, Department of Anthropology, University of Pennsylvania. Ann Arbor, MI: University Microfilms International.

Gopher, A.; Tsuk, T.; Shalev, S.; Gophna, R.
1991 Earliest Gold Artifacts in the Levant. *Current Anthropology* 31:436–443.

Hauptmann, A., and Weisgerber, G.
1990 Periods of Ore Exploitation and Metal Production in the Area of Feinan, Wadi Arabah, Jordan. *Annual Report of the Department of Antiquities, Jordan:* 61–66.

Hauptmann, A.; Lutz, J.; Pernicka, E.; and Yalcin, U.
1993 Zur Technologie der frühesten Kupferverhuttung im ostlichen Mittelmeeraum. Pp. 541–572 in *Between the Rivers and Over the Mountains. Archaeologica Anatolica at Mesopotamica Alba Palmieri Dedicata,* eds. M. Frangipane, H. Hauptmann, M. Liverani, P. Matthiae, M. Mellink. Rome: Dipartimento di Scienze Storiche Archeoloigiche e Antropologiche dell'Antichità Università di Roma "La Sapienze."

Hiebert, F.; Smart, D.; Doonan, O.; and Gantos, A.
1998 From Mountain Top to Ocean Bottom: A Comprehensive Approach to Archaeological Research. Pp. 177–184 in *Ocean Pulse: A Critical Diagnosis,* eds. J. Tanacredi and J. Loret. New York: Plenum.

Hiebert, F.; Smart, D.; Gantos, A.; and Doonan, O.
in press Foreland and Hinterland: Sinop, Turkey. In *Foreland and Hinterland,* ed. J. Fossey. McGill University Monographs in Classical Archaeology and History. Montreal.

Higham, C.
1988 Prehistoric Metallurgy in Southeast Asia: Some New Information from the Excavation of Ban Na Di. Pp. 130–154 in *The Beginning of the Use of Metals and Alloys,* ed. R. Maddin. Cambridge, MA: MIT Press.
1996 *The Bronze Age of Southeast Asia.* Cambridge: Cambridge University Press.

Khakhutaishvili, D. A.
1976 A Contribution of the Kartvelian Tribes to the Mastery of Metallurgy in the Ancient Near East. Pp. 337–348 in *Wirtschaft and Gesellschaft im Vorderasien,* eds. J. Harmatta and G. Komoroćy. Budapest: Akadémiai Kaidó.

Korfmann, M.
1987 Seefahrtbeziehungen zwischen Schwarzem Meer und Ägäis im 2. und 3. Jahrtausend v.u.Z. P. 81 in *Sixth International Congress of Aegean Prehistory. Summary of Papers.* Athens: Ministry of Culture.

Lechtman, H.
1996 Arsenic Bronze: Dirty Copper or Chosen Alloy? A View from the Americas. *Journal of Field Archaeology* 23:477–514.

Lechtman, H., and Klein, S.
1999 The Production of Copper-arsenic Al-

loys (Arsenic Bronze) by Cosmelting: Modern Experiment, Ancient Practice. *Journal of Archaeological Science* 26:497–526.

Levy, T., and Shalev, S.
1989   Prehistoric Metalworking in the Southern Levant: Archaeometallurgical and Social Perspectives. *World Archaeology* 20:352–372.

Maddin, R. (ed.)
1988   *The Beginning of the Use of Metals and Alloys.* Cambridge, MA: MIT Press.

Mair, V. (ed.)
1995   A Collection of Papers on the Mummified Remains Found in the Tarim Basin. *Journal of Indo-European Studies* 23 (3/4).
1998   *The Bronze Age and Early Iron Age Peoples of Eastern Central Asia.* 2 vols. Journal of Indo-European Studies Monograph 26. Institute for the Study of Man.

McConchie, M.
1998   Iron Technology and Ironmaking Communities in Northeastern Anatolia: First Millennium B.C. Ph.D. thesis, Department of Classical Studies and Archaeology, University of Melbourne, Australia.

Merkel, J. F.; Barret, K.; and El-Gayar, E. S.
1998   Egyptian New World Iron Smelting: Archaeometallurgical Evidence from the Hathor Temple at Timna. Paper presented at Second International Conference on Ancient Mining and Metallurgy and Conservation of Metals, Cairo, April 14–16. To appear in *Institute for Archaeo-Metallurgical Studies* 21, Winter 1999.

Merpert, N. I.; Munchaev, R. M.; and Bader, N. O.
1977   The Investigations of Soviet Expedition in Iraq, 1974. *Sumer* 33:65–104.

Moorey, P. R. S.
1994   *Ancient Mesopotamian Materials and Industries.* Oxford: Clarendon Press.

Muhly, J. D.
1977   The Copper Ox-hide Ingots and the Bronze Age Metals trade. *Iraq* 39:73–82.
1993   Early Bronze Age Tin and the Taurus. *American Journal of Archaeology* 97:239-253.
1988   The Beginnings of Metallurgy in the Old World. Pp. 2–20 in *The Beginning of the Use of Metals and Alloys,* ed. R. Maddin. Cambridge, MA: MIT Press.

Murowchick, R. E.
1988   The Development of Early Bronze Metallurgy in Vietnam and Kampuchea: A Reexamination of Recent Work. Pp. 182–199 in *The Beginning of the Use of Metals and Alloys,* ed. R. Maddin. Cambridge, MA: MIT Press.
1989   *The Ancient Bronze Metallurgy of Yunnan and its Environs: Development and Implications.* Ph.D. dissertation, Department of Anthropology, Harvard University. Ann Arbor, MI: University Microfilms International.

Pigott, V. C.
1981   *The Adoption of Iron in Western Iran in the Early First Millennium B.C.: An Archaeometallugical Study.* Ph.D. dissertation, Dept. of Anthropology, University of Pennsylvania. Ann Arbor, MI: University Microfilms International.
1982   The Innovation of Iron: Cultural Dynamics in Technological Change. *Expedition* 25(1):20–25.
1989   The Emergence of Iron Use at Hasanlu. *Expedition* 31(2–3):67–79.
1996a   Near Eastern Archaeometallurgy: Modern Research and Future Directions. Pp. 139–176 in *The Study of the Ancient Near East in the Twenty-First Century,* eds. J. S. Cooper and G. M. Schwartz. Winona Lake, WI: Eisenbrauns.
1996b   The Study of Ancient Metallurgical Technology: A Review. Review of D. B. Wagner, *Iron and Steel in Ancient China. Asian Perspectives* 35(1):89–97.
in press Reconstructing the Copper Production Process as Practiced among Prehistoric Mining/Metallurgical Communities in the Khao Wong Prachan Valley of Central Thailand. In *Metals in Antiquity,* Proceedings of the International Symposium, Harvard University, Sept. 10–13, 1997, eds. S. M. M. Young, A. M. Pollard, P. Budd, and R. A. Ixer. Oxford: Archaeopress of British Archaeological Reports.

Pigott, V. C., and Natapintu, S.
1988   Archaeological Investigations into Prehistoric Copper Production: The Thailand Archaeometallurgy Project 1984–1986. Pp. 156–162 in *The Begin-*

ning of the Use of Metals and Alloys, ed. R. Maddin. Cambridge, MA: MIT Press.

Pigott, V. C.; Weiss, A. D.; and Natapintu, S.
1997   The Archaeology of Copper Production: Excavations in the Khao Wong Prachan Valley, Central Thailand. Pp. 119–157 in *South-East Asian Archaeology, 1992: Proceedings of the Fourth International Conference of the European Association of South East Asian Archaeologists, Rome, 28th September–4th October*, eds. R. Ciarla and F. Rispoli. Rome: L'Istituto Italiano per l'Africa e l'Oriente (Is.AeO-Rome).

Rossovsky, L. N.; Mogarovsky, V. V.; and Chymirev, V. M.
1987   The Metallogeny of Tin and Rare Metals in the Eastern Part of the Mediterranean Folded Belt. Pp. 170–177 in *Mineral Deposits of the Tethyan Eurasian Metallogenic Belt between the Alps and the Pamirs (Selected Examples)*, ed. S. Jankovic. UNESCO/ICCP Project No. 169. Belgrade.

Rostoker, W., and Dvorak, J. R.
1991   Some Experiments with Co-smelting to Copper Alloys. *Archeomaterials* 5(1):5–20.

Rostoker W.; Pigott, V. C.; and Dvorak, J. R.
1989   Direct Reduction to Copper Metal by Oxide/Sulfide Mineral Interaction. *Archeomaterials* 3(1):69–87.

Ryzanov, V.
1979   On General Ancient Tin Ore Sources on the Territory of Uzbekistan. *Material for the History of Uzbekhistan* 15:98–104.

Schillinger, A.
1997   *Die früheste Zinnbronzen im Schwarzmeeraum*. MA thesis. Institute für Ur- und Frühgeschichte. University of Tübingen
in press   The Earliest Tin-Bronzes in Transcaucasia and their Chronological Classification. To be published among the proceedings of the International Symposium on Archeometallurgy in Central and Western Asia, April 19–24, 1997. Tehran: Research Centre for Conservation of Cultural Relics (RCCCR).

Shareq, A.; Chmyriov, V. M.; Stazhilo-Alexseev, K. F.; Dronov, V. I.; Gannon, D. J.; Lubemov, G. K.; Kafarshiy, A. Kh.; and Malyarov, E. P.
1977   *Mineral Resources of Afghanistan*, 2nd ed. Afghan Geological and Mines Survey, United Nations Development Support Project, AFG/74/012. Kabul: Ministry of Mines and Industries.

Smith, C. S.
1966   On the Nature of Iron. Pp. 29–40 in *Made of Iron*. Houston: Art Dept., University of St. Thomas.

Sperl, G.
1990   Zur Urgeschichte des Bleies. *Zeitschrift für Metallkunde* 81:799–801.
1997   New Research on the Beginnings of Metallurgy at Catal Höyük, Turkey (7th mill. BC). Paper presented at "Metals in Antiquity" International Symposium, Boston, organized by S. Young and P. Budd, sponsored by Harvard University and University of Bradford, Sept. 10–13.

Stech, T., and Maddin, R.
1988   Reflections on Early Metallurgy in Southeast Asia. Pp. 163–174 in *The Beginning of the Use of Metals and Alloys*, ed. R. Maddin. Cambridge, MA: MIT Press.

Stech, T., and Pigott, V. C.
1986   The Metals Trade in Southwest Asia in the Third Millennium B.C. *Iraq* 48:39–64.

Tylecote, R. F.
1970   Early Metallurgy in the Near East. *Metal and Materials* 4:285–293.
1981   Iron Sands from the Black Sea. *Anatolian Studies* 31:137–139.

van der Merwe, N., and Avery, D. H.
1982   Pathways to Steel. *American Scientist* 70(2):146–155.

Vandiver, P. B.; Yener, K. A.; and May, L.
1993   Third Millennium Tin Processing Debris from Göltepe (Anatolia). Pp. 545–569 in *Materials Issues in Art and Archaeology III*, eds. P. B. Vandiver, J. Druzik, and G. S. Wheeler. Pittsburgh: Materials Research Society.

Varoufakis, G. J.
1982   The Origin of Mycenaean and Geometric Iron on the Greek Mainland and in the Aegean Islands. Pp. 315–322 in *Acta of the International Archaeological Symposium "Early Metallurgy in Cyprus, 4000–500 B.C.," Larnaca,*

Cyprus 1–6 June 1981, eds. J. D. Muhly, R. Maddin, and V. Karageorghis. Nicosia: Pierides Foundation.

Wallace, A. F. C.
1972 Paradigmatic Processes in Culture Change. *American Anthropologist* 74(3):467–478.

Wagner, D. B.
1993 *Iron and Steel in Ancient China.* Leiden: E. J. Brill.

Wayman, M.; Gualtieri, M.; Konzuk, R. A.
1988 Bronze Metallurgy at Roccagloriosa. Pp. 128–132 in *Proceedings of the 26th International Archaeometry Symposium,* eds. R. M. Farquhar, R.G.V. Hancock, and L. A. Pavlish. Toronto: Archaeometry Lab, Dept. of Physics, University of Toronto.

Wertime, T. A.
1964 Man's First Encounters with Metallurgy. *Science* 146:1257–1267.
1973 The Beginnings of Metallurgy: A New Look. *Science* 182:875–887.

White, J. C.
1988 Early East Asian Metallurgy: The Southern Tradition. In *The Beginning of the Use of Metals and Alloys,* ed. R. Maddin. Cambridge, MA: MIT Press.

White, J. C., and Pigott, V. C.
1996 From Community Craft to Regional Specialization: Intensification of Copper Production in Pre-State Thailand. Pp. 151–175 in *Craft Specialization and Social Evolution: In Memory of V. Gordon Childe,* ed. B. Wailes. University Museum Monograph 93. Philadelphia, PA: University Museum Publications.

Willies, L.
1995 Kestel Tin Mine, Turkey. Interim Report 1995. *Bulletin of the Peak District Mines Historical Society* 12(5):1–11.

Yener, K. A., and Vandiver, P. B.
1993a Tin Processing at Göltepe, an Early Bronze Age Site in Anatolia, *American Journal of Archaeology* 97:207–238.
1993b Reply to J. D. Muhly, "Early Bronze Age Tin and the Taurus." *American Journal of Archaeology* 97:255–264.

Yener, K. A.; Ozbal, H.; Minzoni-Deroche, A.; and Askoy, B.
1989 Kestel: An Early Bronze Age Source of Tin Ore in the Taurus Mountains, Turkey. *Science* 244:200–203.

Wagner, D.
1996 *Iron and Steel in Ancient China.* Leiden and New York: E.J. Brill.

Zettler, R. L., and Horne, L. (eds.)
1998 *Treasures from the Royal Tombs at Ur.* Philadelphia: University of Pennsylvania Museum of Archaeology and Anthropology.

1

Copper and Bronze in Cyprus and the Eastern Mediterranean

James D. Muhly

ABSTRACT Bronze metallurgy, it is argued, developed in the north Aegean (Troy, Thermi, Poliochni) as the result of a trade in metals, especially tin, coming across the Black Sea. The development of bronze metallurgy in Cyprus, on the other hand, came about through trade routes connecting Cyprus with Mesopotamia and the Levant. The ultimate source of the tin might have been the same in both cases, but the location of that source (or sources) remains elusive. Afghanistan seems a prime candidate. [Final ms. received 10/96.]

Thanks to archaeological discoveries and technological research over the past twenty years, the eastern Mediterranean basin has emerged as one of the major centers of metallurgical development in the ancient world. The old idea that Mediterranean, especially Aegean, metallurgy was ultimately derived from southern Mesopotamia, through Syrian and Anatolian intermediaries, can now be completely discounted. The copper technology now documented in southeastern Anatolia, at Çayönü Tepesi, is earlier than anything known from Mesopotamia (see Stech, this volume; also Muhly 1989). The beginnings of metallurgy in the Aegean world actually seem to owe their impetus to developments in the north, in southeastern Europe, rather than to any influence coming from the east (Muhly 1985b).

Mesopotamian developments were obviously of critical importance, especially in the fourth and third millennia, but they are not to be seen simply as the source for all Mediterranean and European metallurgical developments. Such simplistic approaches to questions of technological diffusion are gone for good.

The development of metal technology in the Aegean to the end of the Early Bronze Age, ca. 2200 B.C., has been studied in a recent Oxford doctoral dissertation by Veronica McGeehan-Liritzis (1988). The use of copper and bronze in Cyprus down to the end of the Middle Cypriot period has been traced in a recent Pennsylvania dissertation by Judith Weinstein Balthazar (1986; published 1990). Both studies indicate that it was only in the third millennium B.C., in the Early Bronze Age, that the use of metal became sufficiently widespread to acquire both social and economic significance within the contemporary societies of Greece and Cyprus (see also Stos-Gale and Gale 1994; Nakou 1995).

Whereas the copper metallurgy of the earlier Late Neolithic (Greece) or Chalcolithic (Cyprus) periods had probably made exclusive use of native copper, the Early Bronze Age was marked by the introduction of the mining and smelting of copper ores, especially arsenic-bearing copper ores, and the use of an alloy that has come to be widely known as arsenical copper, although some scholars still prefer the term arsenical bronze or even copper-arsenic bronze (for this problem, cf. Lechtman 1981:77–78).

It seems improbable that this alloy was produced through the deliberate addition of metallic arsenic, as there is no evidence that arsenic was ever recognized as a separate metal in the ancient world. It is more probable that arsenic-rich minerals, such as domeykite (Cu_3As), were added to molten copper, producing an arsenical copper alloy. Domeykite is a mineral with a wide distribution, from southern Iran (Talmessi mine near Anarak; Heskel and Lamberg-Karlovsky 1980:231) to South America (Lechtman 1980:303–304), but it is a rather rare mineral (see Pigott, this volume). Referring to the situation in the northern Andes of Peru, H. Lechtman has stated that "for northern metallurgists to have found domeykite would have been to find the proverbial needle in a haystack" (Lechtman 1980:304). Obviously a wide variety of such materials must have been used in the early production of arsenical copper.

The use of arsenical copper had a wide geographical as well as chronological distribution. Geographically it is attested from India to the British Isles; even in the New World arsenical copper served as the basic alloy used by the ancient Incas of the central Andes (Lechtman 1980, 1981, 1991; Shimada and Merkel 1991). Its absence in Southeast and East Asia raises interesting historical and technological problems that cannot be dealt with at this time (cf. Muhly 1988).

Along the North Coast of Peru the use of arsenical copper seems to have begun some time in the mid-first millennium A.D., during the second half of the Early Intermediate Period, ca. A.D. 200–600 (Lechtman 1980:302). In the Old World the use of arsenical copper began at least as early as the mid-fourth millennium B.C. It is best documented at that time by the metal artifacts from the Ghassulian Culture of Chalcolithic Palestine. Arsenical copper seems to have been the dominant alloy used in the hoard of 416 copper-base artifacts from the Nahal Mishmar Cave, dated to the mid-fourth millennium B.C. (Bar-Adon 1980; Moorey 1988), and it was also used in related objects from nearby sites (Levy and Shalev 1989). New analytical work has revealed that many of the objects from the Nahal Mishmar hoard are made of a complex alloy containing high amounts of antimony in addition to arsenic. Copper sources in eastern Anatolia and the Caucasus seem to provide the only copper ores capable of producing such an alloy (Tadmor et al. 1995). Should this prove to be correct, it would show that connections between the eastern Mediterranean and the Caucasus go back into the fourth millennium B.C.

In eastern Anatolia the hoard from Arslantepe, level VIA, demonstrated that arsenical copper was in use there by at least the late fourth millennium B.C. (Caneva et al. 1985) and it remained a dominant alloy of Anatolian copper metallurgy during the following Early Bronze Age (de Jesus 1980:364–368). In view of the predominant use of arsenical copper in late Chalcolithic and Early Bronze Age Anatolia, it is difficult to explain why the first use of this alloy occurs in Cyprus only at the beginning of the Cypriot Early Bronze Age, ca. 2400 B.C. The beginning of the Early Bronze Age in Cyprus is generally held to have come about as a response to strong Anatolian influence, perhaps even as the consequence of an Anatolian colonization of the island (Gjerstad 1980; for a different point of view, cf. Knapp 1990; 1994:414–421). (On the basis of evidence from their excavations at Marki *Alonia*, D. Frankel and J. M. Webb are preparing a new study of the role of Anatolia in the development of the Cypriot Early Bronze Age. For now, see Frankel and Webb 1994.) This would seem to provide a very attractive agreement between archaeology and technology, with Anatolian Red Polished Wares appearing in Cyprus together with "Anatolian" arsenical copper. The problem comes with the so-called Philia Culture, usually seen as representing a transitional stage between the late Chalcolithic and the Early Bronze Age in Cyprus (for recent literature see Muhly 1993:242–243). The Anatolian influences upon the Philia Culture seem to be apparent everywhere, in pottery, metal typology, and even in architecture. To specific examples already discussed by Stuart Swiny (1985a:20–22; 1986:35–40) can be added the copper spiral "hair-ring" from Grave 529 at Kissongerga *Mosphilia* (Peltenburg 1985:235) as well as a copper awl or borer with a bone handle and a fragment of a flat axe or adze, both from the same site (Peltenburg 1985:62). All three of these copper finds were found in association with the Red Polished Wares characteristic of Philia Culture. They were published too recently to be included in the study by Balthazar (1990) mentioned above (for early metal finds from Cyprus, see also Muhly 1991).

Thus, it would all seem to come together very well, save for the fact that the metal artifacts of the Philia Culture appear to be made not of arsenical but rather of unalloyed copper. In contrast, on the south Anatolian coast, at Cilician Tarsus, the site that has traditionally provided the most numerous and most convincing Cypriot connections, some 28% of the roughly contemporary (EB II) metal artifacts contained arsenic. The problem has been set out by Stuart Swiny, who remarks that the situation "would suggest that if the Philia Culture was in fact introduced from Anatolia, its propagators drastically changed their metallurgical practices upon arrival in Cyprus" (1982:71).

It might be argued that this came about through simple necessity, that a technological change was dictated by the lack of arsenic-bearing copper ores in Cyprus, but this is not the case. Although not common, copper ores containing up to 7.64% arsenic have been identified in the Limassol Forest area southwest of the Troodos massif (Swiny 1982:71).

Nor is it certain that all metal artifacts belonging to the Philia Culture were made of unalloyed copper. P. T. Craddock analyzed two examples from J. R. Stewart's excavations at Nicosia *Ayia Paraskevi*, presently in the Birmingham Museum: a knife containing 89.0% copper and 9.0% arsenic and a hook-tanged dagger containing 96.0% copper and 3.90% arsenic (Craddock 1981:78, table 2; Balthazar 1990:103 and 54, table 38). Both objects are from Tomb 12 at Nicosia *Ayia Paraskevi*, usually dated to the Philia Culture (Stewart 1962:384; Balthazar 1990:103). There is also a toggle pin from Tomb 103 at Vasilia *Evreman*, now in the Ashmolean Museum, containing 3.30% arsenic (Craddock 1986:157), but the *Evreman* tombs are almost certainly to be placed in early Middle Cypriot and are therefore unrelated to the Philia Culture (Swiny 1985a:23–24; Muhly 1991:370).

This situation is indicative of the uncertainty presently surrounding the Philia Culture and the beginnings of the Early Bronze Age in Cyprus. Some of the material from the tombs at Nicosia *Avia Paraskevi*, including the two weapons discussed above, could actually come from the beginning of Early Cypriot; they would then be contemporary with some of the tombs from Bellapais *Vounous* (Swiny 1985b:116), which also include a number of objects made of arsenical copper (Swiny 1982:70, table 1; Craddock 1986:157, table 2; Craddock 1981:78, table 1; for discussion of this complex problem see Philip 1991, esp. pp. 74–75). It is quite possible, therefore, that arsenical copper represents a technology that came to Cyprus from Anatolia, although produced making use of local Cypriot arsenic-bearing ores.

From the Early Cypriot period on, Cyprus developed its own distinctive metallurgical traditions. Within the Early Cypriot (EC) the most interesting development, and one that required some sort of contact with the world outside Cyprus, was the introduction of tin and the resulting use of bronze. The tin necessary for making bronze could not have come from Cyprus itself, since tin forms in the crust of the earth in association with granite rock and there is no granite in Cyprus (Muhly 1985a:277). The appearance of bronze in Cyprus may go back to the time of the Philia Culture, but the analytical evidence for this has yet to be published and the objects in question are almost certainly imports from Anatolia (Balthazar 1990:105–106). On the basis of present analytical data, bronze first came into use in Cyprus during the EC III/MC I transitional period. By the Middle Cypriot (MC) period proper, bronze had established itself as a basic alloy, although arsenical copper still continued in use (Swiny 1982).

Of the handful of bronzes from the EC III/MC I transition at least two have traditionally been seen as imports from Minoan Crete (Catling and Karageorghis 1960:110–111, nos. 2 and 3). They are two daggers from Bellapis *Vounous*: one, no. 32 from Tomb 143, excavated by J. Stewart, has 11% tin (Balthazar 1990:135 and 44, table 21) and the other, no. 89 from Tomb 19, excavated by P. Dikaios, has 10% tin (Balthazar 1990:121 and 41, table 18). There are other examples of bronzes from similar contexts, including a pin, no. 81 from Vounous Tomb 33, with 11.7% tin (Balthazar 1990:123–124 and 41, table 17), and a pair of tweezers, no. 84 from Vounous Tomb 313B, with ca. 8% tin (Balthazar 1990:144 and 42, table 19) that should probably also be seen as imports.

This raises the question of sources of tin and possible centers of bronze production in the third millennium B.C. Early Minoan (EM) Crete was certainly making some use of bronze, beginning in the EM II period, perhaps even EM I (Muhly 1991; Watrous 1994:703), but all such production depended of necessity upon imported tin (Branigan 1982:209; Muhly 1991:364–365). There is a fair amount of evidence for contact between Crete and Cyprus in the late third millennium B.C., demonstrated most recently by the fragments of an Early Cypriot III vase discovered among sherds from the excavations of Sir Arthur Evans at Knossos (Catling and MacGillivray 1983), but it is very doubtful that bronze metallurgy was introduced into Cyprus from Crete.

Of particular interest is the situation in the Cyclades and the northern Aegean. During the Early Cycladic period the predominant alloy was arsenical copper; bronze was virtually unknown (Muhly 1985b:126). Analytical work has shown that arsenic-bearing copper ores were being smelted on the island of Kythnos, producing slags with entrapped copper prills that contained up to 7.23% arsenic (Gale, Papastamataki, et al. 1985:89–90; Gale and Stos-Gale 1989:tables 5 and 6). The resulting slag heaps, especially that known as Skouries, on the east coast of the island north of Ayios Ioannis Bay, have produced evidence dating the smelting operations to the Early Cycladic II period (Gale and Stos-Gale 1989:27; Stos-Gale 1989; for archaeology of site see Hadjianastasiou and MacGillivray 1988). A visit to the Skouries site by the present author, together with Philip Betancourt (August 1996), confirmed the extraordinary importance of this remarkable EBA copper smelting site. The technology of copper recovery involved the crushing of the slag to recover the entrapped prills, a technique not well documented at Chalcolithic and Bronze Age sites throughout the eastern Mediterranean, Cyprus, and the Levant (Gale and Stos-Gale 1984:267), but one that is also attested at the Early Minoan III site of Chrysokamino in the Bay of Mirabello, northeastern Crete (report on the 1996 season at this site, by P. Betancourt, J. D. Muhly, and C. Floyd, forthcoming in *Hesperia*). Moreover, the lead isotope composition of the copper in the ores and the slags from Kythnos seems to match that found in the hoard of arsenical copper artifacts bought many years ago by the British Museum and said to come from Kythnos (Gale and Stos-Gale 1984:268; 1989:29 and 31, fig. 8). The ten analyzed artifacts from the hoard, typologically dated to the Early Cycladic period, were all made of arsenical copper, averaging 3.2% arsenic (Muhly 1985b:126; Craddock 1976:98, 106).

It is reasonable to suggest, therefore, that the Kythnos hoard actually originated in Kythnos, and was even made of arsenical copper smelted locally (Gale and Stos-Gale 1984:267–268). Other examples of Early Cycladic metalwork, from Amorgos, Keos, and Syros, also seem to be made of Kythnean arsenical copper (Gale and Stos-Gale 1984:265; 1989:29). This would mean that being made of Kythnean copper need not mean that the objects themselves were made on Kythnos. It has been argued that the Kythnos hoard itself actually was found in the Cave of Zeus on the island of Naxos (Fitton 1989:38 and n. 31). The reasons given for this identification are interesting but not convincing, since the metal artifacts from the Cave of Zeus, excavated by K. Zachos, all seem to be of Final Neolithic date (Zachos 1990). Strangely enough, Fitton, the author of this study of the Kythnos hoard, does not seem even to be aware of the on-going archaeological investigation of the Naxian Cave of Zeus.

The article also contains an analytical appendix on the Kythnos hoard, by S. La Niece (in Fitton 1989:38–39). Although no explanation is provided, the analyses published therein are identical with those published by Craddock in 1976 (Craddock 1976:106). The only differences come in object no. 8 and these are only the result of an editorial error, with part of the data for object no. 9 being published for no. 8 as well. The new analyses of the Kythnos hoard are, therefore, nothing more than a re-publication of the earlier Craddock analyses. Fortunately none of this has any influence on the provenience argument connecting these artifacts with Kythnos copper; that is based entirely upon the lead isotope data.

But Kythnos seems to have nothing to do with the roughly contemporary metal artifacts from the Kastri settlement on the island of Syros. These do not seem to have been made of copper from Kythnos; more importantly, they constitute the only group of metal artifacts from the Early Bronze Age Cyclades containing significant amounts of tin. The fifteen artifacts excavated by E. M. Bossert in 1962 averaged 5.24% tin but also 1.63% arsenic (Bossert 1967; Muhly 1985b:127). In recent years this group of metal objects from Kastri has taken on a particular significance resulting from interest in the so-called Lefkandi I cultural assemblage and in theories proposing the presence of Anatolian colonists in the Cyclades during the Early Cycladic III period (beginning with Rutter 1979; most recently in Doumas 1988). In addition to the spectrographic analyses published by Bossart, the Gales have now published three different analytical discussions of the same Kastri samples, each time making use of a different analytical technique. This means that the same metal samples have now been studied by four different analytical techniques: emission spectroscopy (Bossart 1967:76, tables 1 and 2), x-ray fluorescence (Gale and Stos-Gale 1984:273, table 2), neutron activation analysis (Gale, Stos-Gale, and Gilmore 1985:148, table 2), and electron microprobe (Stos-Gale et al. 1984:41, table 3).

A comparative study of these analytical results raises some very disturbing questions regarding the accuracy of current techniques of metal analysis (for which see the earlier study by W. T. Chase [1974]), especially given the fact that in none of their studies do the Gales refer to or offer explanation for the very different analytical results published by them elsewhere. In fact, in one case (Gale, Stos-Gale, and Gilmore 1985:154, n. 28), they actually make a special point of the accuracy of their own results! The myriad of questions raised by these analytical discrepancies cannot even be touched upon at this time (cf. Muhly 1991). They do highlight once again the need for a comparative study of the accuracy of all analytical techniques now being used to study ancient metal artifacts (the results of just such a study, organized by V. Rychner, University of Neuchatel, were published by J. P. Northover and V. Rychner in 1998).

The Kastri fort on the island of Syros does indeed seem to represent a settlement established by a group of colonists with strong Anatolian connections, colonists also present at the contemporary Early Bronze II/III settlements at Lefkandi (Euboea), Manika (Euboea), and Rafina (Attica). These northwest Anatolian connections are especially strong in the associated ceramics (Rutter 1983), but are also to be seen in the metalwork, as demonstrated at Kastri and also in a study of the fifty Early Helladic (EH) metal artifacts from Manika (McGeehan-Liritzis 1988). The recent study by P. Sotirakopoulou (1993) convincingly relates this collection of "Kastri Group" sites to an EH II/III transitional period, but ignores the metalwork completely.

The parallels are quite convincing, but what are the proper historical, archaeological, and technological conclusions to be drawn from such evidence? The Gales have been quite critical of my earlier attempt to make sense of the Kastri analyses (Gale, Stos-Gale, and Gilmore 1985:154–155). To some extent our disagreement is only a matter of terminology. I would still maintain that arsenical copper can be considered a deliberate alloy in the sense that ancient metalworkers *deliberately* selected certain ores that, on empirical grounds, they knew were capable of producing a metallic copper superior to that derived from other ores, and superior in properties—hardness, castability—that were of central concern to an Early Bronze Age metalworker.

I would also still argue that low tin bronzes, consisting of 1 to 2% tin, are best seen as deriving from the remelting and mixing of scrap metal. I am less sure than I was before regarding the significance of bronzes that also contain appreciable levels of arsenic (i.e., greater than 1%), but I do believe that they must represent the addition of tin to arsenical copper. I would now admit, however, that the Kastri bronzes are Anatolian, specifically Troadic or north Aegean in origin, to be compared with the bronzes from Troy, Thermi, and Poliochni (cf. Begemann et al. 1995).

What distinguishes the metalwork of all these sites is that much of it is made of bronze, having tin in excess of 5.0%. During the Early Bronze Age, this corner of the northwest Aegean is the only region of the eastern Mediterranean that makes significant use of tin (Muhly and Pernicka 1992). Just how early the earliest bronze from Thermi, Troy, and Poliochni might be depends upon the relative stratigraphy and absolute chronology established for these sites, and there is still little agreement on some very basic issues. It is unlikely, however, that there is any bronze from the sites in question earlier than EB II (Stech and Pigott 1986:52–54; Muhly 1985a:283–285; Muhly 1985b:129).

Bronze is also very much in use in central Anatolia in the EB II period, along with arsenical copper. T. Stech and V. Pigott have proposed that this central Anatolian bronze industry made use of tin received from Mesopotamia in exchange for Anatolian silver (Stech and Pigott 1986). They believe, however, that the tin came to central Anatolia not directly, via an overland route across Syria and Cilicia, but by sea, from some Syrian port and thence to various Anatolian sites (Stech and Pigott 1986:54–55). Troy thus became an emporium supplying Mesopotamians with Anatolian silver and Anatolians with tin from Afghanistan (Stech and Pigott 1986:55–58; on the source of tin, see also Rossovsky et al. 1987; Potts 1994:157).

There is much to be said for the arguments advanced by Stech and Pigott, both for their suggestions on trade patterns and for their ideas that tin and bronze were status metals in the mid-third millennium B.C., such that "the use of tin . . . was as much a cultural matter as a technological one" (Stech and Pigott 1986:55). I am not convinced, however, that Afghan tin came to Troy via Mesopotamia. The Sumerians were people of marvelous ingenuity and enterprise, but I cannot imagine them sailing from Byblos to Troy. Nor, in a context roughly one millennium later, do I believe that the ancient Cypriots ever sailed from Enkomi or Kition to Troy (Muhly 1985c:41, n. 138). It seems much more reasonable to assume, following the work of Manfred Korfmann (1987), that Afghan tin came to Troy (and the north Aegean) across the Black Sea, from Kolchis to Troy.

It has long been recognized that Troy must have served as some sort of emporium, but the topographical setting of the site makes sense only within the context of trade through the Dardanelles and across the Black Sea. I fail to see how Troy's special position could be explained within the context of trade with Mesopotamia. The archaeological evidence also indicates that Troy was not the capital city of a land empire. During the Troy I–II phase the city seems to have stood alone, with no other sites in the area (save for near-by Beşiktepe) and no supporting hinterland (Doumas 1988:27; Özdoğan 1993:10–11).

In such a context it would make good sense to see Troy as the emporium from which north Aegean sites such as Thermi and Poliochni received Afghan tin and from which bronze metallurgy was brought by Anatolian colonists to Kastri on Syros and, perhaps, also to Lefkandi, Manika, and Rafina. This attractive reconstruction puts Troy at the center of movements between the north Aegean and the Black Sea. Machteld Mellink (1986) now argues, however, that Troy was definitely not the moving force behind this EB IIIA West Anatolian expansion. For Mellink, Troy actually was uninhabited during the EB II/III transitional period, a period dominated by West Anatolian seafarers who actually destroyed Troy I, putting an end to EB II occupation at that site (Mellink 1986). Troy II, for Mellink an EB IIIA settlement, is thus seen as a "triumphant station built by a successful warrior king of the new breed of West Anatolian navigators and traders" (Mellink 1986:151).

It may yet be possible to reconcile this new version of West Anatolian Early Bronze Age archaeology with the more traditional view of Troy taken by most Aegean archaeologists. Mellink's reconstruction certainly raises serious problems for the sequence at Lerna, where Caskey (1960) saw an unbroken EB II–III occupation, whereas Mellink would, presumably, see a long and crucial hiatus between Lerna III (representing the end of EB II) and Lerna IV, dated by Mellink (1986:144) to EB IIIB. What is of real importance, for the reconstruction offered here, is not so much the exact sequence at Troy itself but the historical role of the Troad and the north Aegean. In this regard Mellink's emphasis on the aggressive role played by West Anatolian seafarers is very much in the spirit of the interpretation presented here.

If bronze production at Troy II, during the EB III period, was made possible by tin imported by West Anatolian seafarers from the eastern end of the Black Sea, it seems necessary to assume that this trade had no effect upon settlements lying along the southern shore of the Black Sea, it being, in other words, a point-to-point trade. The reason for this is that at the EB III site of Ikiztepe, seven kilometers northwest of Bafra and the only site on the southern shore for which we have any analytical data, the extensive metal industry (Bilgi 1984) is said to have been based on arsenical copper. Nevertheless, although the present director of excavations at the site, Önder Bilgi, states that at Ikiztepe "copper is alloyed only with arsenic"(Bilgi 1984:73), he does, in fact, publish EB III artifacts containing significant amounts of tin (Table 1.1).

TABLE 1.1
IKIZTEPE EB III ARTIFACTS WITH
MORE THAN 2.0% TIN

ARTIFACT NO.*	% Cu	% Sn	% As
7	113.6	2.2	2.1
10	93.6	5.8	2.6
17	103.2	4.1	4.51
79	102.5	3.7	2.39
126	93.6	2.2	2.10
241	78.5	4.9	1.41

*Numeral designation of each artifact as in Bilgi 1984

There are even artifacts dating to EB II that contain tin: (1) an awl (Bilgi 1984:no. 77) with 97.5% Cu, 13.7% Sn, and 2.74% As, and (2) a chisel (Bilgi 1984:no. 124) with 96.7% Cu, 9.4% Sn, and 1.39% As.

Although these analyses present obvious problems, since in many examples the compositions total well over 100% (and one, no. 161, has 205.4% Cu and 32.9% Sn), they also bear an uncanny resemblance to the mixed tin-arsenic alloys known from Kastri on Syros. Stech and Pigott (1986:54) maintain that "the exchange of tin was not a widespread phenomenon, not even within the general cultural sphere of Troy, but directed toward that site alone (and perhaps others very close by, like Beşiktepe)" and that "arsenical copper was the dominant metal used at Ikiztepe." Such statements are made in support of their thesis that bronze at EBA Troy was a non–Aegean-Anatolian phenomenon, one best seen in terms of a Trojan–southern Mesopotamian metals trade.

On the contrary, developments at Troy seem to fit better within the archaeological-analytical continuum now being reconstructed for a north Aegean–Black Sea cultural complex. The EB II/III Anatolian intrusion, represented by the Lefkandi I cultural complex and best seen at Kastri on Syros, seems to represent an extension of this northern complex south into the Cyclades and central Greece. Perhaps Troy played only a secondary role in this expansion, but it was definitely an expansion with strong metallugical connotations. The interaction, even intermingling, of tin and arsenic alloying traditions within this cultural continuum remains an unsolved problem. We are only just beginning to appreciate the complexity of how copper alloys developed and were used during the late Chalcolithic and Early Bronze Ages.

What then of Cyprus? If the tin used at Troy, Thermi, and Kastri came across the Black Sea from Kolchis, it is most unlikely that this northern bronze industry had anything to do with the introduction of bronze to Cyprus. Although the beginning of the Early Bronze Age in Cyprus has, as we have seen, long been associated with influences from Anatolia, those influences were coming from Cilicia and from the site of Tarsus in particular (Gjerstad 1980). It would make sense to have tin coming from the same general area, from the Bolkardağ region of the Taurus mountains (Yener and Özbal 1987; Yener et al. 1989). The difficulty with this is that the traces of cassiterite and stanite identified in the Bolkardağ ores to date do not represent a convincing tin source within an Early Bronze context (cf. Muhly et al. 1991). That conclusion has been sharply contested by the team working on the alleged Taurus tin deposits (cf. Yener and Vandiver 1993; Earl and Özbal 1996; Kaptan 1995), but the present author would still insist that no convincing evidence for an Early Bronze Age Anatolian tin source has yet been presented (Muhly 1993).

In my 1991 paper, written together with German and Turkish colleagues, it was proposed that Kestel was being worked as a gold mine, not a tin mine (Muhly et al. 1991:213–214). The important point that needs to be emphasized here is that no one is contesting the evidence for Early Bronze Age mining activity at Kestel. That activity is clearly placed in the third millennium B.C., a date supported both by the pottery found in the mines, especially in Mine 1 (Willies 1995:5), and by a series of radiocarbon dates (Willies 1995:10). Kestel now provides the best available documentation for Early Bronze Age mining technology in Turkey. The problem comes in identifying the material being mined. This is true not only for Kestel; it is a widespread phenomenon (Craddock 1995:10–11; Willies 1995:8–10). The possibility that Kestel was, at least in the beginning of mining activity there, being worked as a gold mine is now being given serious consideration by the excavators themselves (Willies 1995:8–9; Earl and Özbal 1996:295). If Kestel was being exploited for gold, then the Taurus could not have served as a source of tin for Early Bronze Age Cyprus.

Bronze certainly was being used in Cyprus during the entire Middle Cypriot period, especially for a series of fourteen magnificent shaft-hole axes. These axes have often been seen as Syrian imports, but there is no reason why they could not have been of local Cypriot manufacture (cf. Åström 1977:11–13, 37, 41; Buchholz 1979; Swiny 1982:73–74). Presumably all of these axes were made of bronze; the six examples that have been analyzed had amounts of tin ranging from 8.2% to 17.9%.

The beginnings of bronze metallurgy in Cyprus during the Early Cypriot period, and its expansion during the Middle Cypriot, are developments that had nothing to do with Troy, Thermi, and the bronze metallurgy of Early Bronze Age Anatolia. The emergence of a Cypriot bronze industry in the latter part of the third millennium B.C. must, in fact, be related to contemporary developments in Syria, Palestine, and, ultimately, in southern Mesopotamia. Recent analytical evidence has documented the existence of a number of bronze artifacts dating to the Levantine EB IV period, roughly 2200–2000 B.C. (Stech et al. 1985; Merkel and Dever 1989; Philip 1991:67–68). These seem to constitute the earliest bronze from Syria and Palestine, thus making the first use of bronze in the Levant contemporary with that in Cyprus. Bronze metallurgy developed even earlier in Mesopotamia and was well established by the time of the Royal Cemetery of Ur, ca. 2600–2500 B.C. (cf. Stech, this volume; also Lutz and Pernicka 1996:322).

The spread of metal technology in the Levant during the EB IV (also known as MB I) period has been discussed on many occasions, and typological comparisons have been drawn "from at least Baluchistan to Cyprus and the Aegean and south through Palestine into Egypt" (Braidwood and Braidwood 1960:522). The distinctive range of artifacts includes such items as pins (especially toggle pins), needles, chisels, reamers, several different types of axes, daggers, swords, and both poker-butted and socketed spearheads (for examples of all types, see Philip 1989). The spread of this metal industry into Cilicia is documented by the objects from a hoard found at Soloi-Pompeiopolis, most likely in 1902 (Bittel 1940). The 77 metal artifacts making up this hoard, reportedly found together in a clay vase, are typologically identical to groups of EB IV metalwork from Byblos, Til Barsip (Hypogeum), Jericho, Tell Hesi, and a group of sites near Carchemish. Of the ten Soloi artifacts that were analyzed, five were made of bronze, averaging 9.1% tin (Bittel 1940:201). All the analytical evidence available indicates that the use of bronze was an important aspect of all EB IV metal industries.

The origins of this EB IV metal industry cannot be discussed at this time. What I would like to suggest is that the spread of this industry had a profound impact upon the development of the contemporary metal industry in Cyprus, an island that is, after all, positioned midway between Syria and Cilicia. The EB IV burial cave near 'Enan, north of Hazor, excavated in 1982, shows remarkable parallels with Cyprus, both in the style of burial, in a bilobate tomb, and in the collection of metal artifacts that accompanied the burials (Eisenberg 1985). Again, many of these artifacts were made of bronze (Stech et al. 1985).

The source of tin used in that bronze is almost certainly the same as that used in making the bronzes from Soloi-Pompeiopolis and from the Levant. The bronze industry of Cyprus must have made use of the same source or sources of tin. Just where the tin came from remains to be determined, but I am convinced that it must have come into the eastern Mediterranean by a trade route that went overland, from southern Mesopotamia to Syria, one that must ultimately have extended far to the east, perhaps all the way to Afghanistan.

The extent of these eastern tin deposits, located in Central Asia and Afghanistan, has been established by a group of scholars working at the Institute of Archaeology in the Tadjik Academy of the Sciences in Dushanbe, Tadjik SSR (Rossovsky et al. 1987). They conclude that the tin-ore mineralization of the southern Pamirs and the Hindu Kush is very much like that of Burma, Malaysia, and Indonesia (the main tin-producing regions of the world today), whereas that of southwest Afghanistan has much in common with the famous tin-ore region of Hedzu, South China (Rossovsky et al. 1987:176). They also call attention to the existence of a large tin deposit near Hazara, Pakistan, containing reserves estimated at 500,000 tons of ore containing 7.2–8.0% tin (Rossovsky et al. 1987:174).

It is within such regions that we must look for the sources of the tin that supplied the bronze industries of Mesopotamia, Syria, the Levant, Cyprus, and Anatolia. Much of the trade involving this tin seems to have been a seaborne trade, either across the Indian Ocean and the Arabian Sea into the Persian Gulf, tin that Gudea of Lagash knew as tin from Meluhha (Weisgerber 1986; Potts 1994:153–159), or across the Black Sea into the north Aegean. The voyage of Jason and the Argonauts, a story well known to Homer in the eighth century B.C. (Odyssey 12.70; cf. Kirk 1962:40; Heubeck and Hoekstra 1989:121) might well have as its historical background a metals trade that went back at least as early as the third millenium B.C.

REFERENCES CITED

Åström, P.
1977 *The Pera Bronzes*. Scripta minora, Royal Society of Letters at Lund, 1977–1978:4.

Balthazar, J. W.
1990 *Copper and Bronze Working in Early Through Middle Bronze Age Cyprus*. Studies in Mediterranean Archaeology and Literature, Pocket-book 84. Jorsered: Paul Åströms Förlag.

Bar-Adon, P.
1980 *The Cave of the Treasure: The Finds from the Caves at Nahal 'Mishmar*. Jerusalem: Israel Exploration Society.

Begemann, F.; Pernicka, E.; and Schmitt-Strecker, S.
1995 Thermi on Lesbos: A Case Study of Changing Trade Patterns. *Oxford Journal of Archaeology* 14:123–136.

Bilgi, Ö.
1984 Metal Objects from Ikiztepe—Turkey. *Beiträge zur allgemeinen und vergleichenden Archäeologie* 6:31–96.

Bittel, K.
1940 Depotfund von Soloi-Pompeiopolis. *Zeitschrift für Assyriologie* 46:183–205.

Bossert, E. M.
1967 Kastri auf Syros. Vorbericht über eine Untersuchung der prähistorischen Siedlung. *Archaiologikon Deltion* 22A:53–76.

Braidwood, R., and Braidwood, L.
1960 *Excavations on the Plain of Antioch*. Vol. 1: *The Earlier Assemblages*. Oriental Institute Publications 61. Chicago: University of Chicago Press.

Branigan, K.
1982 Minoan Metallurgy and Cypriot Copper. Pp. 203–211 in *Early Metallurgy in Cyprus: 4000–500 BC*, eds. J. D. Muhly, R. Maddin, and V. Karageorghis. Nicosia: Pierides Foundation.

Buchholz, H.-G.
1979 Bronzene Schaftrohräxte aus Tamassos und Umgebung. Pp. 76–88 in *Studies Presented in Memory of Porphyrios Dikaios*, ed. V. Karageorghis. Nicosia: Lions Club.

Caneva, C.; Frangipane, M.; and Palmieri, A. M.
1985 I metalli di Arslantepe nel quadro dei più antichi sviluppi della metallurgia vicono-orientale. *Quaderni de La Ricerca Scientifica* 112:115–137.

Caskey, J. L.
1960 The Early Helladic Period in the Argolid. *Hesperia* 29:285–303.

Catling, H. W., and Karageorghis, V.
1960   Minoika in Cyprus. *Annual of the British School at Athens* 55:109–127.

Catling, H. W., and MacGillivray, J. A.
1983   An Early Cypriot III Vase from the Palace at Knossos. *Annual of the British School at Athens* 78:1–8.

Chase, W. T.
1974   Comparative Analysis of Archaeological Bronzes. Pp. 148–185 in *Archaeological Chemistry*, Vol. I, ed. C. W. Beck. Advances in Chemistry Series No. 138. Washington, DC: American Chemical Society.

Craddock, P. T.
1976   The Composition of the Copper Alloys Used by the Greek, Etruscan, and Roman Civilizations. Part I. *Journal of Archaeological Science* 3:93–113.
1981   Report on the Composition of Metal Tools and Weapons from Ayia Paraskevi, Vounous, and Evreti, Cyprus, in the Birmingham Museum. Pp. 77–78 in *A Catalogue of Cypriot Antiquities in Birmingham Museum and Art Gallery*, ed. E. Peltenburg. Birmingham: Birmingham Museum and Art Gallery.
1986   Report on the Composition of Bronzes Excavated from the Middle Cypriot Site at Episkopi Phaneromeni and Some Comparative Cypriot Bronze Age Metalwork. Pp. 153–158 in *The Kent State University Expedition to Episkopi Phaneromeni*, Part II, by S. Swiny. Studies in Mediterranean Archaeology 74(2). Nicosia: Paul Åströms Förlag.
1995   *Early Metal Mining and Production*. Edinburgh: Edinburgh University Press.

Doumas, C. G.
1988   EBA in the Cyclades: Continuity or Discontinuity. Pp. 21–29 in *Problems in Greek Prehistory. Papers Presented at the Centenary Conference of the British School at Athens, Manchester, April 1986*, eds. E. B. French and K. A. Wardle. Bristol: Bristol Classical Press.

Earl, B., and Özbal, H.
1996   Early Bronze Age Tin Processing at Kestel/Göltepe, Anatolia. *Archaeometry* 38:289–303.

Eisenberg, E.
1985   A Burial Cave of the Early Bronze Age IV (MB I) near 'Enan. *'Atiqot* (Eng. ser.) 17:59–74.

Fitton, J. L.
1989   *Esse Quam Videre*. A Reconsideration of the Kythnos Hoard of Early Cycladic Tools. *American Journal of Archaeology* 93:31–39.

Frankel, D., and Webb, J. M.
1994   Hobs and Hearths in Bronze Age Cyprus. *Opuscula Atheniensia* 20:51–56.

Gale, N. H.; Papastamataki, A.; Stos-Gale, Z. A.; and Leonis, K.
1985   Copper Sources and Copper Metallurgy in the Aegean Bronze Age. Pp. 81–101 in *Furnaces and Smelting Technology in Antiquity*, eds. P. T. Craddock and M. J. Hughes. British Museum. Occasional Paper No. 48. London.

Gale, N. H., and Stos-Gale, Z. A.
1984   Cycladic Metallurgy in the Prehistoric Cyclades. Pp. 255–276 in *Contributions to a Workshop on Cycladic Chronology*, eds. J. A. MacGillivray and R.L.N. Barber. Department of Classical Archaeology, University of Edinburgh.
1989   Some Aspects of Early Cycladic Copper Metallurgy. Pp. 21–37 in *Miniera y Metalurgia en las Antiguas Civilizaciones Mediterraneas y Europeas*, eds. C. Domergue and J. M. Blázques Martínez. Madrid: Ministerio de Cultura.

Gale, N. H.; Stos-Gale, Z. A.; and Gilmore, G. R.
1985   Alloy Types and Copper Sources of Anatolian Copper Alloy Artifacts. *Anatolian Studies* 35:143–173.

Gjerstad, E.
1980   The Origin and Chronology of the Early Bronze Age in Cyprus. Report, *Department of Antiquities, Cyprus* 1980:1–16.

Hadjianastasiou, O., and MacGillivray, S.
1988   An Early Bronze Age Copper Smelting Site on the Aegean Island of Kythnos. Part Two: The Archaeological Evidence. Pp. 31–34 in *Aspects of Ancient Mining and Metallurgy: Acta of British School at Athens Centenary Conference at Bangor, 1986*, ed. J. E. Jones. Bangor: University College of North Wales.

Heskel, D., and Lamberg-Karlovsky, C. C.
1980   An Alternative Sequence for the Development of Metallurgy: Tepe Yahya,

Iran. Pp. 229–265 in *The Coming of the Age of Iron*, eds. T. A. Wertime and J. D. Muhly. New Haven, CT: Yale University Press.

Heubeck, A., and Hoekstra, A.
1989　*A Commentary on Homer's Odyssey.* Vol. II: *Books XI–XVI.* Oxford: Clarendon Press.

de Jesus, P.
1980　*The Development of Prehistoric Mining and Metallurgy in Anatolia.* BAR International Series 74. Oxford: British Archaeological Reports.

Kaptan, E.
1995　Tin and Ancient Tin Mining in Turkey. *Anatolica* 21:197–203.

Kirk, G. S.
1962　*The Songs of Homer.* Cambridge: Cambridge University Press.

Knapp, A. B.
1990　Production, Location, and Integration in Bronze Age Cyprus. *Current Anthropology* 31:147–176.
1994　The Prehistory of Cyprus: Problems and Prospects. *Journal of World Prehistory* 8:377–453.

Korfmann, M.
1987　Seefahrtbeziehungen zwischen Schwarzem Meer und Ägäis im 2. und 3. Jahrtausend v. u. Z.? P. 81 in *Sixth International Congress of Aegean Prehistory. Summary of Papers.* Athens: Ministry of Culture.

Lechtman, H.
1980　The Central Andes: Metallurgy without Iron. Pp. 267–334 in *The Coming of the Age of Iron*, eds. T. A. Wertime and J. D. Muhly. New Haven, CT: Yale University Press.
1981　Copper-arsenic Bronzes from the North Coast of Peru. *Annals of the New York Academy of Sciences* 376:77–122.
1991　The Production of Copper-arsenic Alloys in the Central Andes: Highland Ores and Coastal Smelters? *Journal of Field Archaeology* 18:43–76.

Levy, T. E., and Shalev, S.
1989　Prehistoric Metalworking in the Southern Levant: Archaeometallurgical and Social Perspectives. *World Archaeology* 20(3):352–372.

Lutz, J., and Pernicka, E.
1996　Energy Dispersive X-ray Fluorescence Analysis of Ancient Copper Alloys: Emperical Values for Precision and Accuracy. *Archaeometry* 38:313–323.

McGeehan-Liritzis, V.
1988　*The Role and Development of Metallurgy in the Late Neolithic and Early Bronze Age of Greece.* Unpubl. D. Phil. thesis, University of Oxford.

Mellink, M. J.
1986　The Early Bronze Age in West Anatolia: Aegean and Asiatic Correlations. Pp. 139–152 in *The End of the Early Bronze Age in the Aegean*, ed. G. Cadogan. Leiden: E. J. Brill.

Merkel, J. F., and Dever, W. G.
1989　Metalworking Technology at the End of the Early Bronze Age in the Southern Levant. *Newsletter, Institute for Archaeo-Metallurgical Studies* 14:1–4.

Moorey, P. R. S.
1988　The Chalcolithic Hoard from Nahal Mishmar, Israel, in Context. *World Archaeology* 20(2):171–189.

Muhly, J. D.
1985a　Sources of Tin and the Beginnings of Bronze Metallurgy. *American Journal of Archaeology* 89:275–291.
1985b　Beyond Typology: Aegean Metallurgy in its Historical Context. Pp. 109–141 in *Contributions to Aegean Archaeology: Studies in Honor of William McDonald*, eds. N. C. Wilkie and W.D.E. Coulson. Minneapolis, MN: Center for Ancient Studies.
1985c　The Late Bronze Age in Cyprus: A 25 Year Retrospect. Pp. 20–46 in *Archaeology in Cyprus 1960–1985*, ed. V. Karageorghis. Nicosia: Zavellis Press.
1988　The Beginnings of Metallurgy in the Old World. Pp. 2–20 in *The Beginning of the Use of Metals and Alloys*, ed. R. Maddin. Cambridge, MA: MIT Press.
1989　Çayönü Tepesi and the Beginnings of Metallurgy in the Old World. Pp. 1–11 in *Archäometallurgie der Alten Welt*, eds. A. Hauptmann, E. Pernicka, and G. A. Wagner. Bochum: Deutsches Bergbau-Museum.
1991　Copper in Cyprus: The Earliest Phase. Pp. 357–374 in *Découverte du métal*, eds. J.-P. Mohen and C. Eluère. Paris: Picard Editeur.

1993 Early Bronze Age Tin and the Taurus. *American Journal of Archaeology* 97:239–253.

Muhly, J. D.; Begemann, F.; Öztunalı, O.; Pernicka, E.; Schmitt-Strecker, S.; and Wagner, G. A.
1991 The Bronze Metallurgy of Anatolia and the Question of Local Tin Sources. Pp. 209–220 in *Archaeometry '90. International Archaeometry Symposium, Heidelberg, April 1990*, eds. E. Pernicka and G. A. Wagner. Basel: Birkhäuser.

Muhly, J. D., and Pernicka, E.
1992 Early Trojan Metallurgy and Metals Trade. Pp. 309–319 in *Heinrich Schliemann. Grundlagen und Ergebnisse moderner Archäologie 100 Jahre nach Schliemanns Tod*, ed. J. Herrmann. Berlin: Akademie Verlag.

Northover, J. P., and Rychner, V.
1998 Bronze Analysis: Experience of a Comparative Programme. Pp. 19–40 in *Atelier du bronzier en Europe du XXe au VIIIe siécle avant notre ére. Vol. 1: Les analyses de composition du mètal...*, ed. C. Mordant, M. Pernot, and V. Rychner. Paris: CTHS.

Nakou, G.
1995 The Cutting Edge: A New Look at Early Aegean Metallurgy. *Journal of Mediterranean Archaeology* 8:1–32.

Özdoğan, M.
1993 The Second Millennium of the Marmara Region. The Perspective of a Prehistorian on a Controversial Historical Issue. *Istanbuler Mitteilungen* 43:151–163.

Peltenburg, E. J.
1985 Lemba Archaeological Project, Cyprus 1983: Preliminary Report. *Levant* 17:53–64.

Philip, G.
1989 *Metal Weapons of the Early and Middle Bronze Ages in Syria-Palestine*. BAR International Series 526 (i & ii). Oxford: British Archaeological Reports.
1991 Cypriot Bronzework in the Levantine World: Conservatism, Innovation and Social Change. *Journal of Mediterranean Archaeology* 4:59–107.

Potts, T.
1994 *Mesopotamia and the East. An Archaeological and Historical Study of Foreign Relations ca. 3400–2000 BC*. Oxford University Committee for Archaeology, Monograph 37. Oxford: Oxbow Books.

Rossovsky, L. N.; Mogarovsky, V. V.; and Chymirev, V. M.
1987 The Metallogeny of Tin and Rare Metals in the Eastern Part of the Mediterranean Folded Belt. Pp. 170–177 in *Mineral Deposits of the Tethyan Eurasian Metallogenic Belt between the Alps and the Pamirs (Selected Examples)*, ed. S. Jankovic. Belgrade: UNESCO/ICCP Project No. 169.

Rutter, J. B.
1979 *Ceramic Change in the Aegean Early Bronze Age: The Kastri Group, Lefkandi I and Lerna IV: A Theory Concerning the Origin of Early Helladic III Ceramics*. Los Angeles: Institute of Archaeology.
1983 Fine Gray-burnished Pottery of the Early Helladic III Period: The Ancestry of Gray-Minyan. *Hesperia* 52:327–355.

Shimada, I., and Merkel, J. F.
1991 Copper-alloy Metallurgy in Ancient Peru. *Scientific American* 265(1):80–86.

Sotirakopoulou, P.
1993 The Chronology of the 'Kastri Group' Reconsidered. *Annual of the British School at Athens* 88:5–20.

Stech, T.; Muhly, J. D.; Maddin, R.
1985 Metallurgical Studies on Artifacts from the Tomb near 'Enan. *'Atiqot* (Engl. ser.) 17:75–82.

Stech, T., and Pigott, V. C.
1986 The Metals Trade in Southwest Asia in the Third Millennium B.C. *Iraq* 48:39–64.

Stewart, J. R.
1962 The Early Cypriote Bronze Age. Pp. 205–401 in *Swedish Cyprus Expedition*, Vol. 4, Part IA, eds. P. Dikaios and J. R. Stewart. Lund: Swedish Cyprus Expedition.

Stos-Gale, Z. A.
1989 Cycladic Copper Metallurgy. Pp. 279–291 in *Archäometallurgie der Alten Welt*, eds. A. Hauptmann, E. Pernicka, and G. A. Wagner. Bochum: Deutsches Bergbau-Museum.

Stos-Gale, Z. A.; Gale, N. H.; and Gilmore, G. R.
1984 Early Bronze Age Trojan Metal

Sources and Anatolians in the Cyclades. *Oxford Journal of Archaeology* 3(3):23–43.

Stos-Gale, Z., and Gale, N.
1994 Metals. Pp. 92–121 in *Provenience Studies and Bronze Age Cyprus. Production, Exchange and Politico-economic Change*, eds. A. B. Knapp and J. F. Cherry. Monographs in World Archaeology No. 21. Madison, WI: Prehistory Press.

Swiny, S.
1982 Correlations between the Composition and Function of Bronze Age Metal Types in Cyprus. Pp. 69–80 in *Early Metallurgy in Cyprus, 4000–500 BC*, eds. J. D. Muhly, R. Maddin, and V. Karageorghis. Nicosia: Pierides Foundation.
1985a The Cyprus American Archaeological Research Institute Excavations at Sotira *Kaminoudhia* and the Origins of the Philia Culture. Pp. 13–26 in *Proceedings of the Second International Congress of Cypriot Studies*, Vol. 1, eds. T. Papadopoullou and S. A. Chadzestylle. Nicosia: Society of Cypriot Studies.
1985b Sotira-*Kaminoudhia* and the Chalcolithic/Early Bronze Age Transition in Cyprus. Pp. 115–124 in *Archaeology in Cyprus, 1960–1985*, ed. V. Karageorghis. Nicosia: A. G. Leventis Foundation.
1986 The Philia Culture and its Foreign Relations. Pp. 29–44 in *Cyprus between Orient and Occident*, ed. V. Karageorghis. Nicosia: Department of Antiquities.

Tadmor, M.; Kedem, D.; Begemann, F.; Hauptmann, A.; Pernicka, E.; and Schmitt-Strecker, S.
1995 The Nahal Mishmar Hoard from the Judean Desert: Technology, Composition and Provenence. *'Atiqot* 27:95–148.

Watrous, L. V.
1994 Review of Aegean Prehistory III: Crete from Earliest Prehistory through the Protopalatial Period. *American Journal of Archaeology* 98:695–753.

Weisgerber, G.
1986 Dilmun—A Trading Entrepôt: Evidence from Historical and Archaeological Sources. Pp. 135–142 in *Bahrain through the Ages: The Archaeology*, eds. S. Haya Ali al Khalifa and M. Rice. Bahrain.

Willies, L.
1995 Kestel Tin Mine, Turkey. Interim Report 1995. *Bulletin of the Peak District Mines Historical Society* 12(5):1–11.

Yener, K. A., and Özbal, H.
1987 Tin in the Turkish Taurus Mountains: The Bolkardağ Mining District. *Antiquity* 61:220–226.

Yener, K. A.; Özbal, H.; Kaptan, E.; Pehlivan, A. N.; and Goodway, M.
1989 Kestel: An Early Bronze Age Source of Tin Ore in the Taurus Mountains, Turkey. *Science* 224:200–203.

Yener, K. A., and Vandiver, P. B.
1993 Tin Processing at Göltepe, an Early Bronze Age Site in Anatolia. *American Journal of Archaeology* 97:207–238.

Zachos, K. L.
1990 The Neolithic Period in Naxos. Pp. 29–32 in *Cycladic Culture. Naxos in the 3rd Millennium BC*, ed. L. Marangou. Athens: Goulandris Museum.

2

The Coming of Iron in the Eastern Mediterranean

Thirty Years of Archaeological and Technological Research

Jane C. Waldbaum

ABSTRACT Until the mid 1960s, theories on the advent of iron in the eastern Mediterranean were based on a number of largely unexamined assumptions, all predicated on the belief that iron was inherently superior to and more desirable than bronze as a utilitarian and military material. This article will review some of the major archaeological and technological research of the past three decades on the early production and use of iron and will show how this new research has affected our thinking on these previously held assumptions. It will examine some of the newer hypotheses currently being explored, and then point out some of the major unanswered questions that remain. [Final ms. received 8/96.]

Until the mid 1960s, thinking on the advent of iron was based on a number of largely unexamined assumptions, all predicated on the belief that iron was inherently superior to and more desirable than bronze as a utilitarian and military material: (1) that the beginning of the "Iron Age," when iron was introduced on a fairly wide scale, represented technological "progress" through the substitution of a "superior" material (iron) for an "inferior" one (bronze); (2) that the earliest iron used was meteoritic in origin; (3) that the earliest production of smelted iron took place under closely guarded "monopolistic" conditions, such as the supposed control by the Hittites of Anatolia over the supply of iron in the Bronze Age Mediterranean; (4) that iron only became more widely available when this monopoly was broken; and (5) that iron was disseminated into certain areas by invaders provided with the new, improved material—e.g., invaders from the north or "Dorians" in Greece, and invaders from the west or Philistines in Palestine.

Over the past thirty years or so, interest in the problem of iron and the reasons for its introduction to and spread through the eastern Mediterranean has been steadily growing. A number of studies have addressed themselves to the older assumptions, often drastically revising them, and have also advanced new theories and raised important new questions as to the relationships between the introduction of new materials and the cultural conditions that led to their adoption.

These studies have tended to take one of several approaches. The first (primarily archaeological) focuses on the objects: the types, functions, geographic and chronological distribution, and context of iron artifacts are documented and often compared to their counterparts in bronze. This has yielded valuable information on the appearance and uses of iron and on when, where, and for what purposes it began to replace bronze (see, e.g., Waldbaum 1978, 1980, 1982; Snodgrass 1980, 1982; McNutt 1990).

Perhaps ancillary to the archaeological approach is the philological, that is, the analysis of ancient texts referring to iron and other metals with a view to illuminating such matters as the relative values, uses, and systems of exchange of metals, which cannot always be determined by study of the objects alone. The relationship between the literary and the archaeological sources, however, while sometimes complementary, is often ambiguous and sometimes contradictory (e.g., Košak 1982, 1986; Limet 1984, 1986; Maxwell-Hyslop 1972, 1980; Heltzer 1977, 1978; Bjorkman 1973; Fensham 1969; Vaiman 1982; Zaccagnini 1990).

A third major avenue of research is technological. Through chemical and metallographic analysis of metal artifacts their relative effectiveness can be assessed, the introduction of new technologies can be observed, and patterns of application of these new technologies over time and geographic range can be documented. In the eastern Mediterranean, these studies often have been hampered by the lack of material available for destructive analysis and by the poor state of preservation of the remains. With few exceptions, researchers have not been able to sample large bodies of objects, nor indeed, to obtain material from all relevant areas. Nevertheless, analytical research, in particular that done by

R. Maddin, J. Muhly, T. Stech, V. Pigott, and others, has begun to elucidate some of the problems involved, at least in Cyprus (Maddin 1982; Stech et al. 1985; Åström et al. 1986), Palestine (Notis, Pigott, et al. 1986; Notis, McGovern, et al. 1986; Stech-Wheeler et al. 1981; Muhly et al. 1990), and to some extent, Hittite Asia Minor (Muhly et al. 1985). In addition, technological expeditions have also been mounted in the field to explore the distribution and availability of ore sources, fuels, and other ecological factors relevant to the production of metals, while examination of ancient mining operations and metalworking installations presents a clearer picture of available resources and production capabilities (e.g., Rothenberg 1988, 1990 and papers by Constantinou, Weisgerber, Koucky and Steinberg, Bachmann, Rothenberg, in Muhly et al. 1982). While this area of investigation is currently providing some of the most fruitful results, its application to early metallurgy in the eastern Mediterranean is still in its early stages and there are many questions left to be answered.

This article will review some of the major research on iron over the past three decades, show how this has affected our thinking on previous assumptions, examine briefly some of the newer hypotheses currently being explored, and then point out some of the major unanswered questions.

"SUPERIORITY" OF IRON TO BRONZE

One of the first developments to have a major impact on thinking about early iron was the demonstration, in literature accessible to non-scientists, that wrought iron, the most common product of direct iron smelting and forging, is not superior to hardened bronze until it has been steeled (carburized) and heat-treated (quenched and tempered) (Smith 1967: 40, fig. 39) or at least coldworked by hammering (Rehder 1992:44). Thus, the mere ability to smelt and forge iron did not by itself confer a strategic military edge over opponents possessing bronze alone until techniques of steeling had been mastered and consistently applied. This fact, while long known to metallurgists, was not until recently generally understood by archaeologists and historians, leading to a number of serious misconceptions in the earlier literature on the subject. The technological inferiority of early wrought iron is now generally accepted (see, e.g., Waldbaum 1978:68–69; 1980:87–88; Snodgrass 1980:337–338; McNutt 1990:113–114 and articles by Hartmann and Champion in Stig Sorensen and Thomas 1989), and has generated new theories for the adoption of iron as well as stimulating scientific research on the introduction of steeling techniques.

IRON IN THE BRONZE AGE: METEORITIC AND SMELTED

It has been shown that both meteoritic and smelted iron artifacts were known and used in most regions of the eastern Mediterranean during all of the Bronze Age and even before (Fig. 2.1). Iron appears sporadically over a wide span of time and space, ranging from a few specimens in fifth and fourth millennium contexts in northern Mesopotamia, Iran, and Egypt; to somewhat more from third millennium or Early Bronze Age (EBA) sites in Mesopotamia, Egypt, and Anatolia; to a few from the Middle Bronze Age (MBA; ca. 2000–1600 B.C.) in Egypt, Anatolia, Cyprus, and Crete; and considerably more from a wider variety of sites in the Late Bronze Age (LBA; ca. 1600–1200 B.C.) in Egypt, Anatolia, Syria, Palestine, Cyprus, Greece, Crete, and the Aegean Islands (Waldbaum 1980:79, table 3.4).[1]

By 1980, the total number of iron objects reported from Bronze Age contexts amounted to somewhat over 100, to which may be added another 50 or so that have come to my attention since then (see Appendix A). With the exception of a possible MBA tool blade from Pella in Jordan (App. A, no. 1), a "nail" from Middle Helladic (MH) Asine in Greece (App. A, no. 18), decorative studs on a box from Acemhöyük (App. A, no. 34), a blade from LB Beşiktepe in northwestern Turkey (App. A, no. 33), pieces of slag from LB Kāmid el-Lōz in Lebanon (App. A, no. 15) and Late Helladic (LH) III Tiryns in Greece (App. A, no. 26), and some small "tubes," "rods," and a rivet from Timna (App. A, nos. 7, 8, 12), all the recent additions to the repertoire are either jewelry pieces or fragments that cannot be identified.

In addition to the newly identified examples of early iron, several pieces, earlier said to be iron, have been shown not to be on recent examination. These include an LH laminated ring from Dendra in Greece (Varoufakis 1982:315), a so-called iron meteorite from Hagia Triada in Crete (Varoufakis 1982:317), and three LBA (Hittite period) artifacts from Alaca Hüyük in Turkey—a nail, a needle, and a bracelet (Muhly et al. 1985:71).

Even as late as 1980 little technological research had been carried out on this material, in part because its poor state of preservation precluded meaningful analysis, but also because of the difficulty in obtaining permission to sample these rare objects. Some of these early iron artifacts were analyzed for nickel under the assumption that significant nickel content indicated meteoritic origin. By this criterion, objects from fifth millennium Iran (Sialk); fourth

Figure 2.1 Map of the Near East and Eastern Mediterranean showing some Bronze Age and Early Iron Age sites at which iron has been found. (Prepared by the University of Wisconsin-Milwaukee Cartographic Services.)

millennium Egypt (Gerzeh); third millennium Mesopotamia (Uruk, Ur), Anatolia (Alaca Hüyük), and Egypt (Dier el-Bahari); and second millennium Syria (Ugarit) and the Tomb of Tut-ankh-amen in Egypt were deemed to be of meteoritic origin (Waldbaum 1980:72–79). Few of these were also analyzed metallographically to reveal the structure characteristic of meteoritic iron. More recently, Varoufakis has examined several of the iron objects from Bronze Age Greece. Although he was not able to perform metallographic analyses, he did take samples for atomic absorption analyses (Varoufakis 1981, 1982). Of the eight rings he examined, ranging from the fifteenth to the thirteenth century B.C. in date, seven contained nickel in quantities from 1.48 to 10.77%; two also contained cobalt (ring bezel from Kakovatos, Pylos, 2.25% Co; ring from Mycenae Tomb 68, 1.29% Co; App. A, no. 22), and one, from Dendra, was not iron at all but made of a copper hoop and lead bezel with an original silver layer that had corroded away. He also examined a small plate from LH IIIB Volos, and two knives from twelfth century B.C. (LH IIIC) Perati, and showed that they contained negligible quantities of nickel (Varoufakis 1982:315–316). Finally, a piece of iron wire wrapped around a gold ring from Late Cycladic (LC) IIB Kition (App. A, no. 35) contained only trace quantities of nickel (Karageorghis 1974:89, n. 2).

Although it is usually assumed that early iron artifacts containing nickel in quantities over one percent were made of meteoritic iron (Waldbaum 1980:69), Varoufakis considers whether these artifacts could have been produced by smelting nickel-rich iron ores such as the laterite deposits (which also contain cobalt) found at Atalanti and on Euboea in Greece (Varoufakis 1981:31; 1982:317 and see also Photos 1989). The point has also been raised by Piaskowski (1982), who performed metallographic analyses on a series of nickel-rich iron objects from Iron Age Europe and showed that they had, in fact, been smelted. Piaskowski proposed that such iron was produced by the smelting of "a mixture of an iron ore and a complex iron-nickel-cobalt-arsenic ore, most probably chloanthite" (Piaskowski 1982:242).

There are several objections (or perhaps cautions) to be noted here before we abandon the notion of Bronze Age use of meteoritic iron: first, Hittite, Egyptian, and Mesopotamian literary texts use terms for iron that seem to suggest knowledge of a meteoritic source for the material (Hittite "black iron of heaven"; Egyptian "iron of heaven"; Sumerian "iron from heaven" [Bjorkman 1973:113–114; Limet 1984:191]). Second, nickel-rich iron ores are very difficult to work (Piaskowski 1982:242; Photos 1989:418) and would seem an odd choice for inexperienced smiths. On the other hand, recent evidence does show that Bronze Age ironworkers did not always choose the most appropriate or easily worked ores (Muhly et al. 1985:77). Third, although there are very few analyses to go on, use of nickel-rich iron in the eastern Mediterranean seems to die out after the thirteenth century, at least in Greece (Varoufakis 1982:317). This could mean either that metalworkers were abandoning meteoritic iron as they became more familiar with smelting techniques or that they had learned to choose more easily worked ores with greater consistency. Only metallographic analyses of the known nickel-rich artifacts will resolve the issue.[2]

A number of tested objects from the Bronze Age, however, showed no nickel and hence (presumably) were the product of smelting operations, though whether they resulted from deliberate smelting of iron ores or were byproducts of copper or lead smelting is still unknown.[3] Pickles (1988:4–5) has suggested that early iron artifacts that do not contain nickel were made of native or telluric iron rather than smelted. Native iron is, however, extremely rare, found now primarily on Disco Island, Greenland (Wertime 1980:11) and also does contain some nickel and cobalt (Piaskowski 1982:238). Since it is relatively easy to produce small quantities of iron accidentally through the smelting of copper or lead (Wertime 1980:12–17; Pickles 1988:10; McNutt 1990:111–112), this seems more likely than use of native iron, which in any case should be sought among the objects containing small amounts of nickel and cobalt. Analysis of several hundred Late Bronze Age copper artifacts deposited as votive objects in the Egyptian temple at Timna in the Wadi Arabah of southern Palestine, in fact, showed "substantial quantities of metallic iron in the copper" (Rothenberg 1988:12), while analyzed iron artifacts from the site contained significant amounts of copper (Gale et al. 1990:186). Research at Timna involving experimental copper smelting operations suggests that iron can indeed become incorporated into the copper during smelting, either from iron oxides used as flux or from the ore itself (Craddock 1988:178–179; Gale et al. 1990:183–185). Furthermore, metallic iron could be recovered from iron-rich copper and forged into usable forms (Gale et al. 1990:185). The Timna analytical team concluded that the iron objects found at Timna "had been produced locally as an adventitious byproduct of copper smelting" (Gale et al. 1990:189). They strongly suggest "that iron smelting was first discovered in a copper smelting furnace" (Rothenberg 1988:12).

Examples of early iron containing no nickel came from Samarra (ca. 5000 B.C.); third millennium Mesopotamia (Tell Asmar and Chagar Bazar); Anatolia (Alaca Hüyük) and Egypt (Giza and Abydos); MBA Buhen (Nubia), Lapithos (Cyprus), and Pella (Jordan) (Waldbaum 1980:72–79; Smith et al. 1984 for Pella); and second millennium Palestine (Timna; Gale et al. 1990:186). Since some of the objects of smelted iron came from the same regions (Mesopotamia, Anatolia, Egypt) and even the same sites (Alaca Hüyük) as objects whose iron contained nickel, it does not seem possible to make regional distinctions with regard to available technologies.

Metallographic analyses have been performed on a few specimens of Bronze Age iron. An object identified by the excavator as a small blade or point was found in a MBA tomb at Pella in Jordan (App. A, no. 1). The piece was analyzed and shown to have been a steel containing ca. 0.8% carbon, which had probably been quenched. Not enough of the original outer layers of the object was preserved to determine whether it had also been tempered, nor could it be ascertained whether the steel was produced deliberately or accidentally. Nevertheless, the piece, if properly identified as ancient, would represent the earliest example of steel that has been recognized to date (Smith et al. 1984:234–235).[4] Since this is the only object to have been analyzed metallographically so far for which so early a date has been proposed, it must be viewed in isolation and cannot be used to claim priority in "invention" of steel technology for MBA Palestine. It is, at any rate, a manufactured piece and not meteoritic. Only time, and more analyses on more securely dated pieces, will show whether the object represents a true technological advance or must remain an anomaly.

A few artifacts of Hittite Empire date from Boğazköy and Alaca Hüyük have been sampled and analyzed metallographically (Muhly et al. 1985:76–79). One of these, a nail, was too corroded to yield information; two, a lugged axe and a knife blade, had been carburized, the former probably by accident, the latter by uncertain means; and one, a socketed point, had not been carburized (Muhly et al. 1985:78–79). On the basis of these, admittedly few, samples the analysts concluded that Hittite smiths could not yet fully control their product since they did not consistently carburize functional artifacts to produce steel (Muhly et al. 1985:79–80). For now, these are important as examples of smelted iron.

An iron ring from a Late Bronze Age context at Kāmid el-Lōz (ancient Kumidi) in Lebanon was found to be mildly carburized. It was one of only two finished iron objects from this period at the site. The other, a pin fragment, was too corroded to analyze (Frisch et al. 1985:146–148, and see below, App. A, nos. 13, 14).

A few fragmentary iron objects from Timna were examined metallographically. These included a small rod (Rothenberg 1988:148, no. 21; App. A, no. 8); a gilded earring (Rothenberg 1988:148, no. 22; App A, no. 9); a small tube embedded in a sea shell (Rothenberg 1988:148, no. 23; App. A, no. 7), and a tube with an iron rivet (Rothenberg 1988:168, no. 497; App. A, no. 12). All of these were completely rusted or mineralized and provided no information about the techniques used to work them (Tylecote 1988:186, 190).

Direct evidence for smelting in Bronze Age contexts is rare but not entirely lacking. I have already mentioned a piece of iron-arsenic slag found in a LH IIIB context, possibly a metal workshop, at Tiryns (Kilian 1983:304, 306, fig. 31; App. A, no. 26). To this may be added a fragment of speiss, one of several collected from a domestic context at Boğazköy, and interpreted as probably resulting from the smelting of an ore containing iron, arsenic, and sulfur. Muhly and colleagues point out that smelted iron containing arsenic deriving from such an ore would have been brittle and difficult to forge, perhaps showing that the smelters did not fully understand the nature of the ores they were using and did not always select appropriate ores (Muhly et al. 1985:77).

Speiss was also found in the Late Bronze Age palace workshop area at Kāmid el-Lōz (App. A, no. 15; Hachmann 1986:27 for the workshop), where it was interpreted as a possible byproduct of copper- or lead-working (Frisch et al. 1985:144–146, 158). Also found at Kāmid el-Lōz were lumps of hematite iron ore and the small pieces of finished iron already mentioned (App. A, nos. 13, 14; Frisch et al. 1985:77–78, 95–96, 105–108, 111). While the preponderance of the evidence from the Kāmid el-Lōz workshops was for copper production, and there was no direct evidence for iron smelting or forging, the excavators and analysts believe that some kind of iron metallurgy did take place there, perhaps utilizing the same furnaces as the copper operations and producing little or no slag from the iron-rich hematite ore (Frisch et al. 1985:178–180). Whether or not this hypothesis is valid must await further scientific discussion of the published evidence.

Another iron smelting operation is reported for the thirteenth century level at Tel Yin'am in Palestine (Liebowitz 1981:82–84; 1993; Liebowitz and Folk 1984). The nature of this installation has been called into question (Stech-Wheeler et al. 1981:261; Rothenberg 1983), though more recent investigators are somewhat more accepting of it as an early and somewhat tentative experiment with iron smelting (Muhly et al. 1990:164).

The evidence for Bronze Age iron production thus accumulated is still slim; it does show, however, that smelting took place in at least two different regions in the Late Bronze Age, and possibly in four.

Iron in the Bronze Age remained rare and expensive and, judging from context, its use was confined primarily to ornamental, ritual, and ceremonial functions rather than military or utilitarian ones, even in areas where smelting appears to have been known (Waldbaum 1978:17–23; 1980:69–82). This suggests either that these artifacts were not suitable for utilitarian purposes or that techniques for producing iron in quantities sufficient for such purposes had not yet been mastered.

THE "HITTITE MONOPOLY" ONCE AGAIN

Another area of research has been on the question of the supposed Hittite "monopoly" over the production and distribution of iron in the Late Bronze Age. I have elsewhere stated my objections to this hypothesis, noting the variety of actual iron objects and texts referring to iron objects available outside of Hittite Anatolia, as well as the essentially local nature of many of the artifacts from such places as Greece and Egypt (Waldbaum 1978:21; 1980:81). A number of iron artifacts, including utilitarian ones, have, indeed, been found in Hittite contexts. There are, however, at least a few objects of a functional nature outside of Hittite Anatolia (viz., the MB steel "blade" from Pella, the MH nail from Asine, and a previously recorded tool point from an LH jeweler's workshop in Kadmeian Thebes [Waldbaum 1980:77]). In no area, however, including Hittite Anatolia, does iron begin to approach bronze as the material of choice for artifacts of daily use, nor does it appear that iron was being exploited as a strategic material, an unlikely prospect in any case, until the techniques for consistent production of steel were mastered.

I have demonstrated above that smelted iron had been produced, at least in small quantities, much earlier than the time of the Hittite Empire, and that in the Late Bronze Age, or Hittite Empire Period, there is evidence for smelting not only at Boğazköy, the Hittite capital, but also in Greece and possibly Syria-Palestine. The few analyzed artifacts confirm that the Hittites were able to smelt, but not always to control the quality of finished products.

Hittite texts are often cited as evidence for Hittite expertise in iron production. Košak (1982, 1986) has collected a number of texts that testify to a far greater quantity and variety of artifacts than can be seen in the archaeological record. Some texts also speak of a specialized guild of blacksmiths as distinct from workers in other metals (Košak 1986:126; Zaccagnini 1990:500). Most of these documents are either palace inventories or ritual and festival texts, and Košak concludes that the items mentioned "served ornamental, prestigious and ceremonial purposes" (Košak 1986:134; but see Siegelová 1984:71–168 for a more utilitarian interpretation of the texts). That is, iron artifacts mentioned in the texts, while apparently more abundant, were no different in function from most of those found archaeologically. Texts referring to iron, sometimes in substantial quantities, are known from other Near Eastern kingdoms of the Bronze Age (Waldbaum 1978:17–18; 1980:75–76, 80; Limet 1984, 1986; Zaccagnini 1990: 500), and while these are somewhat fewer than their Hittite counterparts, they seem to reflect a similar attitude towards iron.

In sum, while the Hittites may have made somewhat greater use of iron than their neighbors, nevertheless, like other Bronze Age cultures, they seemed to esteem it more for its prestige value than for its utilitarian effectiveness. Although they could smelt it, it does not seem to have been readily available among the Hittites for routine purposes; nor does it seem to have been guarded as a military secret. In other words, there are no apparent technological, strategic, or economic reasons to postulate a Hittite monopoly on iron in the Late Bronze Age. My earlier skepticism about the existence of a Hittite monopoly still stands, and is supported by the results of both technological and philological investigations.

IRON AFTER CA. 1200 B.C.

I have previously shown that finds of iron artifacts throughout the eastern Mediterranean become gradually more abundant in relation to bronze for utilitarian and military purposes and appear in a greater variety of types as time goes on. Although the rates at which it was substituted for bronze vary somewhat from region to region, by the tenth century B.C. iron could be said to be in "common use" in most of the eastern Mediterranean (Waldbaum 1978:24–58).

In the past fifteen years a number of new sites producing metalwork have been published and the metal artifacts from them have been examined in some detail. This is not the place to catalogue all the individual objects discovered since the mid 1970s. Rather, I will discuss the data from a few significant, recently published sites from Palestine, Cyprus, and Greece, all of which yielded metal objects from the period in question and for all of which the excavators and other scholars responsible for publication provided scientific studies of the metals as well as full archaeological descriptions of quantities, types, and contexts for the objects included. Study of the material from these sites—the Baq'ah burial caves of Transjordan, Palaepaphos-*Skales* in Cyprus, Lefkandi in Euboea, and Nichoria in the southwest Peloponnese—provides much new information on many aspects of the problems involved in dealing with this period. Unfortunately, there is still little evidence for this period from well-documented sites in Egypt, Syria, or Anatolia,[5] nor is there much new to add to the picture for Crete and the Aegean Islands other than Euboea.[6]

TRANSJORDAN: BAQ'AH VALLEY BURIAL CAVE A4

Burial Cave A4 of the Baq'ah Valley in Transjordan, dating to the Iron IA period, twelfth century B.C., contained the remains of at least 233 burials including individuals of all ages and both sexes. Among the finds are some 134 metal objects, of which 98 or 73% are copper-base and ±36 or 27% are iron. All the metal artifacts are jewelry or dress fastenings, including remains of about 21 copper-base anklets or bracelets and 30 iron anklets or bracelets of which 8 are complete; 35 copper-base rings and 6 iron rings of which 3 are complete; 35 copper-base earrings; 5 copper-base toggle pins (Fig. 2.2), and 2 small "metal" squares of indeterminate use. In contrast with earlier Cave B3 at the same site, no weapons or tools were found in either metal (McGovern 1986:258–267, figs. 82–86).[7] The burials in this cave were not wealthy. The amount of metal deposited averages less than one piece per individual and no artifacts in luxury materials such as gold or ivory were found. There were also few signs of foreign contact, making it likely that the metal artifacts found in the cave were of local manufacture (McGovern 1986:338).

Five of the iron anklets or bracelets were analyzed (App. B, no. 27). According to proton-induced x-ray emission spectrometry (PIXE) analysis, none contain nickel and one has a high cobalt level (ca. 0.40%).

Figure 2.3   Mild steel anklet or bracelet from Baq'ah Cave A4, no. A4.226; University Museum no. 81-6-55, ca. 1200–1050 B.C. (McGovern 1986:pl. 29b). (Photo courtesy P.E. McGovern.)

Four were found on metallographic examination to be mild steel (Fig. 2.3), although without signs of quenching and tempering (Notis, Pigott, et al. 1986:272–278).

The finds from Cave A4 boost the numbers and proportions of iron in relation to bronze for twelfth century Palestine considerably beyond what has been reported earlier (Waldbaum 1978:27, table III.1, 36, table III.10, 39–42). These figures will be further augmented when the finds from two new sites are published. The first is a similar Early Iron Age burial cave at Pella in the Jordan Valley. This cave was discovered in 1987 and contained the remains of over 100 burials and a "large group of iron artifacts" (number unspecified). According to McGovern (1988:52), "[t]he vast majority of the iron artifacts from the Pella tomb were anklets/bracelets and rings of the same types as those from the Baq'ah tomb." The second is a cave tomb at the site of Khirbet Nisya, about eight miles north of Jerusalem. Although pottery from the tomb was poorly preserved, it appears to date to the Iron I period. Remains of some 50–55 disarticulated individuals were found. The metal finds consisted primarily of jewelry in both bronze and iron: 33 bronze items and 18 iron, including 10 pieces identified as bracelets, 1 ring, and several fragments. One of the iron bracelets was analyzed by the team of Vincent Pigott at MASCA and Michael Notis at Lehigh University, and found to be "very similar" in structure to those in the Baq'ah Valley tomb. (The site of Khirbet Nisya is being excavated under the direction of David Livingston under the auspices of the Associates for Biblical Research. I owe the information given here to Gary Byers [see also Byers 1995].)

The exclusively ornamental object types represented by the iron finds from Baq'ah Cave A4, Pella, and

Figure 2.2   Bronze toggle pin from Baq'ah Cave A4, no. A4.223; Cu 87.3%, Sn 11.1% (Notis, McGovern, et al. 1986:279, 282, pl. 33m). (Photo courtesy P. E. McGovern.)

Khirbet Nisya suggest usage more in keeping with Bronze Age practice than with that of the true Iron Age. The practice of steeling such items as bracelets is curious, since there is no obvious functional advantage in terms of hardness or strength to be gained by so doing. On the other hand, if the steeling were indeed deliberate, it might have been employed to impart desirable color or surface texture to the objects, producing a shinier, "jinglier" bangle than dull wrought iron (Notis, Pigott, et al. 1986:276). Alternatively, the shiny metal may have had an apotropaic function (Stager 1985:10). In any case, precedents for steeling in Palestine may exist in the form of the MBA "blade" from Pella cited above (Smith et al. 1984)[8] and a recently discovered pick from Mt. Adir in the northern Galilee, dated by the excavator to the twelfth century, but probably belonging to the eleventh (Amihai Mazar, pers. comm. 1988; and see Muhly et al. 1990:160).[9] The pick, upon examination, was found to have been carburized, quenched, and tempered (App. B, no. 28).

CYPRUS: PALAEPAPHOS-*SKALES* CEMETERY

At Palaepaphos in southwestern Cyprus a major cemetery, called the *Skales* Cemetery, was found containing some 55 tombs ranging in date from Cypro-Geometric (CG) IA through CG III, with a few as late as Cypro-Archaic and Hellenistic (Karageorghis 1983). Most of the CG tombs are rectangular chamber tombs with dromoi, and most contained multiple inhumation burials, though two were cremations. Of interest to us here are 29 tombs of the CG I and II periods (late eleventh to early ninth centuries). Nineteen of the tombs belonged exclusively to the CG I period: nine CG IA (late eleventh century); three CG IB (early tenth century); and seven CG IA–B or just CG I. In addition, two tombs had mixed material ranging from CG I into II and one contained three burials with pottery of CG I, II, and III types respectively, of which the individual metal finds could not be sorted out (Karageorghis 1983:290) and hence will not be considered here. Eight more tombs belong to the CG II period exclusively; while two span the CG II–III periods and will not be considered here. Several of these tombs exhibited a fair amount of wealth, containing bronzes, quite a bit of gold, iron, and a little silver (Stech et al. 1985:198, table II).

Since much has been made of this site as being somehow advanced in its use of iron both functionally and technologically (see, e.g., Snodgrass 1982:286; Karageorghis 1982:299; Stech et al. 1985:192), it is of some interest to examine the remains from the earliest of the tombs to see whether innovations can be detected. I am leaving aside for the moment consideration of tombs dating CG IB, CG IA–B, and CG I without specification, since the material from these contexts either does or could postdate the eleventh century. It should be noted too, that in preliminary publications there was a tendency to lump all of the *Skales* ironwork in the eleventh century (e.g., Karageorghis 1982:299; Snodgrass 1982:286). The final publication, however, with full description of the tomb contents and discussion of chronology, makes it clear that this was not at all the case (Karageorghis 1983).

All nine of the tombs belonging exclusively to the CG IA period contained some metal.[10] All had objects of bronze; five had iron as well, and eight had gold. Of a total of 79 metal objects belonging to this period, 55, or 69.6%, are bronze; 10, or 12.7%, are iron; and 14, or 17.7%, are gold. These figures are, in fact, quite close to those previously observed for Cyprus in general in the eleventh century (Waldbaum 1978:45, fig. IV.5c). In terms of function, there is 1 iron weapon (a dagger); 4 tools (3 knives and a spindle); 4 ornaments (1 fibula, 3 pins); and an attachment. This compares with 5 bronze weapons (spears); 5 tools (needles); 37 ornaments (fibulae, rings, pins); 6 bowls, a tripod, and a tripod-cauldron. All the gold objects are ornaments. Two of the iron knives, the dagger, and the attachment come from a single tomb (89), which contained the remains of at least five burials and also yielded 2 spears, 6 fibulae, 4 needles and a ring in bronze, 3 gold objects, and a large quantity of pottery (Karageorghis 1983:324).[11] Unfortunately, none of the iron artifacts from this earliest group were among those analyzed (Stech et al. 1985:193–195). The dagger and one-edged knives are not new types in Cyprus since examples of similar artifacts can be found as early as the twelfth and early eleventh centuries at several sites. A number of them have bronze rivets or other fittings, a characteristic which has often been taken as a sign of transition from bronze- to ironworking. Use of bronze fittings such as rivets on otherwise iron artifacts has been shown, however, to continue as a technique long after the period of iron's introduction and is not typologically significant (Waldbaum 1982:327–329 and see pp. 339–341 for list of bimetallic knives, daggers, and other implements from Cyprus).

Much of the rest of the material from *Skales* to be considered here is designated CG IB, CG IA–B, or CG I and is likely to fall in the first half of the tenth century; some, however, comes from two tombs dated CG IIA or CG IB–IIA (late tenth century).[12] Eleven out of twelve of these tombs contained some metal; three had only bronze; one, bronze and iron; one, bronze, iron, and silver; three, bronze, iron, and gold; two, bronze and gold but no iron; and one, only iron (two knives) (Stech et al. 1985:198, table II). There is a total of 103 metal objects of which 72, or 69.9%, are bronze; 20, or 19.4%, are iron; and 11, or 10.7%, are gold or silver (mostly gold). These figures differ somewhat from those published previously for tenth century Cyprus in general (Waldbaum 1978:45, fig. IV.5d), in that there seems to be a somewhat greater proportion of both iron and bronze and a somewhat smaller proportion of gold.

The functions of the artifacts can be categorized by material as follows. In iron there are 3 weapons (2 swords from Tomb 76 and a dagger from Tomb 43), 12 tools (10 knives, an axe, and a sickle), and 5 ornaments (1 fibula, 3 pins, 1 ring). This breakdown compares with 4 bronze weapons (spears), 1 tool (an awl), 43 ornaments (including rings, pins, earrings, but mostly fibulae), and 19 vessels (17 bowls, 2 strainers), plus a "knot," a tripod, and 3 obeloi.

All the gold and the single piece of silver are ornaments. The richest tomb in this group is 76, which contained 14 bronzes, 12 iron artifacts, and 4 gold pieces as well as 170 ceramic vessels for some three burials. Among the iron objects are 2 swords, 6 knives, a sickle, an axe, and 2 pins. The most noteworthy change from the situation in CG IA is the increase in relatively large iron tools as opposed to the insignificance of bronze tools.

A number of the iron artifacts from this group have been analyzed metallographically (see Appendix B). They include 2 knives from Tomb 49 (CG I; App. B, nos. 11, 12), 4 knives from Tomb 76 (CG I; App. B, nos. 15–18), and 2 objects called "awls" by the analysts but identified as pins in the final publication, also from Tomb 76 (App. B, nos. 13, 14; Stech et al. 1985:193–195 and cf. Karageorghis 1983:217, nos. 28a, b). One of the knives from Tomb 49 was found to be mildly carburized, though it could not be determined whether deliberately or not; the other analyzed knife from Tomb 49 showed slight evidence of carburization. One of the knives from Tomb 76 was not carburized; one was mildly carburized; the other two were extensively carburized. One of the two pins ("awls") was mildly carburized and the other was not carburized. None of these artifacts showed evidence of quenching or tempering, though the surfaces are so corroded that evidence for this might have been lost (Stech et al. 1985:196). At any rate, it appears that most of the knives were carburized to a varying extent, suggesting some attempt to manipulate the material to advantage, although full-scale heat treatment does not seem to have been practiced.

GREECE, EUBOEA: LEFKANDI CEMETERIES

At the Greek site of Lefkandi on the large offshore island of Euboea, several cemeteries containing material of the Early Iron Age (SubMycenaean, Protogeometric, and SubProtogeometric) came to light. Burials were for the most part in individual cist or shaft graves. Most were cremations, though a few inhumation burials were found. The wealth of the offerings in individual tombs varied from sparse to quite lavish, with some tombs containing imported materials from the Near East, Cyprus, and Egypt, including gold, faience, ceramic, and bronze vessels in addition to pottery brought in from nearby Attica (Popham et al. 1980; Popham et al. 1982a; Popham et al. 1989, 1993). Although there was a settlement mound nearby, occupation material for these periods is largely missing; however, Late Protogeometric (LPG) pit and leveling debris revealed a considerable quantity of bronze foundry refuse, indicating that casting of sizable objects such as tripods took place here in the later tenth century (Popham et al. 1980:93–97; 1979:pls. 12, 13). The following discussion will focus on the tombs.

There are 44 tombs from the eleventh century: 22 SubMycenaean (SM) or first half of the eleventh century; 12 Early Protogeometric (EPG); and 10 Middle Protogeometric (MPG), excluding the, thus far, unique Heroon (Popham et al. 1982b; Popham et al. 1993). The latter two periods together fall within the second half of the eleventh century. Twenty-six tombs were attributed to the Late Protogeometric (LPG) or tenth century (Popham et al. 1980:418–421; Popham et al. 1982a; Popham et al. 1989). The finds from the Heroon should probably be included in the early part of the tenth century as well.[13]

More than half of the eleventh century tombs contained metal: thirteen of the 22 tombs dated to SM, seven of the twelve EPG tombs, and four of the ten MPG tombs. There were 97 metal artifacts from those tombs, of which 76 (78.4%) are bronze; 16 (16.5%) iron; and 5 (5.1%) gold (Popham et al. 1980:418, table 1).[14]

Breaking this down even further, of those thirteen tombs dated to the SubMycenaean period (early eleventh century), ten contained bronze alone; one, bronze and iron; one, bronze and gold; and one, bronze, gold, and iron. There was a total of 53 metal objects found in the tombs: 45 bronze objects (or 85% of the total), 4 iron (7.5%), and 4 gold (7.5%). All the metal is ornamental in character. The iron artifacts consist of 2 pins, a spiral, and a fragment. In the second half of the eleventh century (EPG and MPG), comparable to CG IA, only three tombs contained bronze alone; five contained bronze and iron; two, iron alone; and one, bronze, iron, and gold. From these tombs comes a total of 44 metal artifacts: 31 bronze (all ornaments); 1 gold (a spiral); and 12 iron (4 fibulae, 6 pins, 1 knife, and 1 dagger, the latter two being the only tools or weapons so far found in eleventh century contexts). In comparative terms, this breakdown by metal type amounts to 70.4% bronze, 27.3% iron, and 2.3% gold. It is interesting to note that the *percentage* of iron present in these late eleventh century tombs (27.3%) is more than twice that cited above (12.7%) for the CG IA tombs at *Skales* in Cyprus, though the percentage of gold is much lower. *Skales* and Lefkandi each produced a single iron dagger for the period, and while there is only one knife from Lefkandi and three from *Skales*, this does not seem a significant difference. At both sites iron was still being used for ornamental purposes in this period.

Of the 26 LPG tombs, fifteen are known to have contained metal:[15] four with bronze only, four with iron only, two with iron and bronze, one with iron and gold, one with bronze and gold, two with iron, bronze, and gold, and one with iron, gold, and lead (Popham et al. 1980:419, table 2; Popham et al. 1982a:236–242; Popham et al. 1989). There are more than 78 metal artifacts, including 17 bronze objects, or 21.8%; more than 31 iron, or 39.7%; more than 28 gold, or 35.9%; and 2 lead, or 2.6%.[16] Relative proportions of metals at this one site are quite different from those previously reported for either Greece or the Aegean Islands in the tenth century (Waldbaum 1978:48, fig. IV.7d, 53, fig. IV.11d), showing considerably less bronze and iron and more gold than the former and less bronze and more iron and gold than the latter. It is interesting to note, in any case, that percentages of iron are greater than in contemporary Cyprus.

In terms of function, the gold and lead pieces are all ornamental, although the number of different forms in gold is greater than in earlier periods. Most of the bronzes are also ornamental, although one tomb (T39) contained a curious set of wheels as well as an Egyptian jug and a fibula (Popham et al. 1982a:219, 237, 239–240). There is a considerable shift in the types of iron produced: the more than 9 pins and a fibula are in keeping with earlier usage. There are in addition, however, 2 swords, a dagger, a spear, a spear butt, 10 arrowheads, an axe, a knife, and a needle. In other words, most of the iron from this period consists of tools and weapons.

The same general picture can be seen in the Heroon, a large, apsidal building, dated to the early tenth century, in which had been buried the cremated remains of a male warrior, an inhumed female skeleton (presumably his consort), and a pit containing the skeletons of four horses (Popham et al. 1982b:171–172; Popham et al. 1993). Included with the female burial were 2 gilt coils near the head, a necklace consisting of a gold pendant and 39 gold beads, 2 large sheet gold disks on the breast with a lunate sheet below them, 1 gold and 1 electrum finger ring, and 10 pins—5 bronze, 2 gilt iron with gold caps, 2 iron with bone heads, and 1 fragmentary plain iron. An iron knife with ivory pommel lay near the head (Popham et al. 1993:20–21). The cremation had been placed in a bronze amphora decorated with figurative scenes and stopped at the mouth with a bronze bowl (Catling in Popham et al. 1993:81–96, pls. 18–21). This burial was also accompanied by an iron sword, razor, and spearhead (Popham et al. 1982b:172–173; Popham et al. 1993:19). Remains of 2 iron bits were found with the horse burials and 4 fragments of bronze, 1 of iron, and 2 of lead were found in the fill of the building (Popham et al. 1993:71–72, 76–77, pls. 32, 34). While some of the bronze and gold items are believed to be heirlooms (Snodgrass 1983:82; Catling in Popham et al. 1993:86–87), the sword, razor, spearhead, and knife at least may be added to the general picture of increasing use of iron for utilitarian purposes that is seen in the other tenth century burials.

Unfortunately, although extensive analyses were done on the copper-base objects from Lefkandi (Jones in Popham et al. 1980:447–459), no similar work was done on the iron finds, so we cannot compare the level of technological expertise with that from Cyprus.

GREECE: NICHORIA

A final site that must be looked at is Nichoria in the southwestern Peloponnese. Unlike the other sites discussed, this one provided evidence for stratified occupation rather than burials in the Dark Ages. Unfortunately, however, the site is geographically and culturally somewhat out of the mainstream and the evidence for close dating of the remains for this period is slim (McDonald et al. 1983:3, 318–322) and has been questioned (Snodgrass 1984:152–153; Morris 1989:511–512). The periods we are interested in encompass Dark Age (DA) I, dated ca. 1075–975, and DA II, dated ca. 975–850. Unfortunately again, DA I, of most relevance to us, is least well known, since stratigraphic preservation of remains from this period was poor, and very few finds could therefore be definitely associated with it (McDonald et al. 1983:319). In any event, only 7 bronzes and a single iron artifact could be attributed to DA I: 1 ring, 2 awls or gravers, 2 needles, and 2 statuettes in bronze; and a "nail" in iron. Seventeen bronzes and 3 iron pieces are designated DA I–II, the bronze including a variety of ornaments, small tools, and scrap, the iron 2 knives and a piece of scrap. For DA II, which apparently descends into the ninth century, there are 32 bronze and 11 iron artifacts. Of the bronze, 9 pieces are ornaments, 1 is a weapon, 9 are tools or parts of tools, 3 are vessel rivets, and the rest various "fittings," "trappings," fragments, and scrap. The iron includes 1 pin, an axe, and a blade fragment, a socketed tool, 4 knives, a nail, and 2 pieces of scrap (Catling in McDonald et al. 1983: 274–276, tables 5.1–4, 283–285). Although the absolute quantities are not great, the variety of types in both iron and bronze may well be reasonably representative of what was available in daily use at this small site. While the picture of relative proportions of materials in use may not be wholly accurate, we do, at least, see a definite increase in iron use and iron diversity from DA I to DA II.

As with Lefkandi, extensive series of analyses were performed on the copper-base metals, but not on the iron (Rapp and Aschenbrenner 1978:166–181).

DISCUSSION

Although the material for the new sites in Greece, Cyprus, and Transjordan has been fully and carefully published and adds greatly to our knowledge of the situation in the period of transition from bronze- to iron-using, it still does not provide a fully balanced picture of the development of iron and its changing relationship to bronze in the first three centuries of the Iron Age. For three of the four sites considered here—the Transjordan cave, Palaepaphos-*Skales,* and Lefkandi—the metal finds come only from tombs. Cave A4 in the Baq'ah is—by definition—a burial cave, and the relevant material dates only to the twelfth century B.C.; the material from Palaepaphos-*Skales* is from a cemetery whose life span only begins in the late eleventh century, and continues into later periods. While there is an Iron Age settlement at Lefkandi, stratified habitation levels are lacking between ca. 1100 and the late eighth century, and all the metal artifacts studied come only from the cemeteries, which provide evidence for continuous occupation during that time. Nichoria does have the virtue of being a stratified Early Iron Age site. The site is provincial, however, and may not be entirely typical of developments elsewhere in Greece. Furthermore, the absolute chronology for the period in question is somewhat shaky.

The three cemetery sites, while they provide very useful and interesting material, cannot be considered to provide an accurate picture of the available metal repertoire. Cemetery material is usually rather specialized and selective in what it contains, and virtually never includes a completely representative sampling of available metal types. Burial goods in general tend to be skewed towards valuables—jewelry, valued personal weapons or knives, treasured foreign material such as gold, ivory, amber, or particularly well-made vessels in pottery or metal—but they are conspicuously lacking in such everyday items as tools (other than knives, which may also serve as personal weapons). Furthermore, metal of any kind, being generally more expensive than other materials, tends to be concentrated in tombs, and may appear to be more abundant than it would if more representative, stratified sites were included.[17] This observation is often lost sight of in comparing the material available for study from such areas as Cyprus, where most of the post–twelfth century material is from cemeteries, and that from Palestine in the same period, where there is a greater variety of sites (see Waldbaum 1978:74–78 for range of stratified sites versus cemeteries). In Palestine after the twelfth century, the repertoire of artifact types classified as tools is much broader than in Cyprus, but the apparent proportion of iron tools to those in bronze is lower than in Cyprus (Waldbaum 1978:40, table IV.2, 41, fig. IV.2b, 46, table IV.6, 47, fig. IV.6b). This makes perfect sense since the great majority of iron "tools" in Early Iron Age Cypriot tombs are knives, which often accompany their owners after death. The metal repertoires and relative proportions of bronze to iron in a contemporary Cypriot household and/or farmstead might very well differ considerably.

Another point to consider is that amounts and types of objects in tombs tend to vary with local burial practices and with the nature of the associated community. If local custom dictates against lavish burial, the amount and variety of metal in a given tomb will tend to be low, regardless of the amount of wealth available in the community. Similarly, if the burial ground belongs to a community of farmers, one would not expect graves containing warriors' panoplies. While these observations may seem self-evident, they are often forgotten in the enthusiasm for archaeological "firsts"—and we are often left to compare funerary apples and oranges to very little point. At the three tomb sites considered here, for example, we have at Lefkandi a coastal site with individual cremation graves containing, for their time and place, a fair amount of wealth; at Palaepaphos-*Skales*, a community that favors inhumation in multiple graves containing from two to five burials and rather lavish grave goods, and in Baq'ah Cave A4, a single burial cave containing a large number of burials (perhaps those of the whole community or a large kin group) but very little evidence of wealth. We are still very much in need of excavated settlement material, particularly from Iron Age Cyprus, before we can fully understand patterns of metal use and technological innovation.

DISSEMINATION OF IRON: DORIANS AND PHILISTINES?

The pattern of gradual increase in iron in the eastern Mediterranean over a period of several hundred years has been established for some time (Waldbaum 1968, 1978). The question of why this increase took place when it did and where it did is still controversial. Despite the lack of evidence for a Hittite "monopoly" on iron, the idea persists that the dissemination of iron after ca. 1200 B.C. was somehow owing to the breakup of this monopoly upon the collapse of the Hittite empire. Furthermore, iron was and is perceived to have been spread by peoples invading new territories who had somehow acquired the mastery of iron technology en route from their homelands to their new conquests. These include, of course, the Dorians in Greece and the Philistines in Palestine.

The notion of iron-bearing Dorians has long since been laid to rest. Snodgrass has demonstrated convincingly that the supposed northern homeland of the Dorians lagged behind Greece by some three hundred years in the adoption of iron, and that the coming of the Greek Iron Age is thus in no way dependent on technologically "superior" northern invaders (Snodgrass 1965).

The concept of Philistine mastery of iron, and even a Philistine "monopoly" over the material is more difficult to dispel. Despite the arguments that can be mustered against it (e.g., Waldbaum 1978:42; 1980:84–85), the idea still lingers, and it is therefore worth reviewing the evidence pro and con.

The main basis for the "Philistine monopoly" remains the Biblical text (I Samuel 13:19–22) in which the Israelites of the time of Saul (late eleventh century) possess no weapons and are forced to go to the Philistines to have their agricultural tools repaired because "there was no smith to be found in all the land of Israel." This has frequently been interpreted to mean that there were no *black*smiths in Israel, and that the tools and weapons in question must have been of iron, though the passage cited contains no reference to iron nor is the word translated as "smith" modified to indicate the material being worked. A fairly typical example of the reasoning behind this interpretation may be found in an article by Muhly:

> ... the Biblical passage makes no mention of iron—nor of an ironworking monopoly for that matter. But the Israelites obviously had metal farm implements that needed repairing. On the basis of the surviving artifactual evidence, it is reasonable to assume that, by the end of the 11th century B.C. (the time of the reign of Saul), these farm implements were made of iron. (Muhly 1982:52–53)

And further:

> Iron artifacts from the 11th century are to be found at Israelite as well as Philistine sites, but *all of the weapons*—swords, daggers, spearheads—are found at Philistine sites. (Muhly 1982:52)

In other words, it is first assumed that iron is the preferred material for both tools and weapons by late eleventh century, presumably because of its supposed superiority to bronze (though Muhly and his colleagues elsewhere present evidence that this was not always so in Early Iron Age Palestine [Stech-Wheeler et al. 1981:257–259]); and then asserted that iron weapons are only to be found at Philistine-occupied sites at this time. But is this, in fact, the case?

I have shown elsewhere that the number of different types of tools found in bronze is greater than that for iron in both the twelfth and the eleventh century in Palestine and about the same in the tenth century (Waldbaum 1978:40, table IV.2). In fact, for the eleventh century there are some 14 different tool types in bronze, including such heavy tools as adzes, axe-adzes, plowshares, ox-goads, knives, and chisels, while there are only some eight varieties in iron including axes, chisels, knives, plowshares, and sickles. For weapons and armor, the types available in both bronze and iron in the eleventh century are about the same—swords, daggers, arrow-, lance-, and spearheads, and spear butt—with only armor scale being found in bronze alone. In other words, iron is by no means being used exclusively, or even predominantly for tools or weapons in eleventh century Palestine. As far as distribution goes, while it is true that few (but not no) iron weapons can be found at eleventh century Israelite sites, not many more can be found at Philistine sites (Waldbaum 1978:24–25; McNutt 1990:200–201 and see Muhly et al. 1990:166, 168–169 for an eleventh century iron arrowhead from the Israelite site of Kinneret). Furthermore, most of the weapons at all sites in this period are of bronze, which is by far the predominant material for both tools and weapons in the eleventh century (Waldbaum 1978:41, fig. IV.2a, b). This apparent preference for bronze is supported by analytical evidence showing that iron implements from Philistine sites, in contexts dating from the twelfth century B.C. to as late as the late eighth century B.C., were not regularly carburized (Stech-Wheeler et al. 1981:257–258; Muhly et al. 1990:170 and cf. Muhly 1982:53). While these investigators make every attempt to rationalize a Philistine iron-based technology, and with it a military sway over their enemies the Israelites, even to the point of producing analyses on late and somewhat irrelevant artifacts, several conclusions seem clear: (1) the Philistines do not seem to have been regular producers of steel; (2) bronze remained the metal of preference throughout Palestine in the eleventh century, perhaps because it was still more reliably "superior" to the unsteeled iron that was most commonly produced at the same time; and (3) there is no reason to presume a Philistine monopoly on iron, since its possession would confer no particular advantage on its holders.

The question of whether the Philistines were responsible for the introduction of iron into Palestine in the early twelfth century also deserves some consideration. Several sites with twelfth century Philistine occupation have produced some iron artifacts. These include a knife with ivory ring handle and an amorphous lump from Tell Qasile, Stratum XII (Mazar 1985:6–9, fig. 2.1, photo 3); a similar ivory ring handle with traces of an iron blade was found in the remains of a twelfth century shrine at Tell Miqne-Ekron. (A complete example [Fig. 2.4] and another ivory ring handle without blade came from an eleventh century Philistine temple at Miqne; a fourth, similar, handle was found in a later context [Dothan 1989:199; 1990:31, 33; Dothan and Gitin 1993:1053–1055; 1992:417–419].)[18] In a twelfth century tomb at Azor was found a bracelet, in another at Tell 'Aitun, a ring, and at Tell el-Far'ah (South), bracelets, rings, and a dagger (Waldbaum 1978:24; Dothan 1982:92, table 1; McNutt 1990:199, table 6).

Some of the earliest iron—and even steel—in the region comes from sites with which the Philistines are not associated. In fact, McNutt (1990:198–200) shows that significantly more iron artifacts come from "non-Philistine" twelfth century sites than from sites associated with Philistine settlement.[19] The carburized and quench-hardened pick from Mt. Adir in the Galilee, of twelfth or eleventh century date, is one conspicuous example (Davis et al. 1985); the steel bracelets from Cave A4 in the Baq'ah in Transjordan are another (Notis, Pigott, et al. 1986:262–267, 272–278) and there are also unanalyzed twelfth century iron artifacts from such sites as Madeba and Tell es-Sa'idiyeh in Transjordan (Waldbaum 1978:24; Pritchard 1980:20, 23) and Megiddo in the north (Waldbaum 1978:24) where there is little or no significant contact with the Philistines (see also Dothan 1982:92, table 1 for other

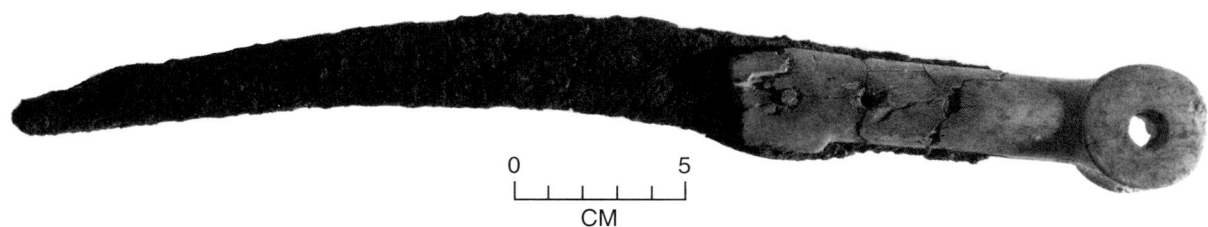

Figure 2.4   Iron knife with ivory ring-shaped handle and bronze rivets from Tel Miqne-Ekron, Stratum V, eleventh century B.C. (Dothan 1989:154–163, figs. 2, 3). (Photo courtesy Tel Miqne-Ekron Excavations; photo by Ilan Stzulman.)

sites). While the general atmosphere of the upheavals associated with the "Sea Peoples," and movements of peoples from the Aegean in the wake of these disturbances, may have in some way contributed to the initial spread of iron production in Palestine, and other parts of the eastern Mediterranean such as Cyprus (cf. Stech-Wheeler et al. 1981:266–267; Snodgrass 1982:293; Mazar 1985:9), these factors do not seem to have contributed significantly to developments east of the Jordan River (Notis, Pigott et al. 1986:277; McGovern 1986:340; McGovern 1987). In any event, it seems unlikely that the Philistines were responsible for the appearance of all early iron, and particularly steel, in Palestine.

WHY THE IRON AGE BEGAN: SOME RECENT HYPOTHESES

The revision of old assumptions and the better understanding of the historical background against which the adoption of iron took place has led to the proposal of new hypotheses to help explain the replacement of bronze by iron.

The first, and until recently, most popular hypothesis suggests that iron began to be exploited out of economic necessity. The political upheavals in the eastern Mediterranean ca. 1200 B.C. caused widespread disruption in trade routes and cut off access to raw materials, especially tin, which was assumed to have traveled long distances to the Mediterranean. The subsequent diminution in the metals trade would have forced the development of the more widely available iron ores (Waldbaum 1968:171–173; 1978:71–73; Snodgrass 1971:237–239; 1989:29). This hypothesis, while attractive for a number of reasons, is now being questioned in the light of new evidence.

In the first place, sources of tin have been identified in the Central Taurus mountain region of southeastern Turkey (Yener 1986a:183; 1986b:472; Yener and Özbal 1987; Willies 1990; Yener et al. 1991:546, 548). Thus far, there is evidence for exploitation of tin in this region only in the Early Bronze Age (third millennium B.C.) at the mining and smelting sites of Kestel and Göltepe (Yener et al. 1989; Willies 1990; Yener and Vandiver 1993a, b). However, the location of tin within the eastern Mediterranean, where it had not previously been recognized, raises the possibility, at least, of easier access to tin in the Bronze Age than had hitherto been believed.[20]

Even if tin did exist in usable form in the eastern Mediterranean, however, it need not necessarily have continued to be available during the period of political dislocation that ushered in the Iron Age. It remains possible that supplies were diminished by a disruption in trade, even if that trade were somewhat more local than has previously been assumed. There is, however, a growing body of evidence suggesting that tin continued to be available throughout the period in question. Recent analyses of bronze objects dating to the twelfth century B.C. and thereafter, from sites as widely scattered as Lefkandi in Euboea (Jones in Popham et al. 1980:447–459), Nichoria in southwestern Greece (Rapp and Aschenbrenner 1978:167–175), Kouklia in Cyprus (Pickles 1988:15–19), the Baq'ah Valley in Transjordan (Notis, McGovern, et al. 1986:278–283), Lachish (David Ussishkin, pers. comm. 1987), Tell Qasile (Mazar 1985:3), Beth Shan (Bonn et al. 1993), and elsewhere (Muhly et al. 1990:161) have shown no diminution of tin contents.[21] In fact, most of the bronzes tested from these sites contain tin well within normal range, and some even showed abnormally high tin contents, with average percentages of tin greater than those for the preceding Late Bronze Age at the same sites (Fig. 2.5, and cf. Fig. 2.2) (see Waldbaum 1989 for detailed discussion). While we still do not have enough analytical data from a large number of sites to be able to say with confidence that there was no tin shortage in the eastern Mediterranean generally, the evidence is beginning to accumulate in favor of this view. Other explanations must be sought to account for the increasing exploitation of iron after ca. 1200 B.C.

A second hypothesis, based on technological evidence, suggests that the discovery and consistent ap-

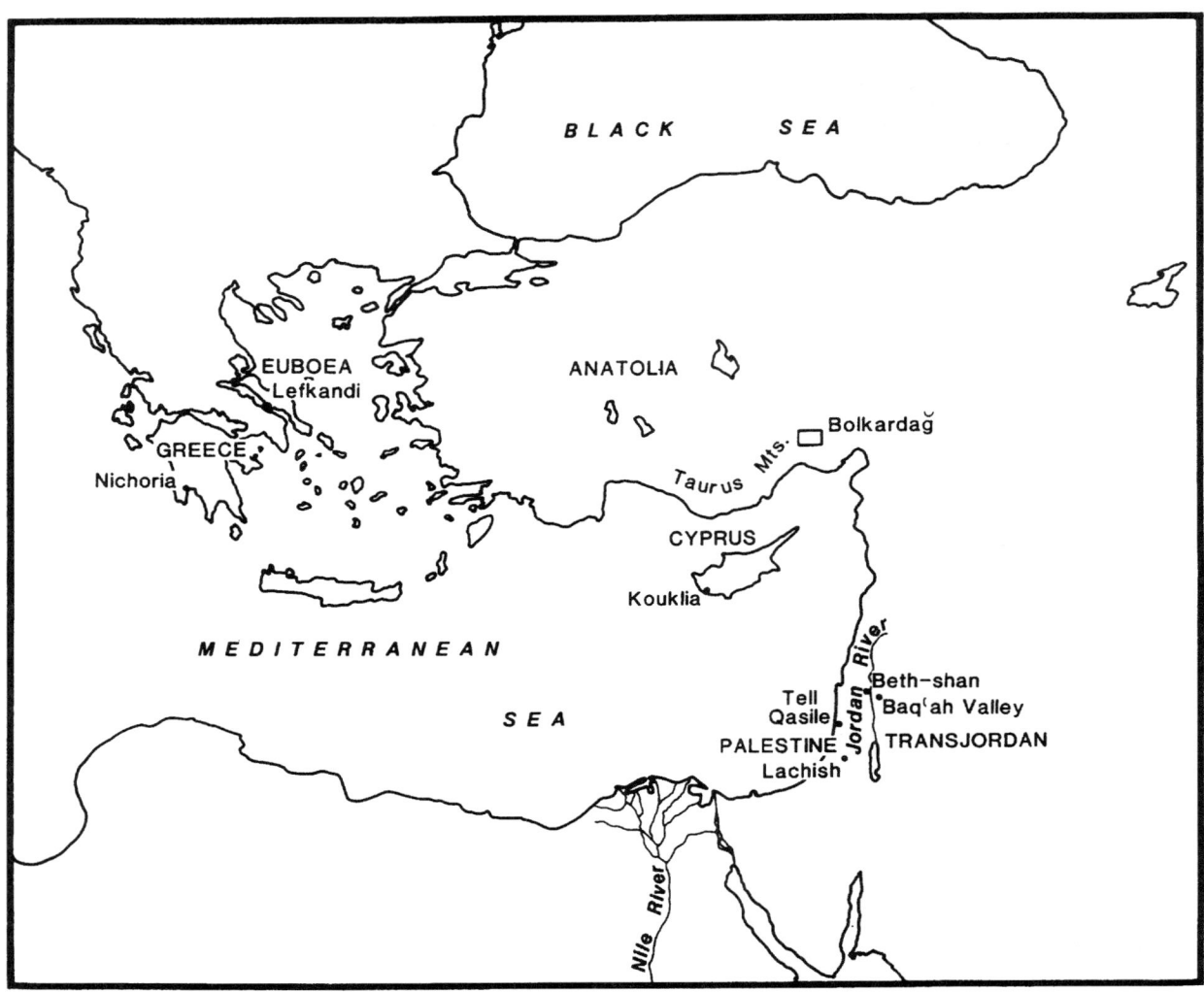

Figure 2.5 Map of the Eastern Mediterranean showing Early Iron Age sites at which normal or high-tin bronzes have been found. (Prepared by the University of Wisconsin-Milwaukee Cartographic Services.)

plication of the proper techniques for hardening iron took place in Cyprus. Iron was first adopted as a viable utilitarian material there and ultimately spread to the rest of the eastern Mediterranean (Maddin 1982:311; Pickles 1988:24). A proper consideration of this theory depends, of course, on analyses of iron artifacts to see, if possible, where steeling was practiced, when it was practiced, and what it was used for. If we are to isolate a single area as a possible originator of a new technology it is also necessary to compare data from all the areas in question and to ensure that the artifacts subjected to analysis can be well dated so that reliable comparisons may be made.

Unfortunately, the principal criterion cannot as yet be met. While several programs of analyses have been undertaken in recent years, so far, for a variety of reasons, they have been carried out only on material from Palestine and Cyprus. Analogous material from the Aegean, in particular, as well as from Anatolia and Syria in the critical transitional years from ca. 1200 to 900 B.C. has not been studied and cannot be added to the equation. It is possible to compare the situations in Palestine and Cyprus to some extent, bearing in mind that most of the Cypriot examples are from funerary contexts while most of the Palestinian ones are from stratified occupation remains.

CYPRUS

Metallographic analyses of several iron artifacts from Palaepaphos-*Skales* Tombs 49 and 76, dating to CG I (from the late eleventh into the tenth century), have been discussed above. Of these, about half, including four out of six knives tested, showed signs of carburization, although it was not clear in every case whether this had been accomplished deliberately or accidentally (Stech et al. 1985:193–196).

The analyses of the *Skales* iron can also be compared to analyses of material from several other sites in Cyprus. The earliest are two twelfth century objects from Hala Sultan Tekke—a knife handle and remnants of wire from a potsherd, neither of which showed signs of deliberate carburization (App. B, nos. 1, 2). Since neither item need have been steeled to function it is not possible to know whether the absence of carburization is deliberate or a function of early date (Åström et al. 1986:37). Artifacts from the Swedish excavations at Idalion, Lapithos, and Amathus as well as from Karageorghis' excavation at Kition have been examined (Åström et al. 1986; Maddin 1982; Maddin et al. forthcoming). Interestingly, the earliest artifacts, two knives and a dagger from LC IIIB (early eleventh century) occupation levels at Idalion, showed the most sophisticated treatment (App. B, nos. 3–5). These objects all showed extensive carburization, the two knives had been quenched, and at least one of them had been tempered. The dagger also showed evidence of rapid cooling (Åström et al. 1986:30–31). The results for the knives confirm those for analyses done earlier by Tholander (1971:17–22).

From Lapithos, an obelos from CG IA Tomb 417 (App. B, no. 6) had not been carburized; and a knife from contemporary Tomb 420 (App. B, no. 7) showed moderate carburization but no evidence of quenching or tempering (Åström et al. 1986:33). Several artifacts from CG II (tenth to early ninth century) had all been moderately to extensively carburized with the exception of another obelos. The carburized objects include a sword from Tomb 409 (App. B, no. 23) and four knives, two from Tomb 409 (App. B, nos. 24, 25), and one each from Tombs 411 and 429 (Åström et al. 1986:34). Several of the carburized objects had also been coldworked, but none showed evidence for quenching and/or tempering. Thus, in tenth and ninth century Lapithos there seems to be evidence for fairly consistent use of carburization and coldworking to strengthen objects that could benefit by such practices, though not of the quenching and heat-treating techniques that would impart still greater strength (for the beneficial effects of coldworking, see Rehder 1992). According to the analysts, the two obeloi would not have needed great strength to fulfill their function, so the absence of carburization may well be by choice (Åström et al. 1986:33–34).

Four artifacts from Amathus were also sampled. Two were from CG I contexts: an uncarburized obelos fragment (?) from Tomb 21.1 (App. B, no. 19), and a knife from Tomb 25.4 (App. B, no. 20), which was found to be heavily carburized and coldworked. A CG IIA (tenth century) knife from Tomb 19 (App. B, no. 21) was moderately carburized and possibly tempered; and another knife from CG II Tomb 21.2 (App. B, no. 26), tenth/ninth century in date, was carburized and coldworked (Åström et al. 1986:34–36). The pattern here is very similar to that from Lapithos.

Three artifacts from eleventh/tenth century Kition were sampled, a dagger tip, a knife blade, and an awl (App. B, nos. 8–10). The knife and the awl had been mildly carburized, the dagger moderately to extensively. Again, no evidence was seen for quenching or heat treatment (Maddin et al. forthcoming). It is interesting to note that several samples of iron artifacts from Phoenician contexts at Kition, postdating the three cited above, showed few or no signs of carburization, suggesting a technological change that apparently persisted into the Cypro-Archaic period, at least at Kition (Maddin et al. forthcoming).[22]

The analyses from Idalion, Lapithos, Amathus, and Kition, together with those from the *Skales* Cemetery, seem to show that by the late eleventh and tenth centuries in Cyprus carburization was practiced on a fairly regular basis in producing artifacts such as knives and weapons that could benefit from the additional strength and hardness. Only at Idalion, however, is there as yet any evidence for the use of quenching and tempering. These observations may, of course, be fortuitous, dependent as they are on the preservation of the surface layers of artifacts that are all too often fully corroded. Only future studies on better preserved objects will be able to clarify this problem.

PALESTINE

For Palestine, the picture is similar, though naturally it differs in detail. There is evidence for carburizing in the twelfth century in the bracelets from Cave A4 in the Baq'ah (App. B, no. 27), and perhaps for carburizing, quenching, and tempering in the pick from Mt. Adir in the Galilee (App. B, no. 28), if the early date is accepted. To these may be added a fragmentary and highly corroded knife from Tell Qasile Stratum XII, or late twelfth century, which showed some evidence of carburization, though possibly produced by accident (App. B, no. 29). These examples are roughly contemporary to the two from Hala Sultan Tekke, the earliest analyzed material from Cyprus, but there are so few of them in either case that any conclusions about precedence in steeling technology would be premature.

Only four objects from eleventh century Palestine have been analyzed so far, a knife from Tell el-Far'ah (South), Tomb 562 (App. B, no. 30), an axe from Tell Qiri, probably dating to the late eleventh century, (App. B, no. 31), an arrowhead from Kinneret, Stratum VI (App. B, no. 32), and a knife from Tel Miqne-Ekron, Stratum V (App. B, no. 33). All four of these artifacts had been carburized. The evidence improves with the tenth century. Eleven artifacts from Taanach were examined, including a sword blade, an arrowhead, two plowshares, a sickle or scythe blade, two armor scales, a chisel, two unfinished tools, and a plowshare scraper (App. B, nos. 37–47). Of these, five—the sword, one of the armor scales, one of the plowshares, the arrowhead, and the chisel—had not been carburized, while six—the scythe, the plowshare scraper, the other armor scale, the other plowshare, and the two unfinished tools—had been deliberately carburized (Stech-Wheeler et al. 1981:249–253). It is difficult to explain why, if carburizing was thoroughly understood, one plowshare and armor scale should be carburized and the others not, nor why such objects as a sword or a chisel should not have been carburized. The analysts do suggest that

> [s]words, arrowheads and armor scales need not be carburized to be adequate weapons; the relative effectiveness of iron or bronze weapons and armor is more or less the same and the greater ductility of wrought iron as opposed to steel would speed the mass-production of armor scales; the manufacturing process will proceed more quickly if the time is not taken for carburization. (Stech-Wheeler et al. 1981:254)

Further, they speculate that the highly corroded object identified as a chisel might have had some other function (Stech-Wheeler et al. 1981:255). It is interesting to note, however, that the same scholars point to a carburized and quenched dagger from Idalion as an example of appropriately applied technology:

> It is then reasonable to assume that the dagger was made of quenched steel. There is no surviving evidence of tempering, but a martensitic blade would...have been very brittle if not tempered. If, however, the dagger was used as a weapon rather than a tool, hardness may have been preferred to durability, and tempering deemed unnecessary. (Åström et al. 1986:31)

Also from tenth century contexts in Palestine were a knife from Ashdod, dated ca. 1000, and a tool blade and an axe from the late tenth century (App. B, nos. 34–36), none of which had been carburized; and from Tell el-Far'ah (South), two knives from Tomb 220, an arrowhead from Tomb 230, and a dagger from Tomb 240, all dated tenth–ninth century and of which only the dagger showed signs of deliberate carburization (App. B, nos. 51–54). A knife from Stratum V at Kinneret had been carburized and quenched, a sickle from Stratum IV had been carburized, but a spearhead tip from Stratum IV had not been carburized (App. B, nos. 48–50).

The evidence presented from Palestine is somewhat ambiguous. From these rather limited studies it appears that carburization was practiced with rather more regularity in the north than in the south, at least in the tenth century, and possibly earlier if the few samples from the twelfth century prove to be part of a larger pattern. It is curious that sites known to have been dominated by the Philistines in the twelfth and eleventh centuries should produce unsteeled iron in the tenth, although I have shown above that the association of Philistines with ironworking may be more tenuous than previously supposed. At any rate, cultural and/or regional differences may play a larger role in technological accomplishment than we can immediately perceive. This can be seen in Cyprus as well, where, for example at Kition, steel production seems to have been practiced in the eleventh/tenth century, then apparently dies out completely under Phoenician domination of the site (Maddin et al. forthcoming; Muhly et al. 1990:171). In this case, at least, technological "progress" did not advance in a straight chronological line.

It is probably too early to make entirely convincing comparisons between the technological achievements of Palestine and Cyprus. The very few analyzed specimens of twelfth or eleventh century date from each region are not sufficient to claim priority in steelmaking techniques for one or the other. By the end of the eleventh century or tenth century steelmaking seems to have been practiced in both regions, though not universally in either. It would be worth exploring in future whether the apparent regional variations noted above reflect genuine differences in practice.

Also missing, of course, are data from other regions, particularly the Aegean. While this area is generally discounted as an early producer of iron, it is paradoxically also credited as in some way providing a "stimulus" for the adoption of iron metallurgy in other areas such as Cyprus and Palestine (e.g., Stech-Wheeler et al. 1981:266–267; Snodgrass 1982:293). The finds from eleventh and tenth century Lefkandi now indicate a fair amount of iron use (and no "metal shortage"), comparable in many respects to what can be seen in Cyprus in the same period. What we are lacking is technological information, and until this is provided through systematic analyses we must remain in the dark as to the true relation among the Aegean, Cyprus, and Palestine in the age of transition from bronze to iron use.

A third hypothesis to explain the replacement of bronze by iron that has gained some credence is ecological. This hypothesis, proposed by the late Theodore Wertime, suggests that pyrotechnological activities making heavy demands on fuel over a long period, in conjunction with other kinds of human activity such as land clearing and agricultural terracing (Stager 1985:5–9), ultimately led to severe deforestation over much of the Mediterranean. Iron smelting, being significantly more fuel-efficient than copper smelting (Horne 1982:12), became economically more feasible despite technological difficulties and the greater labor intensity involved in producing iron (Wertime 1982, 1983). If it could be shown, therefore, that large parts of the Mediterranean were indeed undergoing heavy deforestation around the end of the Bronze Age and the beginning of the Iron Age, then the gradual switch from bronze to iron across a broad geographical range would be a reasonable response to a pressing ecological challenge.

At present, while there appears to be some evidence for deforestation in Greece and Palestine in the

Bronze Age and Early Iron Age, dated evidence on vegetation and land use patterns in the region for the time in question is sparse and cannot always be correlated with the period of change in metal technology (see Waldbaum 1989 for detailed discussion). Furthermore, the fuel efficiency of iron has recently been questioned (Muhly et al. 1990:162–163). More extensive and precisely dated studies from more regions are needed before we can draw any firm conclusions as to the validity of Wertime's hypothesis.

CONCLUSIONS

At present, there is not sufficient evidence to either support or refute conclusively any one of the hypotheses regarding the start of the Iron Age. Any hypothesis, new or old, has to account for the fact that iron was a known metal and smelting a known technique well back into the Bronze Age, though neither the metal nor the technique seems to have been appreciated for its utilitarian potential for a very long time. We still do not really know why the Iron Age began or whether one or more of the current proposals provides a satisfactory solution to the problem. We also do not know whether the proper technology for producing steel began to be consciously applied in just one area, from which it spread to others, or whether it developed empirically in several regions. Analytical data from Cyprus and Palestine are still insufficient and such studies are totally lacking from the Aegean, Anatolia, and Syria. More investigations must be undertaken on material from a broader geographic and chronological range, and an attempt must be made to include more examples of the earliest iron before this issue can be resolved. The possible role of the "Sea Peoples" and/or Aegean peoples in the spread of iron technology as they moved eastward into Cyprus and Palestine after the collapse of Bronze Age civilization in the eastern Mediterranean must be clarified. Did they already possess knowledge of working iron (in the metallurgical sense)? Did they provide some kind of "stimulus" to native metalworking industries as in Cyprus? Or is their presence in the area at the time iron was becoming more accessible mere coincidence? Again, the Aegean provides the missing link and metallurgical analyses the possible hammer with which to forge the chain of evidence.

New finds and new analytical studies are constantly changing our perception of the situation. The next few decades promise exciting new developments and, one hopes, a clearer understanding of the reasons for the adoption of iron in the eastern Mediterranean. For now it is important to persist in asking the right questions.

NOTES

1. Muhly (1992:697), McNutt (1990:99–101), and others have rightly pointed out the pitfalls of relying on tabulations of iron artifacts to provide an accurate picture of quantities of iron available at any given period. It is certainly true that known quantities of iron (or any other excavated material) change with increased archaeological activity as well as with more and better reporting and publication, and that the rapid corrosion of iron leads to its decline and disappearance from the archaeological record over time. It is also true that its often unattractive and formless appearance leads to underreporting in archaeological publications even when it is discovered (cf. Rothenberg 1982:295). These factors, however, do not change substantially in later periods of antiquity (e.g., after the transition from the Bronze Age to the Iron Age). Furthermore, many of these objections, especially with regard to completeness of publication, also pertain to bronze. Nevertheless, numbers and proportions of discovered and reported iron artifacts, in relation to those for artifacts of bronze, do rise significantly during and after this time of transition. Without any claims to absolute accuracy (which I, at least, never made, cf. Waldbaum 1978:12; 1980:69) it is still possible to make a rough assessment of the relative importance, usage, and distribution of iron with respect to bronze by examining the remains.

2. McGhee (1984:5) notes that certain Arctic peoples used meteoritic iron from a known meteorite fall in northwestern Greenland, providing at least one case in which utilization of meteoritic iron can be historically documented. I owe this reference to my colleague, Alice Kehoe.

3. Iron ores are widespread on the earth's surface and were present in most of the regions under consideration here. See Waldbaum (1978:59–66) for brief discussion of the distribution of mineral resources.

4. It should be noted that the tomb in which this object was found had been badly disturbed in antiquity; the excavator seems fairly confident that the "blade" belongs to the undisturbed area of the tomb, and that its MB date is secure (Smith et al. 1984:236). The possibility that the blade was intrusive, however, should not be entirely discounted (see, e.g., McGovern 1988:52).

5. Muscarella (1988:177) summarizes the evidence for a four-hundred-year gap in archaeological sites from central Anatolia following the destruction of the Hittite Empire (ca. 1200–800 B.C.). It is just this period, of course, that is crucial to the understanding of the spread of iron in post-Hittite Anatolia.

6. The North Cemetery at Knossos promises to produce important information on Early Iron Age metals on

Crete, but the metal finds from this site, which will be published by Snodgrass, have not yet appeared. Morris (1989:509–510) has a brief summary of some of the iron finds, while the contents of some of the tombs are enumerated by Catling (1979:45–49).

7. It should be noted that the quantities of objects in copper and iron listed in McGovern (1986:figs. 82–86) differ slightly from those given in the text (p. 59).

8. On the basis of the new finds of Early Iron Age iron from Pella, McGovern (1988:52) is inclined to view the Pella "blade" as intrusive in an MB context.

9. It should be noted that the archaeological context of the Mt. Adir pick has never been fully published; and until this is done the precise dating of this tool must remain uncertain.

10. The CG IA tombs containing metal are: 44, 51, 58, 61, 68, 78, 84, 85, 89.

11. These figures also speak against the view that Cyprus entered Snodgrass' "stage 3" in the adoption of iron as early as the eleventh century and about 100 years before Palestine (Muhly et al. 1990:172; Snodgrass 1980:344; 1982:290). This stage is defined as the period when "iron predominates over bronze as the working metal, although it . . . usually does not completely displace bronze" (Snodgrass 1980:337). In the eleventh century *Skales* tombs, at least, bronze still predominates over iron in every functional category as well as in absolute numbers. What the figures from Cyprus would look like if we had excavated habitation sites to compare with those from Palestine remains unknown.

12. The tenth century tombs being considered are: 43, 45, 48, 48A, 49, 50, 76, 88, 91, 92 (CG IB, CG IA–B, CG I); 53, 82 (CG IIA or CG IB–IIA).

13. Catling and Lemos (1990:4, 7, 95) assign the pottery from the use phase of the Heroon to MPG but favor an absolute date in the early tenth century.

14. The tombs included in this count are: EPG: S8, S10, S16, S20, S31, S32, S46; MPG: S51, P14, P16, T12B.

15. While Popham et al. (1980) and Popham et al. (1982a) provide catalogues of tomb contents for every tomb included in each publication, Popham et al. (1989) do not, preferring only to describe highlights. Similarly, the date of each tomb in this group is not indicated so that the actual number of tombs that fall within our range may be somewhat larger. The finds from the Heroon will be considered separately. The tombs included here are: T12A, T14, T17, T26, T39, T44, T48, T49, T54, T63; P3, P22, P23, P24, P31.

16. Popham et al. (1989:118) mention simply "gold rings, gilt coils and a group of iron pins" from the MPG/LPG tomb T49 without giving exact numbers. Numbers for bronze should probably be raised somewhat too, since discussion of fibulae on p. 120 implies that more were found than the four described.

17. For an interesting discussion of the effects of selective deposition of metal artifacts in tombs and a new, if not totally convincing proposal for the start of the Iron Age in Greece, see Morris (1989).

18. Although only two of the Miqne knife handles had iron blades or traces of blades preserved, all are being used to assess the amount of iron from the site. It is wise to keep in mind, however, that others of similar shape but with bronze blades are known from elsewhere in Cyprus and Palestine (Waldbaum 1982:332). The complete eleventh century knife is being analyzed by M. Notis and D. Belcher of Lehigh University.

19. Muhly (1992:697) criticizes McNutt's figures, saying that "the figures used by M. do not include any of the iron finds from the old excavations at Tell el-Far'ah (South) or the new excavations at Tel Miqne/Ekron and Ashkelon. Inclusion of such material would dramatically change the statistics." In fact, McNutt's tables 6 and 8 do include iron from Tell el-Far'ah (South) in her consideration of the twelfth and eleventh centuries (McNutt 1990:199, 201). As for Miqne, the fragmentary iron knife with ivory handle from the twelfth century, and the complete knife with similar handle and an iron "ingot," both from an eleventh century temple, are the only certain items of iron to be published from Iron I strata. Two more eleventh century ivory knife handles, with no traces of blade preserved, are assumed to have belonged to iron knives (Dothan 1990:28, 30, 31; 1989:199; Dothan and Gitin 1993:1053–1056). Thus far, the Leon Levy Expedition to Ashkelon, led by Professor Lawrence E. Stager, has not turned up any iron from Iron I Philistine deposits. While the situations at both Miqne and Ashkelon could change with further excavation, the few pieces recovered from Miqne to date are hardly cause for a "dramatic" change in McNutt's statistics. Furthermore, the cultic associations of the Miqne pieces are entirely in keeping with her observations that iron appears more frequently in ritualistic or ceremonial contexts at Philistine sites than at non-Philistine (McNutt 1990:198, 201). McNutt does, however, include an iron tool from Tell Qiri with her twelfth century iron (McNutt 1990:199); it should be taken with the eleventh century material (cf. Maddin et al. 1987:244–245).

20. Publications of the Taurus tin sources have unleashed a firestorm of controversy (see *inter alia* Belli 1991; Hall and Steadman 1991; Muhly et al. 1991; Pernicka et al. 1992; Muhly 1993 arguing against the significance of the Taurus as a tin source and Yener and Goodway 1992; Willies 1992; Yener and Vandiver 1993b in response to the above critiques). At present the weight of the evidence favors the existence of tin mining and processing in the Taurus at the Kestel/Göltepe locale during the Early Bronze Age. Whether this mine, or others in the region, produced sufficient quantities of tin to supply more than local needs, or whether it continued to be productive beyond the EBA, awaits the results of future research by Yener and her colleagues. It is interesting to note as well that there were other limited but significant tin sources elsewhere in the Near East including other areas of Turkey, the Caucasus, the former Yugoslavia, Egypt, and Cyprus in addition to the traditionally suggested sources of Afghanistan, central Europe, Southeast Asia, and Cornwall (Yener and Goodway 1992:80; Yener and Vandiver 1993a:212–213). The whole issue of tin supply and trade is, if anything, more complex now than in the past.

21. A local twelfth–eleventh century B.C. bronze industry whose remains have been found at Tel Dan in the northern Galilee, however, was apparently based primarily on

the remelting of scrap bronze without the addition of tin. Tested artifacts show varying and low amounts of tin, though a few objects had tin contents ranging from about 8 to 11% (Shalev 1993:60, 63–64).

22. It also must be noted that the dates for artifacts given in the manuscript available to me may be revised in the final publication. I owe this information to Tamara Stech (pers. comm. 1988).

APPENDIX A

IRON IN THE BRONZE AGE EASTERN MEDITERRANEAN: ADDENDA TO WALDBAUM 1980

Ref no.	Object	Date of Context	Provenance	Technical data	Publication
			SYRIA-PALESTINE (31 OBJECTS)		
1	blade?	MB	Pella, Jordan, tomb	carburized and quenched	Smith et al. 1984:234–236
2	fragments	LB	Lachish, temple		Ussishkin 1978:20
3	large, amorphous lump	LB	Lachish		David Ussishkin, pers. comm. 1987
4	anklet or bracelet	LB	Baqʻah Valley, Transjordan, Burial Cave B3		McGovern 1986:245, fig. 79.1, pl. 30a
5	"piece"	LB II	Jabal Nuzha, tomb		Dajani 1966:48
6	11 rings	13th c. B.C.	Timna, Hathor Sanctuary		Rothenberg 1988:147–148, figs. 54.1–10, 12, nos. 7–16, 18; and cf. Rothenberg 1982:295
7	2 tubes	13th c. B.C.	Timna, Hathor Sanctuary		Rothenberg 1988:148, figs. 54.11, 17, nos. 17, 23
8	3 rods	13th c. B.C.	Timna, Hathor Sanctuary		Rothenberg 1988:148, figs. 54.13–15, nos. 19–21
9	earring? gilded	13th c. B.C.	Timna, Hathor Sanctuary		Rothenberg 1988:148, fig. 54.16, no. 22
10	2 bracelets	LB 14–12th c. B.C.	Timna, Site 2, Area F		Gale et al. 1990:185
11	wire	13th c. B.C.	Timna, Hathor Sanctuary		Rothenberg 1988:152, fig. 58.24, no. 124
12	rivet	13th c. B.C.	Timna, Hathor Sanctuary		Rothenberg 1988:168, fig. 76.11, no. 497*
13	ring	LB	Kāmid el-Lōz, Lebanon, metalworking area of palace, phase P4a	mildly carburized	Frisch et al. 1985:111, no. 129, 146–148 (analysis), pls. 29.2, 34.6
14	pin fragment	LB	Kāmid el-Lōz, Lebanon, metalworking area of palace, phase P4a		Frisch et al. 1985:111, no. 130, pls. 28.2, 34.5
15	slag, 4 pieces of iron-arsenic speiss	LB	Kāmid el-Lōz, Lebanon, metalworking area of palace, phase P4a1	possibly from copperworking?	Frisch et al. 1985:108, no. 94, 144–146, 158, pl. 34.3–4
16	ring	LB 13th c. B.C.	Tel Nami, tomb		Michal Artzy, pers. comm. 1990
17	arrowhead	under deposit of LB II material	Tel Gezer, Locus 22,020, just outside the city wall		Dever 1993:50**
			GREECE (10 OBJECTS)		
18	nail?	MH	Asine, fill of tomb		Dietz 1980:86
19	ring	LH IIIA/B	Thebes		Varoufakis 1981:25
20	signet ring, bezel covered with gold foil	LH IIIA/B	Mycenae, Tomb 58	2.20% Ni	Athens National Museum 2856; Varoufakis 1982:315, pl. 30.2
21	ring, bezel missing	LH IIIA/B	Mycenae, Tomb 58	4.94% Ni	Athens National Museum 2866; Varoufakis 1981:29; 1982:316, pl. 30.6

APPENDIX A (CONTINUED)

Ref no.	Object	Date of Context	Provenance	Technical data	Publication
			GREECE (CONTINUED)		
22	signet ring with silver bezel, iron hoop	LH IIIA/B	Mycenae, Tomb 68	1.86% Ni, 1.29% Co	Athens National Museum 2986; Varoufakis 1982:315–316, pl. 30.5
23	ring bezel, laminated with lead	LH?	No provenance	0.78% Ni, probably 3.18% Ni originally	Athens National Museum 2347; Varoufakis 1981:29, fig. 2.3
24	ring, laminated with lead	LH?	No provenance	3.28% Ni, probably 10.77% Ni originally	Athens National Museum 2337; Varoufakis 1981:29 OR Athens NM 2377; Varoufakis 1982:316, pl. 30.7
25	ring fragment	LH?	No provenance	0.15% Ni	Athens National Museum, no. unspecified; Varoufakis 1981:29, fig. 2.6
26	slag, iron-arsenic	LH IIIB	Tiryns, workshop?		Kilian 1983:304, 306, fig. 31
27	"piece"	LH IIIB–C	Tiryns, Grave V(D)		Rudolph 1973:40, no. 22, pl. 18.5
			CRETE (3 OBJECTS)		
28	ring, silver and iron	MM II	Archanes, Sanctuary, on hand of a skeleton		Catling 1981:42; Varoufakis 1981:25
29	2 beads	LM IIIA	Archanes, Tholos Tomb A, inside a larnax on chest of burial		Heraklion Museum; Sakellerakis 1970:150, 153, nos. 202–203; Varoufakis 1981:25, fig. 1.3
			ANATOLIA (7 OBJECTS)†		
30	dagger with gold plated hilt	EBA	Alaca Hüyük, Tomb C		Stronach 1957:102, fig. 3.5; Muhly et al. 1985:71
31	ring	Hittite context	Boğazköy, Unterstadt 2, J/20 Haus 19 Raum 1		Boehmer 1979:35, pl. 21, no. 3466
32	3 rings	Hittite (called Phrygian in Boehmer 1972:155–156, pl. 55)	Boğazköy, Unterstadt 1		Boehmer 1979:35, nos. 1628–1630
33	long blade	late 13th/early 12th c. B.C.	Beşiktepe, Troad, found in a pit with transitional Myc IIIB–IIIC wares		Mellink 1984:446
34	studs on an ivory box	"no later than the 18th c. B.C."	Acemhöyük, on a floor of level III		Özgüç 1976:556††
			CYPRUS (1 OBJECT)		
35	wire, wrapped around a gold ring	LC IIB	Kition, Tomb 9		Karageorghis 1974:89
			TOTAL ALL AREAS: 52 OBJECTS		

(Notes appear on next page)

APPENDIX A (CONTINUED)

*This object appears in the catalogue of metal objects as bronze. Metallographic analysis by R. F. Tylecote, however, shows it to be iron (Tylecote 1988:190).

**The thirteenth century date of this deposit, and of the arrowhead, has been called into question (see Finkelstein 1994:278). It should be noted, also, that the article by Waldbaum cited in Dever's note 33 as published in 1991 (Dever 1993:53), and in his bibliography as 1992 (Dever 1993:54) was not at the time published. What Dever saw was, in fact, an earlier manuscript of the present article.

†It has been thought best to omit four objects from Korucutepe, ostensibly dating to the end of the LBA, but quite possibly extending as late as the eleventh century (Van Loon 1978:40; 1980:147–148, 276).

††In addition to iron studs the box was also decorated with bronze studs set in gold and lapis lazuli studs (Özgüç 1976:556).

APPENDIX B

ANALYZED ARTIFACTS FROM CYPRUS AND PALESTINE, CA. 1200–900 B.C.

This appendix does not include all the artifacts from Cyprus and Palestine that have been sampled and analyzed. Only those samples that can be dated by context between the twelfth and tenth centuries B.C., as well as a few for which the dates may descend into the early ninth century, have been included.

Ref no.	Object	Date of Context	Provenance	Technical data	Publication
			CYPRUS		
1	knife handle	12th c. B.C.	Hala Sultan Tekke	not carburized	Blomgren and Tholander 1976:123–126
2	"wire"	12th c. B.C.	Hala Sultan Tekke	not carburized	Åström et al. 1986:37
3	knife blade	LC IIIB	Idalion, no. 106	carburized, quenched, tempered	Tholander 1971:18, 22; Åström et al. 1986:30
4	knife blade	LC IIIB	Idalion, no. 1068	carburized, quenched; no evidence of tempering	Tholander 1971:18, 22; Åström et al. 1986:31
5	dagger	LC IIIB	Idalion, no. 517	carburized and quenched	Åström et al. 1986:31
6	obelos	CG IA	Lapithos, Tomb 417, no. 12a	not carburized	Åström et al. 1986:33
7	knife	CG IA	Lapithos, Tomb 420, no. 46	moderately carburized	Åström et al. 1986:33
8	dagger tip	11th–10th c. B.C.?	Kition, Area II, Room 4, between Floors II and I, no. 569	moderate to extensive carburization	Maddin et al. forthcoming
9	knife blade	11th–10th c. B.C.?	Kition, Area II, Bothros 2, Floor I, no. 829	mildly carburized	Maddin et al. forthcoming
10	awl	11th–10th c. B.C.?	Kition, Area II, Bothros 2, Floor I, no. 1041	mildly carburized	Maddin et al. forthcoming
11	knife blade	CG I	*Skales*, Tomb 49, no. 12	mild carburization, possibly not deliberate	Stech et al. 1985:193
12	knife point	CG I	*Skales*, Tomb 49, uncat.; Larnaca Mus. no. 416/79	very slight carburization	Stech et al. 1985:193
13	pin or awl	CG I	*Skales*, Tomb 76, no. 28a	mild carburization	Stech et al. 1985:195
14	pin or awl	CG I	*Skales*, Tomb 76, no. 28b	not carburized	Stech et al. 1985:195
15	knife edge	CG I	*Skales*, Tomb 76, no. 72	not carburized	Stech et al. 1985:195

APPENDIX B (CONTINUED)

Ref no.	Object	Date of Context	Provenance	Technical data	Publication
\multicolumn{6}{c}{CYPRUS (CONTINUED)}					
16	knife end	CG I	*Skales,* Tomb 76, no. 132	mild carburization	Stech et al. 1985:195
17	knife hilt	CG I	*Skales,* Tomb 76, no. 133	carburized	Stech et al. 1985:195
18	knife point	CG I	*Skales,* Tomb 76, no. 134	thoroughly carburized	Stech et al. 1985:195
19	obelos fragment?	CG I	Amathus, Tomb 21.1, no. 46	not carburized	Åström et al. 1986:34
20	knife point	CG I	Amathus, Tomb 25.4, no. 4	heavily carburized and coldworked	Åström et al. 1986:34
21	knife blade tip	CG IIA	Amathus, Tomb 19, no. 31	moderately carburized and possibly tempered	Åström et al. 1986:34–36
22	knife blade	CG II	*Skales,* Tomb 77, no. 27	carburized	Stech et al. 1985:195
23	sword fragment	CG II	Lapithos, Tomb 409, no. 6	mildly carburized and coldworked	Åström et al. 1986:33
24	knife	CG II	Lapithos, Tomb 409, no. 17a	moderately carburized and possibly coldworked	Åström et al. 1986:33
25	knife	CG II	Lapithos, Tomb 409, no. 17b	carburized and coldworked	Åström et al. 1986:33–34
26	knife blade	CG II	Amathus, Tomb 21.2, no. 19	carburized and coldworked	Åström et al. 1986:36
\multicolumn{6}{c}{PALESTINE}					
27	bracelets or anklets	Iron IA	Baq'ah Cave A4, nos. 55, 77, 147, 226, 202	carburized	Notis, Pigott, et al. 1986:273–275
28	pick	12th or 11th c. B.C.?	Fortress on Mt. Adir, Galilee, below earliest floor	carburized, quenched, tempered	Davis et al. 1985:41–42
29	knife blade	12th c. B.C.	Tell Qasile, Stratum XII	some carburization but possibly not deliberate	Stech-Wheeler et al. 1981:257
30	knife blade	11th c. B.C.	Tell el-Far'ah (South), Tomb 562	carburized	Stech-Wheeler et al. 1981:258
31	axe	late 11th c. B.C.?	Tel Qiri, Area D	carburized	Maddin et al. 1987:244–245
32	arrowhead	11th c. B.C.	Kinneret, Stratum VI	moderately carburized, estimated carbon content 0.4 %	Muhly et al. 1990:166
33	knife blade	11th c. B.C.	Tel Miqne-Ekron, Building 350, south room, Stratum V	corroded through but some remnant structure preserved; "probably a mild steel"	Michael Notis, pers. comm. 1993
34	knife blade	ca. 1000 B.C.	Ashdod	not carburized	Stech-Wheeler et al. 1981:257–258
35	pick or tool blade	late 10th c. B.C.	Ashdod	not carburized	Stech-Wheeler et al. 1981:257–258
36	axe	late 10th c. B.C.	Ashdod	not carburized	Stech-Wheeler et al. 1981:257–258
37	sword blade fragments	10th c. B.C.	Taanach TT 71	slight carburization, possibly not deliberate	Stech-Wheeler et al. 1981:249–250
38	plowshare	10th c. B.C.	Taanach TT 91	not carburized	Stech-Wheeler et al. 1981:250
39	plowshare	10th c. B.C.	Taanach TT 820	carburized	Stech-Wheeler et al. 1981:253
40	sickle or scythe fragment	10th c. B.C.	Taanach TT 322	carburized	Stech-Wheeler et al. 1981:250
41	plowshare scraper	10th c. B.C.	Taanach TT 387	carburized	Stech-Wheeler et al. 1981:250–251

APPENDIX B (CONTINUED)

Ref no.	Object	Date of Context	Provenance	Technical data	Publication
			PALESTINE (CONTINUED)		
42	armor scale	10th c. B.C.	Taanach TT 408	carburized	Stech-Wheeler et al. 1981:251
43	armor scale	10th c. B.C.	Taanach TT 602	not carburized	Stech-Wheeler et al. 1981:253
44	arrowhead	10th c. B.C.	Taanach TT 409	not carburized	Stech-Wheeler et al. 1981:251–252
45	chisel?	10th c. B.C.	Taanach TT 726	not carburized	Stech-Wheeler et al. 1981:252
46	unfinished tool	10th c. B.C.	Taanach TT 1879	carburized	Stech-Wheeler et al. 1981:252
47	unfinished blade	10th c. B.C.	Taanach TT 1880	carburized	Stech-Wheeler et al. 1981:252–253
48	knife blade	10th c. B.C.	Kinneret, Stratum V	thoroughly carburized and quenched	Muhly et al. 1990:166, 169
49	sickle	10th c. B.C.	Kinneret, Stratum IV	thoroughly carburized with "a well developed Widmenstätten structure"; more than 0.8% C	Muhly et al. 1990:167
50	spearhead tip	10th c. B.C.	Kinneret, Stratum IV	no trace of carburization	Muhly et al. 1990:167
51	knife blade	10th–9th c. B.C.	Tell el-Far'ah (South), Tomb 220	not carburized	Stech-Wheeler et al. 1981:258
52	knife blade	10th–9th c. B.C.	Tell el-Far'ah (South), Tomb 220	not carburized	Stech-Wheeler et al. 1981:258
53	arrowhead	10th–9th c. B.C.	Tell el-Far'ah (South), Tomb 230	not carburized	Stech-Wheeler et al. 1981:258
54	dagger blade	10th–9th c. B.C.	Tell el-Far'ah (South), Tomb 240	carburized	Stech-Wheeler et al. 1981:258

REFERENCES CITED

Åström, P.; Maddin, R.; Muhly, J. D.; and Stech, T.
    1986   Iron Artifacts from Swedish Excavations in Cyprus. *Opuscula Atheniensia* 16(3):27–41.

Belli, O.
    1991   The Problem of Tin Deposits in Anatolia and its Need for Tin, According to the Written Sources. Pp. 1–9 in *Anatolian Iron Ages. The Proceedings of the Second Anatolian Iron Ages Colloquium Held at Izmir, 4–8 May 1987*, eds. A. Cilingiroglu and D. H. French. British Institute of Archaeology at Ankara Monograph No. 13. Oxford.

Bjorkman, J. K.
    1973   Meteors and Meteorites in the Ancient Near East. *Meteoritics* 8:91–130.

Blomgren, S., and Tholander, E.
    1976   An Iron Handle from Hala Sultan Tekke in Cyprus Dated to about 1200 B.C. Metallographic Examination and Technical Judgment. Pp. 123–126 in *Hala Sultan Tekke. I: Excavations 1897–1971*, by P. Åström, D. M. Bailey, and V. Karageorghis. Studies in Mediterranean Archaeology 45(1). Göteborg: Paul Åströms Förlag.

Boehmer, R. M.
    1972   *Boğazköy-Ḫattuša. Ergebnisse der Ausgrabungen des deutschen archäologischen Instituts und der deutschen Orient-Gesellschaft herausgegenben von Kurt Bittel. VII: Die Kleinfunde von Boğazköy aus den Grabungskampagnen 1931–1939 und 1952–1969*. Wissenschaftliche Veroffentlichung der deutschen Orient-Gesellschaft 87. Berlin: Gebr. Mann Verlag.
    1979   *Boğazköy-Ḫattuša. Ergebnisse der Ausgrabungen herausgegeben von Kurt Bittel. X: Die Kleinfunde aus der Unterstadt von Boğazköy, Grabungskampagnen 1970–1978*. Berlin: Gebr. Mann Verlag.

Bonn, A. G.; Moyer, H.; and Notis, M. R.
    1993   The Typology and Archaeometallurgy of the Copper-base Artifacts. Addendum: Metallographic Study of the Level VI Copper-base Objects. Pp. 219–220 in *The Late Bronze Egyptian Garrison at Beth Shan: A Study of Levels VII and VIII*, Vol. I, by F. W. James and P. E. McGovern. University Museum Monograph 85. Philadelphia: The University of Pennsylvania Museum.

Byers, G.
    1995   A 12th Century B.C.E. Cave Tomb from Khirbet Nisya. Paper presented at the Annual Meeting of the American Schools of Oriental Research, Philadelphia.

Catling, H. W.
    1979   Knossos, 1978. *Archaeological Reports for 1978–79*, pp. 43–58. The Society for the Promotion of Hellenic Studies and The British School at Athens.
    1981   Archaeology in Greece, 1980–81. *Archaeological Reports for 1980–81*, pp. 3–62. The Society for the Promotion of Hellenic Studies and The British School at Athens.

Catling, R. W. V., and Lemos, I. S.
    1990   *Lefkandi*. II: *The Protogeometric Building at Toumba. Part 1: The Pottery*. The British School of Archaeology at Athens. Oxford: Thames and Hudson.

Craddock, P. T.
    1988   The Composition of the Metal Finds. Pp. 169–181 in *Researches in the Arabah 1959–1984*. I: *The Egyptian Mining Temple at Timna*, by B. Rothenberg. London: Institute for Archaeo-Metallurgical Studies, Institute of Archaeology, University College.

Dajani, R. W.
    1966   Jabal Nuzha Tomb at Amman. *Annual of the Department of Antiquities Jordan* 11:48–49.

Davis, D.; Maddin, R.; Muhly, J. D.; and Stech, T.
    1985   A Steel Pick from Mt. Adir in Palestine. *Journal of Near Eastern Studies* 44:41–51.

Dever, W. G.
    1993   Further Evidence on the Date of the Outer Wall at Gezer. *Bulletin of the American Schools of Oriental Research* 289:33–54.

Dietz, S.
    1980   *Asine*. II: *Results of the Excavations East of the Acropolis 1970–1974. Fasc. 2: The Middle Helladic Cemetery, The Middle Helladic and Early Mycenaean Deposits.*

Skrifter Utgivna av Svenska Institutet i Athen, 4°, XXIV:2. Stockholm: Paul Åströms Förlag.

Dothan, T.
1982 *The Philistines and their Material Culture.* New Haven and London: Yale University Press, and Jerusalem: Israel Exploration Society.
1989 Iron Knives from Tel Miqne-Ekron. *Eretz Israel* 20:154–163 (Hebrew), 199 (Eng. summary).
1990 Ekron of the Philistines, Part I: Where They Came From, How They Settled Down and the Place They Worshipped In. *Biblical Archaeology Review* 16(1):26–36.

Dothan, T., and Gitin, S.
1992 Ekron. Pp. 415–422 in Vol. 2 (D–G) of *The Anchor Bible Dictionary*, ed. D. N. Freedman. New York: Doubleday.
1993 Miqne, Tel (Ekron). Pp. 1051–1059 in *The New Encyclopedia of Archaeological Excavations in the Holy Land*, Vol. 3, ed. E. Stern. Jerusalem: The Israel Exploration Society.

Fensham, C.
1969 Iron in the Ugaritic Texts. *Oriens Antiquus* 8:209–213.

Finkelstein, I.
1994 Penelope's Shroud Unraveled: Iron II Date of Gezer's Outer Wall Established. *Tel Aviv* 21:276–282.

Frisch, B.; Mansfeld, G.; and Thiele, W.-R.
1985 *Kāmid el-Lōz 6. Die Werkstätten der spätbronzezeitlichen Paläste.* Saarbrücker Beiträge zur Altertumskunde, Band 33, eds. R. Hachmann and W. Schmitthenner. Bonn: Dr. Rudolf Habelt GMBH.

Gale, N. H.; Bachmann, H. G.; Rothenberg, B.; Stos-Gale, Z. A.; and Tylecote, R. F.
1990 The Adventitious Production of Iron in the Smelting of Copper. Pp. 182–191 in *Researches in the Arabah 1959–1984.* I: *The Ancient Metallurgy of Copper Archaeology-Experiment-Theory*, ed. B. Rothenberg. London: Institute for Archaeo-Metallurgical Studies, Institute of Archaeology, University College.

Hachmann, R.
1986 *Kāmid el-Lōz 1977–81: Bericht über die Ergebnisse der Ausgrabungen in Kāmid el-Lōz in den Jahren 1977 bis 1981.* Saarbrücker Beiträge zur Altertumskunde, Band 36. Bonn: Dr. Rudolf Habelt GMBH.

Hall, M. E., and Steadman, S. R.
1991 Tin and Anatolia: Another Look. *Journal of Mediterranean Archaeology* 4(1):217–234.

Heltzer, M.
1977 The Metal Trade of Ugarit and the Problem of Transportation of Commercial Goods. *Iraq* 39:203–211.
1978 *Goods, Prices and the Organization of Trade in Ugarit.* (*Marketing and Transportation in the Eastern Mediterranean in the Second Half of the II Millenium B.C.E.*) Wiesbaden: Dr. Ludwig Reichart Verlag.

Horne, L.
1982 Fuel for the Metal Worker: The Role of Charcoal and Charcoal Production in Ancient Metallurgy. *Expedition* 25(1):6–13.

Karageorghis, V.
1974 *Excavations at Kition.* I: *The Tombs (Text).* Nicosia: Republic of Cyprus, Ministry of Communications and Works, Department of Antiquities.
1982 Metallurgy in Cyprus during the 11th Century BC. Pp. 297–301 in *Acta of the International Archaeological Symposium Early Metallurgy in Cyprus, 4000–500 BC, Larnaca, Cyprus 1–6 June 1981*, eds. J. D. Muhly, R. Maddin, and V. Karageorghis. Nicosia: Pierides Foundation.
1983 *Palaepaphos-Skales. An Iron Age Cemetery in Cyprus*, 2 vols. Deutsches archäologisches Institut, Ausgrabungen in Alt-Paphos auf Cypern, Band 3, ed. F. G. Maier. Konstanz: Universitätsverlag.

Kilian, K.
1983 Ausgrabungen in Tiryns 1981. Bericht zu den Grabungen. *Archäologischer Anzeiger* 3:294–328.

Košak, S.
1982 *Hittite Inventory Texts* (CTH 241–250). Heidelberg: Carl Winter Universitätsverlag.
1986 "The Gospel of Iron." Pp. 125–135 in *Kaniššuwar. A Tribute to Hans G. Güterbock on his Seventy-Fifth Birthday, May 27, 1983*, eds. H. A. Hoffner, Jr. and G. M. Beckman. The Oriental Institute of the University of Chicago Assyriological Studies No. 23. Chicago.

Liebowitz, H.
1981 Excavations at Tel Yin'am: The 1976 and 1977 Seasons: Preliminary Report. *Bulletin of the American Schools of Oriental Research* 243:79–94.
1993 Yin'am, Tel. Pp. 1515–1516 in *The New Encyclopedia of Archaeological Excavations in the Holy Land*, Vol. 4, ed. E. Stern. Jerusalem: The Israel Exploration Society.

Liebowitz, H., and Folk, R.
1984 The Dawn of Iron Smelting in Palestine: The Late Bronze Age Smelter at Tel Yin'am, Preliminary Report. *Journal of Field Archaeology* 11:265–280.

Limet, H.
1984 Documents relatifs au fer à Mari. Pp. 191–196 in *Mari. Annales de recherches interdisciplinaires*, Vol. 3. Paris: Editions Recherche sur les Civilizations.
1986 *Textes administratifs relatifs aux métaux*. Archives Royales de Mari XXV. Paris: Editions Recherche sur les Civilisations.

Maddin, R.
1982 Early Iron Technology in Cyprus. Pp. 303–312 in *Acta of the International Archaeological Symposium Early Metallurgy in Cyprus, 4000–500 BC, Larnaca, Cyprus 1–6 June 1981*, eds. J. D. Muhly, R. Maddin, and V. Karageorghis. Nicosia: Pierides Foundation.

Maddin, R.; Muhly, J. D.; and Stech, T.
1987 An Iron Axe from Tell Qiri. Pp. 244–245 in *Tell Qiri: A Village in the Jezreel Valley. Report of the Archaeological Excavations 1975–1977*, by A. Ben-Tor and Y. Portugali. Qedem. Monographs of the Institute of Archaeology Vol. 24. Jerusalem: The Hebrew University of Jerusalem.

Maddin, R.; Stech, T.; and Muhly, J. D.
forth- Metallurgical Studies of Iron Artifacts coming from Kition. In *Excavations at Kition VI*, eds. V. Karageorghis and M. Demas.

Maxwell-Hyslop, K. R.
1972 The Metals *Amūtu* and *Aši'u* in the Kültepe Texts. *Anatolian Studies* 22:159–162.
1980 A Note on the Jewellery Listed in the Inventory of Manninni (CTH 504). *Anatolian Studies* 30:85–90.

Mazar, A.
1985 *Excavations at Tell Qasile. Part Two: The Philistine Sanctuary: Various Finds, The Pottery, Conclusions, Appendixes*. Qedem. Monographs of the Institute of Archaeology. Vol. 20. Jerusalem: The Hebrew University of Jerusalem.

McDonald, W. A.; Coulson, W. D. E.; and Rosser, J. (eds.)
1983 *Excavations at Nichoria in Southwest Greece. III: Dark Age and Byzantine Occupation*. Minneapolis: The University of Minnesota Press.

McGhee, R.
1984 The Timing of the Thule Migration. *Polarforschung* 54:1–7.

McGovern, P. E.
1986 *The Late Bronze and Early Iron Ages of Central Transjordan: The Baq'ah Valley Project, 1977–1981*. University Museum Monograph 65. Philadelphia: The University of Pennsylvania Museum.
1987 Central Transjordan in the Late Bronze and Early Iron Ages: An Alternative Hypothesis of Socio-economic Transformation and Collapse. Pp. 267–273 in *Studies in the History and Archaeology of Jordan*, Vol. 3, ed. A. Hadidi. Department of Antiquities of Jordan, and London: Routledge and Kegan Paul.
1988 The Innovation of Steel in Transjordan. *Journal of Metals* 40(7):50–52.

McNutt, P. M.
1990 *The Forging of Israel: Iron Technology, Symbolism, and Tradition in Ancient Society*. Journal for the Study of the Old Testament suppl. series, Vol. 108; The Social World of Biblical Antiquity Series Vol. 8. Sheffield: Almond Press.

Mellink, M. J.
1984 Archaeology in Asia Minor. *American Journal of Archaeology* 88:441–459.

Morris, I.
1989 Circulation, Deposition and the Formation of the Greek Iron Age. *Man* (n.s.) 23:502–519.

Muhly, J. D.
1982 How Iron Technology Changed the Ancient World—And Gave the Philistines a Military Edge. *Biblical Archaeology Review* 8(6):42–54.

1992 Review of *The Forging of Israel: Iron Technology, Symbolism, and Tradition in Ancient Society*, by P. M. McNutt. *Journal of the American Oriental Society* 112(4):696–702.

1993 Early Bronze Age Tin and the Taurus. *American Journal of Archaeology* 97:239–253.

Muhly, J. D.; Begemann, F.; Öztunalı, Ö.; Pernicka, E.; Schmitt-Strecker, S.; and Wagner, G. A.
1991 The Bronze Metallurgy of Anatolia and the Question of Local Tin Sources. Pp. 209–220 in *Archaeometry '90. International Symposium on Archaeometry 2–6 April 1990, Heidelberg, Germany*, eds. E. Pernicka and G. A. Wagner. Basel, Boston, Berlin: Birkhäuser Verlag.

Muhly, J. D.; Maddin, R.; and Karageorghis, V. (eds.)
1982 *Acta of the International Archaeological Symposium Early Metallurgy in Cyprus, 4000–500 BC, Larnaca, Cyprus 1–6 June 1981*. Nicosia: Pierides Foundation.

Muhly, J. D.; Maddin, R.; and Stech, T.
1990 The Metal Artifacts. Pp. 159–175 in *Kinneret. Ergebnisse der Ausgrabungen auf dem Tell el-'Orēme am See Gennesaret 1982–1985*, by V. Fritz. Wiesbaden: Otto Harrassowitz.

Muhly, J. D.; Maddin, R.; Stech, T.; and Özgen, E.
1985 Iron in Anatolia and the Nature of the Hittite Iron Industry. *Anatolian Studies* 35:67–84.

Muscarella, O. W.
1988 The Background to the Phrygian Bronze Industry. Pp. 177–192 in *Bronzeworking Centres of Western Asia c. 1000–539 B.C.*, ed. J. Curtis. London and New York: Kegan Paul International.

Notis, M. R.; McGovern, P. E.; Moyer, H.; Pigott, V. C.; and Swann, C. P.
1986 The Copper-base Archaeometallurgy. Pp. 178–283 in *The Late Bronze and Early Iron Ages of Central Transjordan: The Baq'ah Valley Project, 1977–1981*, ed. P. E. McGovern. University Museum Monograph 65. Philadelphia: The University of Pennsylvania Museum.

Notis, M. R.; Pigott, V. C.; McGovern, P. E.; Liu, K. H.; and Swann, C. P.
1986 The Metallurgical Technology: The Archaeometallurgy of the Iron IA Steel. Pp. 272–278 in *The Late Bronze and Early Iron Ages of Central Transjordan: The Baq'ah Valley Project, 1977–1981*, ed. P. E. McGovern. University Museum Monograph 65. Philadelphia: The University of Pennsylvania Museum.

Özgüç, N.
1976 An Ivory Box and a Stone Mould from Acemhöyük. *Türk Tarih Kürümü Belleten* 40:555–560.

Pernicka, E.; Wagner, G. A.; Muhly, J. D.; and Öztunalı, Ö.
1992 Comment on the Discussion of Ancient Tin Sources in Anatolia. *Journal of Mediterranean Archaeology* 5(1):91–98.

Photos, E.
1989 The Question of Meteoritic Versus Smelted Nickel-rich Iron: Archaeological Evidence and Experimental Results. *World Archaeology* 20(3):403–421.

Piaskowski, J.
1982 A Study of the Origin of the Ancient High-nickel Iron Generally Regarded as Meteoritic. Pp. 237–243 in *Early Pyrotechnology. The Evolution of the First Fire-using Industries*, eds. T. A. Wertime and S. F. Wertime. Washington, DC: Smithsonian Institution Press.

Pickles, S.
1988 *Metallurgical Changes in Late Bronze Age Cyprus*. University of Edinburgh Department of Archaeology, Occasional Paper No. 17. Edinburgh.

Popham, M. R.; Calligas, P. G.; and Sackett, L. H.
1989 Further Excavation of the Toumba Cemetery at Lefkandi, 1984 and 1986, A Preliminary Report. *Archaeological Reports for 1988–1989*, pp. 117–129. The Society for the Promotion of Hellenic Studies and The British School at Athens.

1993 *Lefkandi. II: The Protogeometric Building at Toumba. Part 2: The Excavation, Architecture and Finds*. The British School of Archaeology at Athens.

Popham, M. R.; Sackett, L. H.; and, with Themelis, P. G.
1979 *Lefkandi. I: The Iron Age* (plates). The British School of Archaeology at Athens Suppl. Vol. 11. London: Thames and Hudson.

1980   *Lefkandi*. I: *The Iron Age* (text). The British School of Archaeology at Athens Suppl. Vol. 12. London: Thames and Hudson.

Popham, M. R.; Touloupa, E.; and Sackett, L. H.
1982a   Further Excavation of the Toumba Cemetery at Lefkandi, 1981. *British School at Athens Annual* 77:213–248.
1982b   The Hero of Lefkandi. *Antiquity* 56:169–174.

Pritchard, J. B.
1980   *The Cemetery at Tell es-Sa'idiyeh, Jordan*. University Museum Monograph 41. Philadelphia: The University of Pennsylvania Museum.

Rapp, G., Jr., and Aschenbrenner, S. E. (eds.)
1978   *Excavations at Nichoria in Southwest Greece*. I: *Site, Environs, and Techniques*. Minneapolis: The University of Minnesota Press.

Rehder, J. E.
1992   Iron Versus Bronze for Edge Tools and Weapons. *Journal of the Minerals, Mining and Materials Society* 44(8):42–46.

Rothenberg, B.
1982   Discussion following Snodgrass. P. 295 in *Acta of the International Archaeological Symposium Early Metallurgy in Cyprus, 4000–500 BC, Larnaca, Cyprus 1–6 June 1981*, eds. J. D. Muhly, R. Maddin, and V. Karageorghis. Nicosia: Pierides Foundation.
1983   Corrections on Timna and Tel Yin'am in the *Bulletin*. *Bulletin of the American Schools of Oriental Research* 252:69–70.
1988   *Researches in the Arabah 1959–1984*. I: *The Egyptian Mining Temple at Timna*. London: Institute for Archaeo-Metallurgical Studies, Institute of Archaeology, University College.

Rothenberg, B. (ed.)
1990   *Researches in the Arabah 1959–1984*. II: *The Ancient Metallurgy of Copper Archaeology-Experiment-Theory*. London: Institute for Archaeo-Metallurgical Studies, Institute of Archaeology, University College.

Rudolph, W.
1973   Die Nekropole am Prophitis Elias bei Tiryns. Pp. 23–127 in *Tiryns: Forschungen und Berichte* VI, ed. U. Jantzen. Deutsches archäologisches Institut, Athen. Mainz: Philipp von Zabern.

Sakellerakis, J. A.
1970   Das Kuppelgrab A von Archanes und das kretisch-mykenische Tieropferritual. *Prähistorische Zeitschrift* 45:135–219.

Shalev, S.
1993   Metal Production and Society at Tel Dan. Pp. 57–65 in *Biblical Archaeology Today, 1990. Proceedings of the Second International Congress on Biblical Archaeology. Pre-Congress Symposium: Population, Production and Power. Jerusalem, June 1990*. Supplement, eds. A. Biran and J. Aviram. Jerusalem: The Israel Exploration Society.

Siegelová, J.
1984   Gewinnung und Verarbeitung von Eisen im Hethitischen Reich im 2. Jahrtausend v. u. Z. *Annals of the Náprstek Museum* 12:71–168.

Smith, C. S.
1967   The Interpretation of Microstructures of Metallic Artifacts. Pp. 20–52 in *Application of Science in Examination of Works of Art*, Proceedings of the Seminar: September 7–16, 1965, conducted by the Research Laboratory, Museum of Fine Arts, Boston.

Smith, R. H.; Maddin, R.; Muhly, J. D.; and Stech, T.
1984   Bronze Age Steel from Pella, Jordan. *Current Anthropology* 25:234–236.

Snodgrass, A. M.
1965   Barbarian Europe and Early Iron Age Greece. *Proceedings of the Prehistoric Society* 31:229–240.
1971   *The Dark Age of Greece. An Archaeological Survey of the Eleventh to the Eighth Centuries BC*. Edinburgh: Edinburgh University Press.
1980   Iron and Early Metallurgy in the Mediterranean. Pp. 335–374 in *The Coming of the Age of Iron*, eds. T. Wertime and J. D. Muhly. New Haven and London: Yale University Press.
1982   Cyprus and the Beginnings of Iron Technology in the Eastern Mediterranean. Pp. 285–294 in *Acta of the International Archaeological Symposium Early Metallurgy in Cyprus, 4000–500 BC, Larnaca, Cyprus 1–6 June 1981*, eds. J. D. Muhly, R. Maddin, and V.

Karageorghis. Nicosia: Pierides Foundation.

1983 The Greek Early Iron Age: A Reappraisal. *Dialogues d'histoire ancienne,* pp. 73–86.

1984 Review of W. A. McDonald, W. D. E. Coulson, and J. Rosser (eds.), *Excavations at Nichoria in Southwest Greece.* III: *Dark Age and Byzantine Occupation. Antiquity* 58:152–153.

1989 The Coming of the Iron Age in Greece: Europe's Earliest Bronze/Iron Transition. Pp. 22–35 in *The Bronze Age–Iron Age Transition in Europe. Aspects of Continuity and Change in European Societies c. 1200 to 500 B.C.,* Part 1, eds. M. L. Stig Sorensen and R. Thomas. BAR International Series 483(i). Oxford: British Archaeological Reports.

Stager, L. E.
1985 The Archaeology of the Family in Ancient Israel. *Bulletin of the American Schools of Oriental Research* 260:1–35.

Stech, T.; Muhly, J. D.; and Maddin, R.
1985 The Analysis of Iron Artifacts from Palaepaphos-*Skales. Report of the Department of Antiquities,* Cyprus, 1985, pp. 192–202.

Stech-Wheeler, T.; Muhly, J. D.; Maxwell-Hyslop, K. R.; and Maddin, R.
1981 Iron at Taanach and Early Iron Metallurgy in the Eastern Mediterranean. *American Journal of Archaeology* 85:245–268.

Stig Sorensen, M. L., and Thomas, R. (eds.)
1989 *The Bronze Age–Iron Age Transition in Europe. Aspects of Continuity and Change in European Societies c. 1200 to 500 B.C.* BAR International Series 483(i). Oxford: British Archaeological Reports.

Stronach, D. B.
1957 The Development and Diffusion of Metal Types in Early Bronze Age Anatolia. *Anatolian Studies* 7:89–125.

Tholander, E.
1971 Evidence of the Use of Carburized Steel and Quench Hardening in Late Bronze Age Cyprus. *Opuscula Atheniensia* 10(3):15–22.

Tylecote, R. F.
1988 Metallurgical Notes on Selected Metal Objects. Pp. 186–190 in *Researches in the Arabah 1959–1984.* I: *The Egyptian Mining Temple at Timna,* ed. B. Rothenberg. London: Institute for Archaeo-Metallurgical Studies, Institute of Archaeology, University College.

Ussishkin, D.
1978 Excavations at Tel Lachish 1973–1977. Preliminary Report. *Tel Aviv* 5:1–97.

Vaiman, A. A.
1982 Eisen in Sumer. Pp. 33–38 in *Vorträge gehalten auf der 28. Rencontre Assyriologique Internationale in Wien. 6–10 Juli 1981.* Archiv für Orientforschung Beiheft 19. Horn.

Van Loon, M. N.
1978 *Korucutepe. Final Report on the Excavations of the Universities of Chicago, California (Los Angeles) and Amsterdam in the Keban Reservoir, Eastern Anatolia 1968–1970,* Vol. 2. Amsterdam, New York, Oxford: North-Holland Publishing Company.

1980 *Korucutepe. Final Report on the Excavations of the Universities of Chicago, California (Los Angeles) and Amsterdam in the Keban Reservoir, Eastern Anatolia 1968–1970,* Vol. 3. Amsterdam, New York, Oxford: North-Holland Publishing Company.

Varoufakis, G. J.
1981 Investigation of some Minoan and Mycenaean Iron Objects. Pp. 25–32 in *Frühes Eisen in Europa. Acta des 3. Symposiums des "Comité pour la sidérurgie ancienne de l'USIPP" Schaffhausen und Zurich 24.–26. Oktober 1979,* ed. H. Haefner. Schaffhausen: Verlag Peter Meili.

1982 The Origin of Mycenaean and Geometric Iron on the Greek Mainland and in the Aegean Islands. Pp. 315–322 in *Acta of the International Archaeological Symposium Early Metallurgy in Cyprus, 4000–500 BC, Larnaca, Cyprus 1–6 June 1981,* eds. J. D. Muhly, R. Maddin, and V. Karageorghis. Nicosia: Pierides Foundation.

Waldbaum, J. C.
1968 *The Use of Iron in the Eastern Mediterranean: 1200–900 B.C.* Unpublished Ph.D. dissertation, Department of the Classics, Harvard University, Cambridge, MA.

1978 *From Bronze to Iron: The Transition from the Bronze Age to the Iron Age in the East-*

1980 The First Archaeological Appearance of Iron and the Transition to the Iron Age. Pp. 69–98 in *The Coming of the Age of Iron,* eds. T. A. Wertime and J. D. Muhly. New Haven and London: Yale University Press.

ern Mediterranean. Studies in Mediterranean Archaeology 54. Göteberg: Paul Åströms Förlag.

1982 Bimetallic Objects from the Eastern Mediterranean and the Question of the Dissemination of Iron. Pp. 325–347 in *Acta of the International Archaeological Symposium Early Metallurgy in Cyprus, 4000–500 BC, Larnaca, Cyprus 1–6 June 1981,* eds. J. D. Muhly, R. Maddin, and V. Karageorghis. Nicosia: Pierides Foundation.

1989 Copper, Iron, Tin, Wood: The Start of the Iron Age in the Eastern Mediterranean. *Archeomaterials* 3:111–122.

Wertime, T. A.
1980 The Pyrotechnologic Background. Pp. 1–24 in *The Coming of the Age of Iron,* eds. T. A. Wertime and J. D. Muhly. New Haven: Yale University Press.

1982 Cypriot Metallurgy Against the Backdrop of Mediterranean Pyrotechnology: Energy Reconsidered. Pp. 351–361 in *Acta of the International Archaeological Symposium Early Metallurgy in Cyprus, 4000–500 BC, Larnaca, Cyprus 1–6 June 1981,* eds. J. D. Muhly, R. Maddin, and V. Karageorghis. Nicosia: Pierides Foundation.

1983 The Furnace Versus the Goat: The Pyrotechnologic Industries and Mediterranean Deforestation in Antiquity. *Journal of Field Archaeology* 10:445–452.

Willies, L.
1990 An Early Bronze Age Tin Mine in Anatolia, Turkey. *Bulletin of the Peak District Mines Historical Society* 11:91–96.

1992 Reply to Pernicka *et al.*: Comment on the Discussion of Ancient Tin Sources in Anatolia. *Journal of Mediterranean Archaeology* 5(1):99–103.

Yener, K. A.
1986a Tin in the Taurus Mountains: The Bolkardağ Mining Survey. *American Journal of Archaeology* 90:183 (abstract).

1986b The Archaeometry of Silver in Anatolia: The Bolkardağ Mining District. *American Journal of Archaeology* 90:469–472.

Yener, K. A., and Goodway, M.
1992 Response to Mark E. Hall and Sharon R. Steadman "Tin and Anatolia: Another Look." *Journal of Mediterranean Archaeology* 5(1):77–90.

Yener, K. A., and Özbal, H.
1987 Tin in the Turkish Taurus Mountains: The Bolkardağ Mining District. *Antiquity* 61:220–226.

Yener, K. A.; Özbal, H.; Kaptan, E.; Pehlivan, A. N.; and Goodway, M.
1989 Kestel: An Early Bronze Age Source of Tin Ore in the Taurus Mountains, Turkey. *Science* 244:200–203.

Yener, K. A.; Sayre, E. V.; Joel, E. C.; Özbal, H.; Barnes, I. L.; and Brill, R. H.
1991 Stable Lead Isotope Studies of Central Taurus Ore Sources and Related Artifacts from Eastern Mediterranean Chalcolithic and Bronze Age Sites. *Journal of Archaeological Science* 18:541–577.

Yener, K. A., and Vandiver, P. B.
1993a Tin Processing at Göltepe, an Early Bronze Age Site in Anatolia. *American Journal of Archaeology* 97:207–238.

1993b Reply to J. D. Muhly, "Early Bronze Age Tin and the Taurus," with an Appendix by Lynn Willies: Early Bronze Age Tin Working at Kestel. *American Journal of Archaeology* 97:255–264.

Zaccagnini, C.
1990 The Transition from Bronze to Iron in the Near East and in the Levant: Marginal Notes. *Journal of the American Oriental Society* 110:493–502.

3

Aspects of Early Metallurgy in Mesopotamia and Anatolia

Tamara Stech

ABSTRACT The Mesopotamian Metals Project (MMP) was undertaken in order to answer basic questions about the nature of metals used in different time periods and different areas of Mesopotamia. Drawing on the collections of the University of Pennsylvania Museum, this project involves the analysis of about 350 copper-base artifacts. Their chronological and geographical range is from third millennium Ur, Kish, Fara, Gawra, and Billa, to second millennium Nippur, Khafajeh, and Billa, to Neo-Assyrian Billa.

Taking the MMP data along with results published by other scholars, sufficient information is available to begin to describe and interpret the course of metallurgical activities in Mesopotamia. One of the important results of this research which is discussed in detail is that we can no longer postulate a smooth linear development from the use of native copper to smelted copper to arsenical copper to tin bronze. In Mesopotamia the "Bronze Age" was in fact the "copper-base" metal age, and the same situation pertained in Bronze Age Anatolia. [Final ms. received 10/96.]

INTRODUCTION

Archaeometallurgical research at the University of Pennsylvania has, in recent years, been directed toward elucidating the role of metals in the technological, economic, commercial, cultural, and symbolic spheres in early Southwest Asia.[1] Because archaeology in this area has had a long history, it depends on many hypotheses that need to be re-examined in the light of current theoretical formulations. For example, ancient metals are generally regarded as "valuable," but the nature of that value is not made explicit. Does value relate to technological function or aesthetic qualities, or to usefulness in commerce or validation of status? The listing of possible lines of inquiry could be extended, but the point is that the Mesopotamian "model" for the development of metallurgy—one which is based on orderly technological evolution—has become a monolith rather than a flexible and revisable hypothesis.

In fact, so little was really known about Mesopotamian metallurgy that as late as 1977 Mallowan (1977:4) was able to say that we still had not determined whether the Sumerians used copper or bronze. This observation led James Muhly to conceive the Mesopotamian Metals Project (MMP), which has been going on at the University of Pennsylvania since 1980, and is now approaching publication.[2] The project includes elemental analyses of about 350 copper-base artifacts, all from known archaeological contexts, metallographic observation of about 150, and scanning electron microscopy of about 75. Lead isotope analysis of 60 is being performed by F. Begemann and S. Schmitt-Strecker in Mainz. The chronological and geographical range is from third millennium Ur, Kish, Fara, Gawra, and Billa to second millennium Nippur, Khafajeh, and Billa to Neo-Assyrian Billa. Taking our results with those compiled by Moorey (1985:51–68; 1994:276–278), as well as those of Berthoud (1979:table 2) and Müller-Karpe (1990), we can begin the task of description and interpretation.

Anatolia is included in this examination because, from the time metal was first used, technological change and adoption of innovations occurred more or less contemporaneously there and in Mesopotamia, with initial priority on the side of Anatolia. The similarity in metallurgical developments stands in contrast to the differences in the relevant natural resources in the two areas, a factor which must have been important in the creation and maintenance of any connection, as was relative ease of communication on the Tigris and Euphrates rivers. Studying the two cases together can show how alike they really are and what factors could have made them that way.

NEOLITHIC ANATOLIA

The factual background to the inquiry starts at Çayönü Tepesi (Fig. 3.1), a site in southeastern Turkey occupied from about 7250 to 6750 B.C. (see, e.g., Çambel and Braidwood 1980; Braidwood et al. 1981).[3] About one hundred artifacts of native copper have been found at the site, along with hundreds of malachite disc beads and much debris from the processing of malachite. Contemporary Cafer Hüyük near Çayönü has apparently yielded no metal (no mention of metal is made in Cauvin 1985), but it should be borne in mind that the horizontal exposure at Çayönü is greater than at any other Neolithic site in Southwest Asia. In addition, while copper artifacts were found in several parts of the site, the largest concentration of them and of malachite as well was in two areas in a single courtyard. The malachite concentrations are also unusually rich in small finds, particularly stone and bone ornaments and small clay artifacts, and noteworthy for the paucity of animal bones and chipped stone artifacts.[4] The working areas were most intensively used in the intermediate phase, which is the least understood of the five strata of Çayönü Phase I (the main period of prehistoric occupation), but metal and mineral processing also took place immediately preceding and following the intermediate phase, albeit on a smaller scale. Therefore, the indications of an unusual interest in native copper at Çayönü might be attributable to the vagaries of excavation, as well as to the large exposure. Recent excavations bear out this observation, since copper beads roughly contemporary with those of Çayönü have been found at Aşıklı Höyük near Aksarary (Esin 1993; Maddin et al. 1999). Possible early copper artifacts found at Nevali Çori (Hauptmann et al. 1993:543–544) are stratigraphically and typologically of the Aceramic Neolithic, but since they have an impure chemical composition suggestive of smelted rather than native copper, such an early chronological assignment may be incorrect.

As far as we know archaeologically, working of native copper did not continue, and metallurgy was not practiced again until the later seventh millennium at Çatal Hüyük in central Anatolia. Some copper artifacts are known from the site (Mellaart 1967:218), as are beads from Level IX (ca. 6400 B.C.) identified as lead. Lead very rarely occurs as a native metal, so it must usually be smelted from the lead sulfide ore galena, which is easier to reduce than copper. Analysis of the beads has, however, shown that they are in fact galena, not metal (Sperl 1990). In addition, slag from

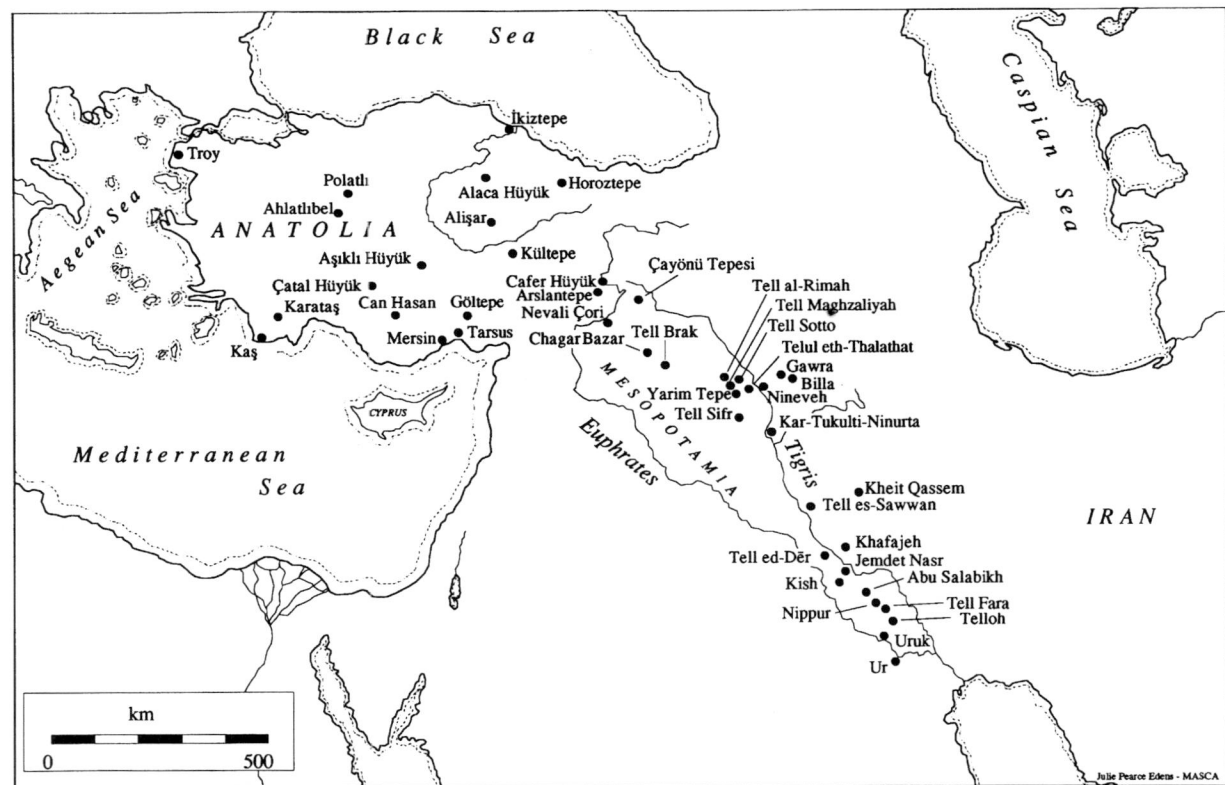

Figure 3.1   *Map of sites mentioned in the text.*

Level VI said to derive from a copper smelting operation (Neuninger et al. 1964) has also been questioned on the basis that it could be a crucible or melting slag (Tylecote 1976:5). And ores found at Çatal Hüyük could actually have been intended for use as minerals, as they were at Çayönü, rather than for smelting to metal.

Çayönü and Çatal Hüyük both appear to stand out in the general context of the Neolithic period in Southwest Asia in terms of relative sophistication in communal organization, symbolic life, and technological achievement, but their distinctiveness is probably due to large exposures and good preservation, particularly at Çatal Hüyük (Voigt 1990:13–14; Maddin et al. 1999). In terms of metalworking it is important to point out that the finds themselves are not numerous; all the copper artifacts from Çayönü together weigh only a few kilos, so the raw materials could have been collected by a single person on one visit to an ore-bearing area. The placement of the Aşıklı beads in burials might suggest that they were objects of value taken out of circulation by burial, but what we know so far of Aşıklı does not indicate a level of social complexity that would call for that kind of validation of different social roles. More reasonably, the beads can be interpreted as possessions of the deceased during life, with which they went to the grave. For Çayönü, Aslı Özdoğan (1995:88) has noted that in the late Cell Building Phase an increase in clay tokens coincides with a decrease in metal artifacts. This fact would suggest that metal was not integral to developing social complexity, an idea supported by the paucity of metallurgical remains attributable to the next three millennia. The macehead from Level 2b at Can Hasan remains remarkable for its date and the apparent nature of the casting technology used to make it (see Raymond 1984:11, top left for an illustration).

NEOLITHIC MESOPOTAMIA

That no site in Mesopotamia has produced as much native copper as Çayönü is not surprising, but there was certainly interest in obtaining the material, and perhaps a system of valuation was developing. The earliest finds are from seventh millennium Tell Maghzaliyah, an awl found on a house floor (Bader et al. 1981:62; Ryndina and Yakhontova 1985), and Tell Sotto, two beads among a necklace of stone and shell beads in an infant burial. The excavators of Tell Sotto note that it is difficult to tell whether the beads are in fact copper ore—malachite—or folded sheet copper (Merpert et al. 1978:47). The distinction is important because the metals of Çayönü have shown that native copper was almost always worked into sheet before being made into an object. Of sixth millennium date are two copper fragments from Telul eth-Thalathat (Fukai and Matsutani 1977:54) which were found in fill. At Tell es-Sawwan three copper beads and a small piece of copper ore were on the floor of a building in Level II; a very small copper knife with a perforation at one end appeared in a grave beneath the floor of Room 142 (Al-A'dami 1968:59). Several buildings in this level, including those in which the copper artifacts were found, have an unusual number of infant burials beneath them and contained no "household debris or agricultural implements"; Al-A'dami (1968:60) interprets them as mortuary structures (cf. Muhly 1989:2–3; Moorey 1994:255).

Yarim Tepe I has yielded 26 finds of metal and malachite, which come from almost every level. Of these, 17 appear to be malachite fragments and one other is a malachite pendant (Merpert et al. 1977:82; 1978:36; Merpert and Munchaev 1987:17). Metallic finds are two rings, from Level XI and X respectively, a copper sheet bead from Level VII, and the lead bracelet found beneath the wall of house no. 35 of Level XII (Merpert et al. 1977:82). Specific contexts for most of these artifacts have not yet been published. Yarim Tepe II shows continued acquisition of ore and metal, with at least nine pieces of malachite and one malachite pendant/bead; a copper bead was also found. The most unusual piece is a copper seal/pendant, located below the floor of Tholos 67, the largest circular building in Yarim Tepe II (Merpert et al. 1978:44, 39; Merpert and Munchaev 1987:27).

To ascertain the number of metallic copper artifacts in Neolithic Mesopotamia, we should—pending further study—exclude the beads since uncertainty has been expressed in several cases about the material of which they were made. On analogy with Çayönü, it is likely that most beads are in fact malachite. Thus, we are left with few native copper artifacts in Neolithic Mesopotamia. The finds which are most likely to be metal are the awl from Tell Maghzaliyah; the knife from Tell es-Sawwan; two rings, a bead, and the lead bracelet from Yarim Tepe I; and the seal/pendant from Yarim Tepe II. It is striking that the lead bracelet and the seal/pendant were found below floor levels, and that the Sawwan knife was in a sub-floor burial in a building with unusual contents. A possible interpretation is that metal in the form of finished artifacts was more highly regarded than copper ore, and that the artifacts may, in some cases, represent early versions of foundation deposits. Malachite and native copper may have come together from their sources, which are presently unknown, but their differences and the rarity of the metal seem to have been appreciated.

POSSIBLE FACTORS IN THE DEVELOPMENT OF SMELTING, FIFTH AND EARLY FOURTH MILLENNIA

Little visible experimentation with metals seems to have taken place during the fifth and early fourth millennia in Anatolia and Mesopotamia, in contrast to the considerable metallurgical activity in Iran (see Pigott, this volume). Although unambiguous evidence for the beginnings of smelting is lacking, since the remains at sites in Iran like Tal-i Iblis and Tepe Ghabristan are only beginning to be understood clearly and such sites are in any case lacking in Anatolia, the fifth and fourth millennia must be the period in which this procedure was being worked out, presumably in areas that had the proper resources for experimentation and cultural reasons for doing so.

The discoveries of lead artifacts at Yarim Tepe I may suggest that lead smelting was known, since, as mentioned above, lead rarely occurs as a native metal. The large number of lead artifacts (38 of 84 metal objects; Moorey 1994:294) in the "Jemdet Nasr" cemetery at Ur (more likely, Jemdet Nasr through Early Dynastic II; see Vértesalij and Kolbus 1985) may also point to lead as the material for which smelting was first developed, for this was also the period in which copper artifacts started appearing in both Mesopotamia and Anatolia in significant quantities. The proportion of silver work represented in the "Sammelfund" of Uruk (Heinrich 1936), probably a hoard buried at the end of Uruk IV (Muhly 1983:354–355), may also be an indication of the importance of lead in the development of smelting, since silver can often be recovered from lead. Also in the Sammelfund is a copper lion figurine which contains about 9% lead and no other alloying elements. This high lead alloy may represent an experiment or an accident, since leading does not seem to become a deliberate part of ancient metallurgy even during the third millennium B.C., a period in which metallurgical techniques expanded to encompass almost all ways of manipulating metals (Craddock 1977:passim; Moorey 1994:294).

COPPER ALLOYS

Because alloys of copper are prominent in the discussion that follows, it is necessary to define them. An alloy results when two or more metals are, intentionally or inadvertently, mixed together in the presence of heat; the product is assumed to have some kind of desirable, enhanced properties. What is of concern in technological history is deliberate alloying, when a person creates the mixture. This may be done by combining metals which have been processed separately to the metallic state, or by co-smelting two ores, or in some cases by mixing an ore with a metal. In addition, the selection of ores which contain several metallic elements should be regarded as a deliberate manipulation to produce a desired effect.

Bronze—by which I mean only the combination of copper and tin—is easier to classify as an alloy than arsenical copper because tin is of limited occurrence in nature, and in Southwest and South Central Asia rarely occurs in conjunction with copper ores. In southwestern Afghanistan, however, tin and copper do co-occur, with tin contents sufficient to yield a low tin ($\pm 2\%$) bronze (Cleuziou and Berthoud 1982:15). Because of this fact and because 2% tin in copper yields a metal that is noticeably stronger and harder than pure copper (R. Maddin, pers. comm.), 2% has been selected as the minimum amount of tin to define "bronze." Below that amount, we cannot determine whether the alloy was produced by smelting a joint copper-tin ore, or by the recycling of bronze scrap, or by deliberate addition (see Cleuziou and Berthoud 1982:15; Craddock 1980:171).

It is more difficult to define arsenical copper as a deliberate alloy because arsenical copper ores do occur in Iran (see, e.g., Bazin and Hübner 1969:passim; Berthoud, pers. comm.), Turkey (de Jesus 1980:90–95), and Cyprus (Zwicker 1982:64–67). In smelting such ores, some arsenic may be lost because of the tendency of arsenic to volatilize, but some is retained, the amount depending on the smelting conditions and the quantity originally present in the ore. The volatility of arsenic would account for the variation in arsenic contents in most groups of analyzed artifacts; i.e., there was little direct control over the amount of arsenic that would be retained. Even if the arsenic was added to copper in the form of orpiment or realgar (arsenic sulfides), the resulting metal would have been the same as if a joint ore had been used. Since an element of selectivity may have been involved in both cases, we still regard it as a deliberate alloy. The designation of the amount of arsenic necessary to constitute an alloy is difficult to determine because quantities present range from parts per million to over 5%. Northover (1989:113) points out that quantities of arsenic less than 2% offer little improvement over pure coppers, but about 1% is used here to facilitate discussion, since the existing analyses generally show only traces of arsenic or amounts over 1%.

MESOPOTAMIAN COPPER-BASE ARTIFACTS, LATE FOURTH–EARLY THIRD MILLENNIUM

The oldest analyzed artifacts from Mesopotamia, other than those in the Pennsylvania project, are a pure copper pin and mirror from Telloh (Parrot 1948:44), an axe from Gawra XII with a trace of tin (Levey and Burke 1959:43), and a pure copper spearhead from Nineveh 4 (Campbell Thompson and Mallowan 1933:145, n. 2). Late fourth millennium artifacts that we have studied include: an arsenical (0.61%) flat adze from Gawra Level XII, a pure copper pin or awl from Gawra XI, and two awls, one with a trace of arsenic, from Gawra IX; seven artifacts from Jemdet Nasr levels at Tell Fara, of which three are coppers with arsenic contents above 1%, and the rest have significant traces (above 0.2%) of arsenic; and three from Jemdet Nasr/Early Dynastic (ED) I contexts at Ur, two of which are arsenical coppers. Arsenical copper is also the material of a hook from Jemdet Nasr (Moorey and Schweizer 1972:180). Although the analyses are too few to determine how extensively arsenical copper was used before Early Dynastic I, it is clear that it was known, and by ED I it appears to have been fairly common. Of 13 artifacts from ED I Kheit Qassem in the Hamrin area, 5 contain over 1% arsenic, while 6 others have 0.2 to 1% (Berthoud 1979:nos. 6601–604, 6607–609, 6611, 6613, 6616–619). The same is true at Kish Cemetery Y, where amounts of arsenic up to 4.8% occur in artifacts. Cemetery Y is also notable because the first verifiable tin bronzes in Mesopotamia are found there.

The beginning of the use of tin bronze has been clarified by the Pennsylvania project, which has shown that some of the earliest tin bronzes are questionable. An artifact called a "mouthpiece" and said to come from below Gawra Level XI (Tobler 1950:pl. 182) is a tin bronze, but it is Achaemenid in type and probably a "kohl tube," an observation confirmed by Roger Moorey in conversation. The pin from Gawra VIII with 5.62% tin (Speiser 1935:102) was never identified by number, so the analysis cannot be checked. Woolley (1934:290, table II) lists high tin (over 8%) bronzes from the "earliest tombs at Ur," but he did not give the numbers of the graves or artifacts, so they are untraceable and therefore undatable.

We have confirmed the presence of bronze in Kish Cemetery Y, in which 3 of 14 artifacts are bronze and 2 others have significant impurities of tin. Moorey and Schweizer (1972:182) had similar findings, that 5 of 9 artifacts contained 1% or more of tin. The occurrence of bronze at Kish, in the graves at least, is in the context of the concentration of copper-base artifacts in the graves of a few individuals, who are distinguished by other features. They usually had a special set of implements, consisting of at least one saw (one with a dagger), chisel, and axe or adze. A similar set in gold came from the later grave of Puabi at Ur (Woolley 1934:pl. 158b) and from one grave at Abu Salabikh (Martin et al. 1985:168–169). Several of these graves at Kish are associated with chariots, and in one group of five graves there are no pottery vessels, their functions presumably having been filled by the extant counterparts in metal.

What is known about Kish in pseudo-historical tradition fits well with a possible interpretation of the Y graves. Kingship descended from heaven to Kish in what we term ED I, and the First Dynasty of Kish was preeminent in Mesopotamia. Although the Y graves may not constitute all the burials made in ED I Kish, they do suggest that, in at least one group that used a common burial ground, there were differences that called for the interment of unusual amounts and kinds of metal with some of the deceased. Such disposal of what might be designated as wealth is known in other instances to accompany the founders of dynasties or communities, in other words to betoken and validate some kind of political change. This is not the place in which to repeat arguments about how the social persona of the deceased might be reflected in death, or if power occasions the acquisition of goods, or if having the goods means having the power. But it is important to note that special burials took place at a dynamic moment in the history of Kish, and that tin bronze first occurs at this time.

ANATOLIAN COPPER-BASE ARTIFACTS, LATE FOURTH–EARLY THIRD MILLENNIUM

The evidence for alloying practices in Anatolia in the later fourth and early third millennia is sparse, with indications that tin bronze was available around that time and coexisted with arsenical copper and to a lesser extent with pure copper (Stech and Pigott 1986:53–54). The remarkable hoard of 22 metal objects found at Arslantepe (Malatya), late Uruk in date, shows the strength of the arsenical copper tradition, with the average arsenic content being about 4% (Palmieri 1981:109–110). This tradition, technologically and to a certain extent typologically, can be seen in the Early Bronze (EB) II and III metalwork from Ikiztepe near the southwestern shore of the Black Sea, a site where arsenical copper was used exclusively when tin bronze was available at some other sites in Anatolia (Kunç 1986).

TIN AND BRONZE IN MESOPOTAMIA AND ANATOLIA, THIRD–SECOND MILLENNIUM

The introduction of tin in bronze in both Anatolia and Mesopotamia did not occasion its entry into general circulation, as Table 3.1 demonstrates. None of the artifacts analyzed from Gawra VII (ED III) were bronze and only a few from Gawra VI (Akkadian), as were some of the analyzed artifacts from Cemetery A at Kish (ED III/Akkadian) (MMP data). In Cemetery A, metal was more evenly distributed than in Cemetery Y. Although again we do not know how representative Cemetery A is of all graves at Kish in this period, we might speculate that the passing of political control to Ur by ED III is reflected in a less restricted access to metal and/or a change in the politico-cultural need to dispose of it visibly.

In ED III, only at Ur in the Royal Cemetery was bronze an important factor in local metallurgy. Of 48 artifacts we have analyzed, 25 were bronze. In the Akkadian period, the proportion changed to 12 of 37 (MMP data). Other analytical programs show the same general results. Craddock (1984) detected 6 bronzes among 15 artifacts from ED III Ur, and 4 among 9 of the Akkadian period; Berthoud (1979:table 2), 8 among 21 of ED III date.

In later periods, tin bronze did not become more common, as analyses of material from Fara, Nippur, and Gawra V of the Ur III period, and of Nippur, Khafajeh, Tell Sifr, and Tell Brak of the Old Babylonian period show.

Thus, in terms of proportions, the major concentrations of tin bronze identified to this point in Mesopotamia in the third and second millennia were in the graves of the Royal Cemetery of Ur and to a much lesser extent in those of ED I Kish. The question is whether bronze was differentially preserved in the funerary context or whether the burial of certain individuals called for bronze. In regard to the latter, it is very difficult to derive any kind of correlation because the number of artifacts analyzed is generally small in comparison with the total found, and the evidence from Ur is archivally so defective that the complete catalogue of finds can probably never be reconstructed.

To look at the problem a different way, we might try to determine why tin could have had a special status even though it is not a metal that was commonly used on its own, although the finds of tin jewelry at Tell ed-Der may indicate that such artifacts have been in the past identified as corroded silver (Van Lerberghe and Maes 1984:100). The simplest reason why tin may have been held in special regard is that possible sources for Mesopotamia—Afghanistan, Central Asia, or Anatolia—are distant. If the tin did come from Afghanistan, its trade or exchange either bypassed or did not interest the settlements of the Iranian plateau (Stech and Pigott 1986:43). The tin could have been obtained directly from Afghanistan or through the intermediary agency of a bronze-using group like the Harappans (Stech and Pigott 1986:44).

The search for the sources of tin for the bronze of Southwest Asia is a topic that has received increased attention because of recent claims that a tin source exists in the Taurus Mountains of south central Anatolia (Yener and Özbal 1987; Yener and Vandiver 1993). The controversy surrounding the possibility that this area was a tin source has generated many articles (see Yener and Vandiver 1993:209, n. 12, 215, ns. 37 and 38) and there are still many unanswered questions about the geology of the region and the production techniques used. Tin, however, seems to have been processed in crucibles at Göltepe and at least some of the processing took place during the third millennium. The extent to which this tin was used in Anatolia during the Early Bronze Age and the patterns of its dissemination remain crucial issues.

Anatolia is the only area in Southwest Asia west of Mesopotamia in which tin bronze occurs with some frequency during the third millennium. (The data below come from the tables in Esin 1969.) It was available in Troy II in non-funerary contexts and in the north Aegean (see Muhly, this volume). In contrast, the few analyzed samples from the contemporary "Yortan" graves of western Anatolia are predominantly arsenical copper. In central Anatolia, tin bronze was certainly present, but it is not as proportionately well represented as in the Troad, being so only in the graves of Alaca Hüyük and Ahlatlıbel in EB II. EB I and II Alişar, and EB II Polatlı and Kültepe have very few bronzes. For EB III, the most significant body of analyzed artifacts is from the graves of Horoztepe, where over half of 56 were bronze and about a quarter arsenical copper. In contrast, at Tarsus in EB II and III, 9 of 55 analyzed artifacts were bronze; a similar situation prevailed at contemporary Mersin.

In the second millennium in central Anatolia, tin bronze was a fairly rare commodity. Middle Bronze Age (MBA) Alişar had 1 bronze among 8 analyzed artifacts; in the Late Bronze Age (LBA), 12 of 51 are bronze. Virtually all copper-base artifacts of both periods contained significant impurities of arsenic. In light of the texts describing tin importation to Anatolia by Assyrian traders, we would expect a well-developed bronze industry in central Anatolia in the Middle Bronze Age. This does not seem to have been the case at Alişar, where an Assyrian trading colony existed, but the evidence is better for Kültepe, ancient Kaneš, the source of the relevant texts. Broken down by phases, the figures for bronze representation at Kültepe are: levels IV and III, 3 of 11 analyzed artifacts; level II, 9 of 15; level Ib, 12 of 28; level Ia, 2 of 4. Arsenical copper remained more common than tin bronze throughout. In Late Bronze Age Alaca Hüyük, 4 of 14 artifacts were

TABLE 3.1
NUMBERS OF BRONZE ARTIFACTS IN ANALYZED COLLECTIONS FROM MESOPOTAMIAN SITES

PERIOD	SITE	NO. OF ARTIFACTS ANALYZED	NO. OF BRONZE ARTIFACTS	REMARKS	REFERENCE
ED III	Gawra VII	15	0		MMP*
ED III	Ur, Royal Cemetery	48	25		MMP
ED III	Ur	15	6		Craddock 1984
ED III	Ur	21	8		Berthoud 1979:table 2
ED III/Akkadian	Cemetery A, Kish	42	7		MMP
Akkadian	Gawra VI	71	6		MMP
Akkadian	Ur, Royal Cemetery	37	12		MMP
Akkadian	Ur	9	4		Craddock 1984
Ur III	Fara + Nippur	6	1	4 of analyzed artifacts are arsenical copper	MMP
Ur III	Gawra V	14	4	all 14 analyzed artifacts contained As	MMP
Old Babylonian	Khafajeh	6	1	the tin bronze contains 1.9% As	MMP
Old Babylonian	Nippur	2	1	a low tin bronze (1.0%)	MMP
Old Babylonian	Tell Sifr	85	8	4 of the 8 are low tin bronzes	Moorey 1985:60
Old Babylonian	Tell Brak	2	1	a low tin bronze	Moorey 1985
Old Babylonian	Gawra IV	4	1	a low tin bronze w/ 1.6% As	MMP
Middle Assyrian	Tell al-Rimah	14	2		McKerrell 1977
Middle Assyrian	Kar-Tukulti-Ninurta	4	0		McKerrell 1977:21, tables 9, 10
Middle Assyrian	Tell Brak	5	1	a low tin bronze	Moorey 1985
Middle Assyrian	Chagar Bazar	5	2	1 bronze, 1 low tin bronze	Moorey 1985:53–54
Kassite	Nippur	4	2		MMP

*MMP: Mesopotamian Metals Project

bronze. The tentative conclusion is that bronze never became the dominant type or alloy of copper during the central Anatolian Bronze Age. If the Taurus Mountains were supplying this area, they were not doing so regularly or abundantly.

The situation at Tarsus in the second millennium may help to explain how the tin trade might have operated. As remarked above, there were 9 good tin bronzes among the 55 EBA artifacts, while 17 others have traces of tin in the 0.1 to 1.0% range. The 8 artifacts analyzed from MBA Tarsus include no real bronzes: all are arsenical coppers. In the Late Bronze I period, 9 of 11 analyzed artifacts from Tarsus are bronze; none from the stratigraphically mixed LB II deposits was analyzed. Although one site cannot provide all the answers, we can speculate that the tin-bearing copper ores from the Taurus might have been used at Tarsus in the EBA. During the MBA, in contrast, these ores may not have been used, and the area would also appear to have been bypassed by Assyrian traders. By the Late Bronze Age, the situation had changed again. Perhaps the real trade in tin at this time was a seaborne phenomenon of the Mediterranean, a part of the metals trade which has tangible expression in the cargo of the fourteenth century shipwreck at Kaş, in which tin oxhide ingots occur (Bass 1986:276–277; Pulak 1988:4, 8–10), and in the oxhide ingots of Crete, Cyprus, Greece, and Sardinia. Tarsus, as a coastal site, could have benefited from participation in the Mediterranean commercial sphere in a way that its central Anatolian neighbors did not. Politically, the early part of the Late Bronze Age saw conflict between the Hittites and the kingdom of Mitanni for control of Kizzuwatna, of which Tarsus must have been an important city, perhaps even the capital. The pottery of this period at Tarsus has affinities with Hittite types, and Syrian connections are also apparent (Goldman 1956:349). Important for the thesis of involvement with the

Mediterranean is the presence of Levantine bichrome ware (Goldman 1956:183, 199–200). By LB II, Kizzuwatna and presumably Tarsus were under Hittite control, but since the types of alloys used in this period are unknown, we cannot extend the interpretation further.

SUMMARY

In both Mesopotamia and Anatolia there are a few early artifacts of native copper. Çayönü stands apart for the number of these, but the discovery of native copper could have been a byproduct of the search for malachite. The several kilos of malachite and the native copper found at the site could have been gathered by one person on a single visit to an ore source. The working of native copper did not lead, in any way that is visible to us, to continued exploitation of copper mines and eventually to smelting. The properties of metal that are different from those of stone and bone seem to have been known to some Neolithic craftsmen, but no incentive to continue experimenting with them appears to have existed.

Although we cannot see an unbroken thread of development, it is possible that lead showed the possibilities of smelting. In Mesopotamia, a relatively intensive use of lead and its byproduct, silver—in the Sammelfund of Uruk and the "Jemdet Nasr" cemetery of Ur—immediately precedes the extensive consumption of copper and its alloys. Although we know too little about fourth millennium metallurgy in Anatolia to see if a similar development took place there, in both Mesopotamia and Anatolia by the early third millennium pure copper, arsenical copper, and tin bronze were in service, the last most frequently in funerary contexts and at Troy. The graves of Kish, Ur, Alaca Hüyük, and Horoztepe, and the citadel of Troy contain most of the tin bronze known from Southwest Asia in the third millennium B.C. In Anatolia in the second millennium, bronze was an important component only in a few special situations—at Middle Bronze Age Kültepe, where Assyrians are known to have brought tin to trade, and at Late Bronze Age Tarsus, a site with connections to the Mediterranean cultural and commercial sphere. In Mesopotamia in the second millennium, tin was being acquired from somewhere to the east, a trade described in Old Assyrian texts, but the available analyses suggest that it was little used, both in Assyria and in the south.

Thus, in inland Southwest Asia, regular bronze consumption in the Bronze Age occurred only in a few places, particularly in graves. The technology for making bronze was known by the early third millennium, as some sources of tin must have been, so part of the reason for the relative rarity of bronze must have been cultural rather than technological. The interpretation of funerary context is therefore essential in understanding how tin and tin bronze functioned in the Bronze Age economy. At both Kish and Ur (Charvát 1982), there is historical justification for associating bronze with the establishment of dynasties, political situations which were reinforced by the burial of considerable amounts of materials like gold, silver, and lapis lazuli. The alloy of tin and copper appears to have been part of this complex of goods. Of the 12 graves in Kish Cemetery Y that are conspicuous for concentrations of metal, bronze with 1% or more tin occurs in 4 (the figure of 1% being used in this case because of the way in which Moorey and Schweizer [1972] reported their data). Bronzes with similar amounts of tin are known in three other graves (120, 469, 479). The evidence for the association of bronze with other indications of status is not, it must be admitted, compelling, for the total sample comprises only 17 artifacts. For Ur, the attempt to relate bronze to high-status graves, as defined by Susan Pollock (1983, 1985, 1991), is frustrating because we have not analyzed sufficient material from the relevant graves. Pollock notes that there is a steady decline in the proportion of graves containing pottery as rank increases, and that the number of copper/bronze vessels also increases with other indications of greater rank. Because of poor preservation, we do not have sufficient analyzable metal vessels from the Royal Cemetery, but the majority of the copper-base vessels analyzed by Michael Müller-Karpe (1990:164, table 1) are bronze. This preliminary observation provides the best link so far between tin bronze and status. It would be useful to know how the vessels from the Royal Cemetery were made, to see if there is any technological reason for them to be bronze. Bronze is easier to cast than copper, but copper can be worked more easily. If the vessels were worked, then bronze would not have been the better choice of material from the technical point of view. Unfortunately, the vessels in the University of Pennsylvania Museum are completely corroded and cannot be studied metallographically to determine method of manufacture.

Although there is no textual evidence on the individuals buried at Alaca Hüyük and Horoztepe, their graves have been interpreted as those of early dynasts, perhaps Hittites or related groups. Trojan bronze is the most notable exception, perhaps because of the nature of Troy in this period. It was not a settlement with a diverse economic base, as far as we can tell, but rather a citadel which could have accommodated only a few families. Presumably it existed and flourished because it derived benefits from trade and/or communication between the Aegean and the Black Sea. The inhabitants appear to have been an elite group, already singled out by special circumstances in life. How they fared in death is not known, since the graves of Early Bronze Age Troy have not yet been found; and how they related to and were supported by the Trojan hinterlands is similarly obscure. These gaps in knowledge may be filled by the current excavations at both Troy and nearby Beşiktepe. The greater proportion of bronzes among the Trojan metals than in central Anatolia might indicate that Troy was the place where the tin or bronze entered Anatolia, hence by sea, either from the north or south, the former route crossing the Black Sea and bringing tin from Central Asia or Afghanistan (Muhly et al. 1991:216, 218–219). The alternative explanation is that the tin trade to Anatolia went overland, as it did in the second millennium, so the messages about meaning imparted by the Mesopotamians might not have been clearly received at Troy.

The evidence for a decline in bronze consumption in the second millennium may relate more to the funerary context than to decrease in wealth. Tin might have continued to be regarded as a material with special status and was therefore used only in certain situations. This perception appears to have been maintained throughout the second millennium in inland Southwest Asia, as the conjunction of rich graves and bronze at Kültepe may indicate. Unfortunately, we cannot reconstruct the complete contexts in which bronze appeared at Kültepe, since half the analyzed artifacts are unpublished. The true "Bronze Age," in which tin was used for technological reasons, as the case of Tarsus might indicate, may have taken place in the circum-Mediterranean area, but analyses of the relevant material are too few to allow more than speculation. Conspicuous consumption of metal in the graves of early dynasts has, however, been documented in the Shaft Graves of Mycenae and the "burials with bronzes" of Knossos.

Pigott and I have advanced the hypothesis that tin was obtained from Afghanistan first by the Sumerians, and later the Akkadians and Assyrians (Stech and Pigott 1986). They valued it as a rare material from a distant region which came along with prized goods such as lapis lazuli and gold. The presumed conjunction of the three would thus have given tin its worth in Mesopotamian society. It was a commodity which could be used to indicate status and to obtain other valued goods. Therefore, tin was not regarded strictly as a technological enhancement to the properties of copper, but a cultural support to the bases of power, both political and economic. (See Kristiansen 1987 for a similar argument relating to Bronze Age Northern Europe.)

Under this hypothesis, Sumerians would have brought tin to Anatolia, perhaps as early as the Early Bronze Age, presumably to obtain silver, which is not well represented in the Afghan metallic repertoire, and perhaps gold. The selection of tin as the medium of exchange requires explanation, since it would not seem to be the material of choice in transactions involving luxury materials because its properties are mostly invisible. Lapis lazuli does not have this disadvantage, but it is virtually lacking in third millennium Anatolia, so tin might have been chosen as the vehicle to acquire silver in order to preserve supplies of lapis relative to those of tin. The Anatolians, however, as the residents of a metal-rich land may have had a preference for using metals rather than stones to express rank. In either case, Mesopotamian ideas about the desirability of including tin in the form of bronze in the burials of special individuals must have been communicated to and adopted by their Anatolian trading partners. In accepting tin as a material which conveyed status and which was part of a group of goods used to legitimize political control, Anatolians also accepted Mesopotamian standards of value. By doing so, they may have been expressing the desire to emulate the success of Sumerian and Akkadian rulers in acquiring power and wealth. Mesopotamian intervention in Anatolia to obtain materials which solidified political and economic control may have ultimately created the context in which the Middle Bronze cities of central Anatolia developed.

The transformation of the village-based societies of Early Bronze Age Anatolia into the Middle Bronze Age urban concentrations is a phenomenon which has required Indo-European invasions to explain. Perhaps we can now invoke—in a tentative way—the desire to centralize the population in a manner which allowed for greater control of larger numbers of people and hence their economic production, following the Mesopotamian model. The lack of massive destruction levels at the end of the Early Bronze Age has been an impediment to the invasion theory, but a major disjunction clearly occurred. The remains in the cemetery at Karataş in southwestern Turkey can be interpreted to support the theory of increasing centralization during the latter part of EB II and into EB III.[5] When Mesopotamians reappeared in Anatolia in the second millennium, they brought with them tin, and exchanged it for silver, but they did so in an urban setting which was congenial to their semi-permanent settlement.

Anatolians still wanted the commodity being offered, and seem to have used it more extensively, at least at Kültepe, than did the Mesopotamians who brought it in trade. For the Assyrians, tin apparently was the means to acquire other desired goods. Anatolians wished to obtain a metal which they did not have in abundance—tin—and exchanged it for one which they did—silver. Mesopotamians wanted silver, and found tin the most effective means of obtaining it. In the process of interaction and exchange, the Mesopotamians communicated more than commodities. The acceptance of symbols validating rank may imply a broader acceptance of the political system being symbolized, and thus created the context for the ideological transformation of the administrative structures of central Anatolia. Perhaps this transformation is the ultimate explanation of why Mursili I raided Babylon in 1595 B.C. in an aggressive rather than passive attempt to participate in the perceived economic and political power of Mesopotamian rulers.

NOTES

1. The mandate of the IREX conference in Georgia in 1988 was for the then-Soviet and American participants to communicate their own current research. The present article is substantially what was presented at that conference, with some bibliographic additions, so it should not be construed as an attempt to review completely early metallurgy in Mesopotamia and Anatolia.

2. The Mesopotamian Metals Project is the joint work of Stuart Fleming, James D. Muhly, Vincent C. Pigott, and me. The elemental analyses were obtained using the PIXE technique (proton-induced X-ray emission) by Charles Swann, University of Delaware, and Fleming.

3. F. Begemann, U. Esin, A. Hauptmann, R. Maddin, J. Muhly, E. Pernicka, S. Schmitt-Strecker, and I have been privileged to conduct the analyses of the metal artifacts from Çayönü, due to the generosity of the excavators, Halet Çambel and Mehmet Özdoğan of the University of Istanbul and Robert J. Braidwood of the University of Chicago. Elemental analyses were done by PIXE by Swann and Fleming, to whom we are grateful for their contributions. The American Philosophical Society has supported several aspects of the research on Çayönü, through grants to Maddin for elemental analysis and to Stech to study the total corpus of metals and minerals.

4. This information was kindly supplied by Aslı Özdoğan in July 1988.

5. The Karataş cemetery was used from the end of EB I through most of EB II (ca. 2900–2500/2400 B.C.). Over time, the burials became less rich in terms of grave goods and special treatments of the burial jars, and multiple burials became more common. Although the site was inhabited into EB III, there are no known graves of this period. One possible explanation is that centralization of political control was increasing, with the elites of each community toward the end of EB II being buried in special cemeteries near the seat of the regional leader. By EB III, perhaps most individuals were buried in these locations. There are, of course, other possible explanations for this situation.

REFERENCES CITED

Al-A'dami, K. A.
1968   Excavations at Tell es-Sawwan (Second Season). *Sumer* 24:57–94.

Bader, N. O.; Merpert, N. I.; and Munchaev, R. M.
1981   Soviet Expedition's Surveys in the Sinjar Valley. *Sumer* 37:55–95.

Bass, G. F.
1986   A Bronze Age Shipwreck at Ulu Burun (Kaş): 1984 Campaign. *American Journal of Archaeology* 90:269–296.

Bazin, D., and Hübner, H.
1969   *Copper Deposits in Iran*. Geological Survey of Iran, Report 13. Tehran.

Berthoud, T.
1979   *Etude par l'analyse de traces et la modelisation de la filiation entre minérai de cuivre et objets archéologiques du Moyen-Orient (IVème et IIIème millénaires avant notre ère)*. Doctoral thesis, Université Pierre et Marie Curie, Paris.

Braidwood, R. J.; Çambel, H.; and Schirmer, W.
1981   Beginnings of Village-farming Communities in Southeastern Turkey: Çayönü Tepesi, 1978 and 1979. *Journal of Field Archaeology* 8:249–258.

Çambel, H., and Braidwood, R. J.
1980   *The Joint Istanbul-Chicago Universities' Prehistoric Research in Southeastern Anatolia*. Istanbul University, Faculty of Letters, Publication No. 2589. Istanbul.

Campbell Thompson, R., and Mallowan, M. E. L.
1933   The British Museum Excavations at Nineveh, 1931–32. *Liverpool Annals of Art and Archaeology* 20:70–186.

Cauvin, J.
1985   Le Néolithique de Cafer Höyük (Turquie): Bilan provisoire après quatre campagnes (1979–1983). *Cahiers de l'Euphrate* 4:123–133.

Charvát, P.
1982   Early Ur—War Chiefs and Kings of Early Dynastic III. *Altorientalische Forschungen* 9:43–59.

Cleuziou, S., and Berthoud, Th.
1982   Early Tin in the Near East: A Reassessment in the Light of New Evidence from Afghanistan. *Expedition* 24(3):14–19.

Craddock, P. T.
1977   The Composition of the Copper Alloys used by the Greek, Etruscan and Roman Civilizations. 2: The Archaic, Classical and Hellenistic Greeks. *Journal of Archaeological Science* 4:103–113.

1980   The Composition of Copper Production at the Ancient Smelting Camps in the Wadi Timna, Israel. Pp. 165–173 in *Scientific Studies in Early Mining and Extractive Metallurgy,* ed. P. T. Craddock. British Museum Occasional Paper No. 20. London.
1984   Tin and Tin Solder in Sumer: Preliminary Comments. *MASCA Journal* 3(1):7–9.

Esin, U.
1969   *Kuantatif Spektral Analiz Yardımıyla Anadolu'da Baslangıcından Asur Kolonileri Çagina Kadar Bakır ve Tunç Madenciligi.* Istanbul Üniversitesi Edebiyat Fakültesi Yayınları No. 1427. Istanbul.
1993   Copper Beads of Aşıklı. Pp. 179–183 in *Aspects of Art and Iconography: Anatolia and its Neighbors,* eds. M. J. Mellink, E. Porada, and T. Ozguc. Ankara: Türk Tarih Kürümü.

Fukai, S., and Matsutani, T.
1977   Excavations at Telul 1976. *Sumer* 33:48–64.

Goldman, H.
1956   *Excavations at Gözlü Küle, Tarsus,* Vol. 2. Princeton: Princeton University Press.

Hauptmann, A.; Lutz, J.; Pernicka, E.; and Yalcin, U.
1993   Zur Technologie der frühesten Kupferverhuttung im östlichen Mittelmeerraum. Pp. 541–572 in *Between the Rivers and Over the Mountains. Archaeologica Anatolica et Mesopotamica Alba Palmieri Dedicata,* eds. M. Frangipane et al. Rome: Università di Roma "La Sapienza."

Heinrich, E.
1936   *Kleinfunde aus den archaischen Tempelschichten in Uruk.* Berlin: Deutsche Forschungsgemeinschaft.

de Jesus, P. S.
1980   *The Development of Prehistoric Mining and Metallurgy in Anatolia.* BAR International Series 74. Oxford: British Archaeological Reports.

Kristiansen, K.
1987   From Stone to Bronze: The Evolution of Social Complexity in Northern Europe, 2300–1200 BC. Pp. 30–51 in *Specialization, Exchange, and Complex Societies,* eds. E. M. Brumfiel and T. K. Earle. Cambridge: Cambridge University Press.

Kunç, S.
1986   Analyses of Ikiztepe Metal Artifacts. *Anatolian Studies* 36:99–101.

Levey, M., and Burke, J. E.
1959   A Study of Ancient Mesopotamian Bronze. *Chymia* 5:37–50.

Maddin, R.; Muhly, J. D.; and Stech, T.
1999   Early Metalworking at Çayönü. In *The Beginnings of Metallurgy,* eds. A. Hauptmann, E. Pernicka, T. Rehren, and Ü. Yalcin. Der Anschnitt, Beiheft 9.

Mallowan, M. E. L.
1977   Recollections of C. Leonard Woolley. *Expedition* 20(1):3–4.

Martin, H. P.; Moon, J.; and Postgate, J. N.
1985   *Abu Salabikh Excavations.* Vol. 2: *Graves 1 to 99.* London: British School of Archaeology in Iraq.

McKerrell, H.
1977   The Use of Tin-bronze in Britain and the Comparative Relationship in the Near East. Pp. 7–24 in *The Search for Ancient Tin,* eds. A. D. Franklin, J. S. Olin, and T. A. Wertime. Washington, DC: Smithsonian Institution.

Mellaart, J.
1967   *Çatal Hüyük. A Neolithic Town in Anatolia.* London: Thames and Hudson.

Merpert, N. Ya., and Munchaev, R. M.
1977   The Most Ancient Metallurgy of Mesopotamia. *Sovetskaya Arkheologiya* 1977(3):154–163.
1987   The Earliest Levels at Yarim Tepe I and Yarim Tepe II in Northern Iraq. *Iraq* 49:1–36.

Merpert, N. I.; Munchaev, R. M.; and Bader, N. O.
1977   The Investigations of Soviet Expedition in Iraq, 1974. *Sumer* 33:65–104.
1978   Soviet Investigation in the Sinjar Plain. *Sumer* 34:27–70.

Moorey, P. R. S.
1985   *Materials and Manufacture in Mesopotamia: The Evidence of Archaeology and Art. Metals and Metalwork, Glazed Materials and Glass.* BAR International Series 237. Oxford: British Archaeological Reports.
1994   *Ancient Mesopotamian Materials and Industries.* Oxford: Clarendon Press.

Moorey, P. R. S., and Schweizer, F.
- 1972 Copper and Copper Alloys in Ancient Iraq, Syria and Palestine and New Analyses. *Archaeometry* 14:177–198.

Muhly, J. D.
- 1983 Kupfer. B. Archäologisch. *Reallexikon der Assyriologie und vorderasiatischen Archäologie* 6(5):348–364.
- 1989 Çayönü Tepesi and the Beginnings of Metallurgy in the Old World. Pp. 1–12 in *Old World Archaeometallurgy*, eds. A. Hauptmann, E. Pernicka, and G. A. Wagner. Bochum: Deutsches Bergbau-Museum.

Muhly, J. D.; Begemann, F.; Öztunalı, O.; Pernicka, E.; Schmitt-Strecker, S.; and Wagner, G. A.
- 1991 The Bronze Metallurgy of Anatolia and the Question of Local Tin Sources. Pp. 209–220 in *Archaeometry '90, International Archaeometry Symposium, Heidelberg, April 1990*, eds. E. Pernicka and G. A. Wagner. Basil: Birkhaüser.

Müller-Karpe, M.
- 1990 Metallgefässe das dritten Jahrtausends in Mesopotamien. *Archäologisches Korrespondenzblatt* 20:161–176.

Neuninger, H.; Pittioni, R.; and Siegl, W.
- 1964 Frühkeramikzeitliche Kupfergewinnung in Anatolien. *Archaeologia Austriaca* 35:98–110.

Northover, J. P.
- 1989 Properties and Uses of Arsenic Copper Alloys. Pp. 111–118 in *Old World Archaeometallurgy*, eds. A. Hauptmann, E. Pernicka, and G. A. Wagner. Bochum: Deutsches Bergbau-Museum.

Özdoğan, A.
- 1995 Life at Çayönü during the Pre-Pottery Neolithic. Pp. 79–100 in *Readings in Prehistory. Studies Presented to Halet Çambel*. Istanbul: Graphis Yayinlari.

Palmieri, A.
- 1981 Excavations at Arslantepe (Malatya). *Anatolian Studies* 31:101–119.

Parrot, A.
- 1948 *Tello*. Paris: Editions Albin Michel.

Pollock, S.
- 1983 *The Symbolism of Prestige: An Archaeological Example from the Royal Cemetery of Ur*. Ph.D. dissertation, Department of Anthropology, University of Michigan.
- 1985 Chronology of the Royal Cemetery of Ur. *Iraq* 47:129–158.
- 1991 Of Priestessses, Princes and Poor Relations: The Dead in the Royal Cemetery of Ur. *Cambridge Journal of Archaeology* 1:171–189.

Pulak, C.
- 1988 The Bronze Age Shipwreck at Ulu Burun, Turkey: 1985 Campaign. *American Journal of Archaeology* 92:1–37.

Raymond, R.
- 1984 *Out of the Fiery Furnace: The Impact of Metals on the History of Mankind*. Melbourne.

Ryndina, N. V., and Yakhontova, L. K.
- 1985 The Earliest Copper Artifact from Mesopotamia. *Sovietskaya Arkheologija* 1985(2):155–165.

Speiser, E. A.
- 1935 *Excavations at Tepe Gawra. I: Levels I–VIII*. Philadelphia: University of Pennsylvania Press.

Sperl, G.
- 1990 Zur Urgeschichte des Bleies. *Zeitschrift für Metallkunde* 81:799–801.

Stech, T., and Pigott, V. C.
- 1986 The Metals Trade in Southwest Asia in the Third Millennium B.C. *Iraq* 48:39–64.

Tobler, A. J.
- 1950 *Excavations at Tepe Gawra* II. Philadelphia: University of Pennsylvania Museum.

Tylecote, R. F.
- 1976 *A History of Metallurgy*. London: The Metals Society.

Van Lerberghe, K., and Maes, L.
- 1984 Contribution à l'étude des métaux de Tell ed-Der. Pp. 97–143 in *Tell ed-Der* IV, ed. L. De Meyer. Leuven: Uitgeverij Peeters.

Vértesalij, P. P., and Kolbus, S.
- 1985 Review of Protodynastic Development in Babylonia. *Mesopotamia* 20:53–109.

Voigt, M. M.
- 1990 Reconstructing Neolithic Societies. *Archeomaterials* 4:1–14.

Woolley, C. L.
- 1934 *Ur Excavations*. Vol. 2: *The Royal Ceme-*

tery. London: British Museum.

Yener, K. A., and Özbal, H.
1987 Tin in the Turkish Taurus Mountains: The Bolkardağ Mining District. *Antiquity* 61:220–226.

Yener, K. A., and Vandiver, P. B.
1993 Tin Processing at Göltepe, an Early Bronze Age Site in Anatolia. *American Journal of Archaeology* 97:207–238.

Zwicker, U.
1982 Bronze Age Metallurgy at Ambelikou-Aletri and Arsenical Copper in a Crucible from Episcopi-Phaneromeni. Pp. 63–68 in *Early Metallurgy in Cyprus: 4000–500 B.C.*, eds. J. D. Muhly, R. Maddin, and V. Karageorghis. Nicosia: Pierides Foundation.

4

The Development of Metal Production on the Iranian Plateau

An Archaeometallurgical Perspective

Vincent C. Pigott

ABSTRACT As an ore-rich metallogenic zone, the Iranian Plateau was, from the Neolithic period on, a major center for the development of a variety of metallurgical technologies. This chapter comprises an overview of the archaeometallurgical (i.e., archaeological and analytical) data which form the foundation for the complex story of how the technologies of copper, its alloys, and iron evolved over the span of some six millennia from the Neolithic period through the Chalcolithic into the early Iron Age.

In the Neolithic period, the only metal recognized and worked was native copper. With the advent of the Chalcolithic, the dependence on the massive native copper and copper arsenide deposits on the Plateau appears to have continued, but smelting technologies developed that enabled the reduction of rich oxides and arsenical ores of copper, perhaps in combination. Tin bronze made an initial appearance at the close of this period and occurred with increasing regularity through the third millennium. During the early second millennium bronze technology became more widespread. Iron arrived on the scene at the beginning of the first millennium, as bronze technology continued to flourish. [Final ms. received 12/98.]

During the past three decades considerable research has been focused on the early manipulation of copper, its alloys, and iron on the Iranian Plateau over a period of seven millennia, from ca. 6500 into the first millennium B.C. Field surveys directed toward archaeometallurgical inquiry and laboratory analyses of ores and metals, together with substantial archaeological excavation, have contributed a significant body of information on the development of early metalworking technology in this ore-rich metallogenic zone.

The corpus of information forms a basis for the continuing efforts to understand the processes underlying the adoption and subsequent development of particular technological innovations and their sociocultural and politico-economic significance. This chapter, as a first step toward the above goal, is an attempt to bring together the data supporting our current understanding of copper-base and iron metallurgy and metal use on the Plateau and in some adjacent regions.[1] The developments of the metallurgies of the other base metals (e.g., antimony, lead) and the precious metals (e.g., silver, gold) remain inadequately investigated.

THE NEOLITHIC (CA. 7500–5500 B.C.) AND CHALCOLITHIC PERIODS (CA. 5500–3200 B.C.)

Current evidence suggests that on the Iranian Plateau native copper was used initially in the Neolithic and continued in use through the Chalcolithic period (Heskel 1982:384–388). The ninth millennium B.C. copper mineral pendant from Zawi Chemi in the Zagros Mountains (Solecki 1969:361) indicates an early familiarity with copper ore, but not with metalworking. Among the earliest excavated metal artifacts that are arguably native copper in composition are a rolled bead (mid-seventh millennium) from the site of Ali Kosh on the Deh Luran Plain of northern Khuzistan (Smith 1969), several awls (mid-sixth millennium) from Tepe Zagheh on the Qazvin Plain (Shahmirzadi 1979), copper fragments (fifth millennium) from Chogha Sefid (Hole 1977:245), and pins, projectile points, awls, and spiral coils (mid-fifth millennium) from Tepe Sialk near Kashan (Halm 1939; see also Coghlan 1942:22–34; Lamberg-Karlovsky 1967:145–146). (See Fig. 4.1 for locations of sites and copper mines on the Iranian Plateau.)

Analyses of the Ali Kosh bead (Fig. 4.2) and a Sialk pin indicate that they are made of coldworked native copper (Smith 1968:239–240; Smith 1965:28–30; see also Wertime 1964:1260; Knauth 1974:40). Smith (1968) analyzed two samples of native copper from the ore body at Talmessi and both were found to contain native silver, with one sample containing 0.08 wt.% As (by spectrographic analysis). He also demonstrated that small lumps of native copper can be deformed extensively without annealing. Smith made a replica of the Ali Kosh copper bead from native copper and the two beads were markedly similar in macrostructure (Fig. 4.3). From Tepe Yahya near Kerman the earliest metal (mid-fifth millennium context) consisted of two copper awls, one of which was shown to have been shaped from native copper (Heskel and Lamberg-Karlovsky 1980:232; see also Heskel 1983:364, table 1 for a somewhat dated list of the earliest metal artifacts in Southwest Asia, as well as a discussion of the influence which contexts—social, political and economic—had on the development of early metal use; see also Stech 1990 and Voigt 1990 for useful discussion of Neolithic metallurgy and its context).

Metalworking during the Neolithic must have been haphazard. The shapes are simple and it is likely that native copper was being worked in the manner of stone, that is, through "selective surface collecting, and then quarrying, hammering and polishing, flaking and drilling" (Moorey 1982:83). Innovation and the development of techniques may have been stimulated initially by the desire to achieve decorative effects (Smith 1976). Initial encounters with copper ores appear to have occurred on this basis. The Shanidar pendant may serve as one such example and the malachite discs and rectangles (inlay?) at eighth millennium Cayönü in Anatolia also support this supposition (Stech 1990; see also Glumac 1985 on malachite bead manufacture at Neolithic Divostin, in the former Yugoslavia).

It is during the Chalcolithic that the use of copper expanded dramatically. The Chalcolithic repertoire of arsenical copper artifacts is modestly diversified and includes needles, pins, tanged dagger blades, chisels, and shaft-hole and flat axes. At the close of the period the repertoire has diversified considerably to include seals, midribbed daggers, shaft-hole maceheads, spiral-headed pins, coiled bracelets, earrings, and finger rings (Moorey 1982:86). Items in lead, silver, and gold also occur for the first time.

Artifacts analyzed from the period have arsenic present in varying amounts, but there are no indications that arsenical copper was being produced by intentional alloying. It is likely that at this early period arsenic-rich native copper and ores from Plateau sources were being exploited (although see discussion below regarding copper deposits with arsenical mineralizations on the Plateau). Unfortunately, the microstructure of copper that forms following the melting of native copper is indistinguishable from that of an artifact cast from smelted copper (Maddin et al. 1980). Regardless of whether the copper is melted or smelted, the presence of arsenic in copper conveys improved mechanical properties, particularly in the annealed state, quite similar to those conveyed by tin in copper (Maréchal 1958; cf. Coghlan 1975:83). Moorey (1994:250) summarized the fundamental role which arsenic played in ancient copper-base alloys, namely, that it acts as a deoxidant (cf. Northover 1989:111–112); and that it also promotes fluidity in the molten metal which facilitates casting. Moorey cites Northover (1989:113) who notes that "small quantities of arsenic, say up to about 2%, offer very little improvement over pure coppers and it is only at about 4% and over that a good balance between strength and toughness, with properties approaching those of medium tin bronzes, can be obtained." Northover (1989:113) also points out that "another constraint on Cu-As alloys is that with ancient technology it was difficult to create alloys with over 8% arsenic, whatever route was taken for their manufacture." While it cannot be pursued in detail here, Paul Budd et al. (1992) describe three techniques by which arsenical copper may have been smelted in antiquity in the British Isles. These techniques would also have been applicable in the ancient Iranian context, and will be briefly reviewed in discussion below (see also Pigott in press; Pigott et al. in press).

Crucibles are known from a variety of Chalcolithic contexts on the Plateau, while archaeologically documented, in situ, metal smelting furnaces are known only from the Bronze Age site of Shahdad in southern Iran, as discussed below (Hakemi 1992; Vatandoost-Haghighi 1978). More furnaces may yet be excavated at other sites, but current evidence supports the contention that it was the crucible that played a primary role in copper production. This may be true not only for melting native copper but also for smelting oxidic ores (Tylecote 1974) as well as co-smelting a self-reducing mixture of oxidic and sulfidic ores (Rostoker et al. 1989; Rostoker and Dvorak 1991; see also Yener and Vandiver 1993; Vandiver et al. 1993). Crucible smelting is a technologically simple process and may have characterized the production of copper and its alloys on the Iranian Plateau well into the Bronze Age.

THE EVIDENCE FOR CHALCOLITHIC CRUCIBLE SMELTING ON THE PLATEAU

At Chalcolithic Tal-i Iblis (ca. 5500–3500 B.C.) near Kerman large quantities of crucible fragments oc-

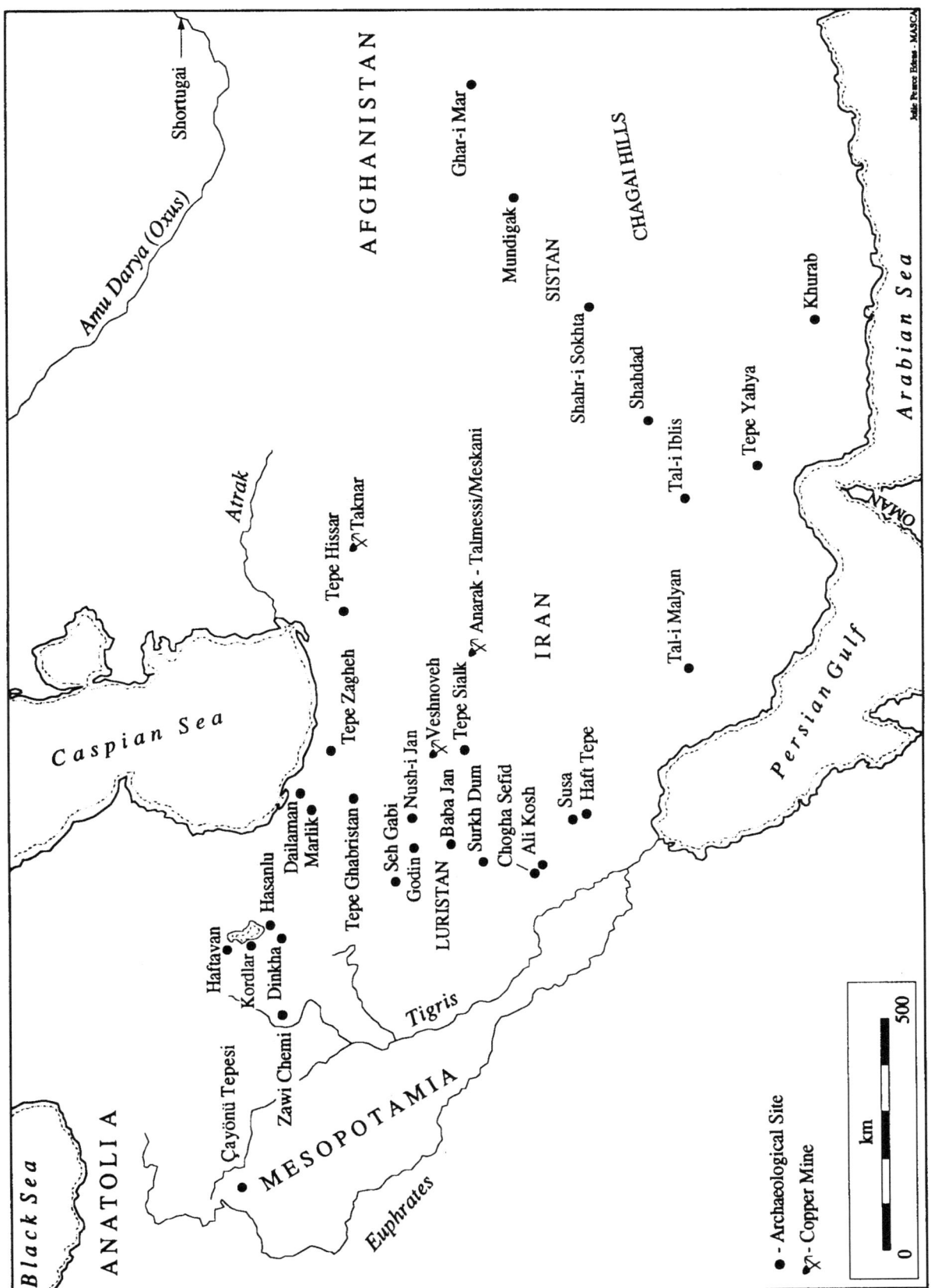

Figure 4.1 *The Iranian Plateau, locating copper mines and sites mentioned in the text. Map by J. P. Edens.*

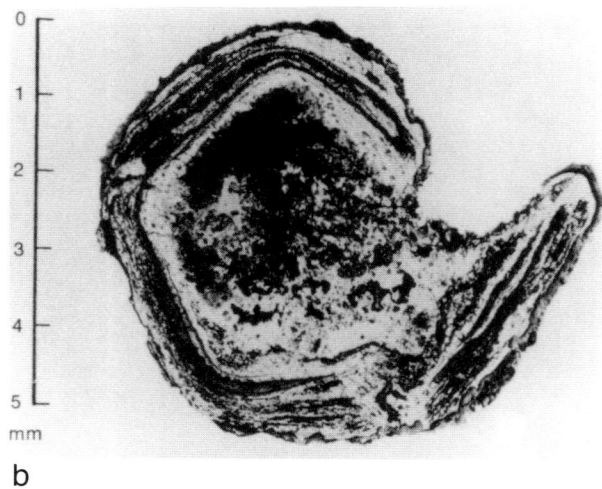

Figure 4.2   (a) Native copper rolled bead from Ali Kosh (mid-seventh millennium). L. ca. 12 mm. From Smith 1969. Courtesy of Museum of Anthropology Publications, University of Michigan. (b) Polished cross section of Ali Kosh bead. The metal is corroded but the resulting cuprite and malachite have preserved the original shape. From Smith 1967:28. Courtesy Museum of Fine Arts, Boston. Reproduced with permission. ©1999 Museum of Fine Arts, Boston. All rights reserved.

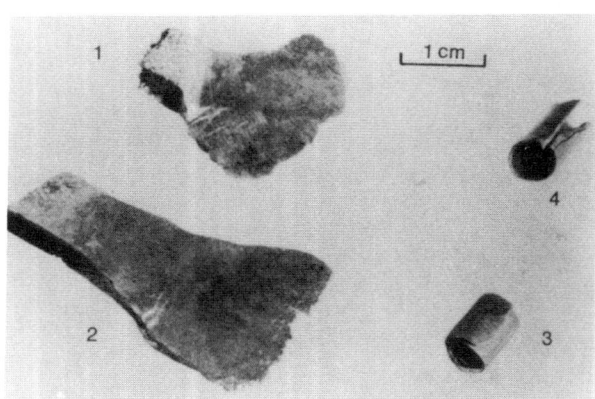

Figure 4.3   Examples of native copper, hammered and rolled by Cyril Stanley Smith. (1) Samples of copper from Talmessi and (2) from North Michigan, hammered without annealing, show that native copper can be extensively deformed without annealing. Rolled cylinders (3) and (4) are attempts to replicate the Ali Kosh bead: (4) was rolled after annealing while (3) was coiled in the work-hardened condition and shows characteristic abrupt bends. From Smith 1967:28. Courtesy Museum of Fine Arts, Boston. Reproduced with permission. ©1999 Museum of Fine Arts, Boston. All rights reserved.

curred with some slag, ore, and copper-base artifacts, the last comprising mostly pins, awls (?), and rings (i.e., small artifacts produced by simple methods) (Caldwell 1967; see also Caldwell 1968; Caldwell and Shahmirzadi 1966; Dougherty and Caldwell 1967). The ores encountered at the site were primarily malachite, with some azurite and one piece of chalcocite (Dougherty and Caldwell 1967:19–20).

There is excavated evidence which strongly suggests the practice of crucible smelting. More than 300 crucible fragments with slagged interiors were found, a number in a dumping area ca. 60 cm thick by 100 m long (see description of crucible fragments, Dougherty and Caldwell 1967:18, and full reconstruction, p. 185) (Fig. 4.4). Analysis of one fragment indicated temperature tolerances just below 1000°C, sufficient for reduction of ores but not the melting of native copper (Dougherty and Caldwell 1967:19). Spectrochemical analysis of copper stains on the fragment's interior yielded, besides copper, "significantly higher concentrations of cobalt, nickel, phosphorous and tin, than the surrounding ceramic or the surface dross." The presence together of trace elements Co and Ni is interesting in the light of the discussion below concerning the deposits of Cu-Ni arsenides at Anarak-Talmessi/Meskani.

Both Dougherty and Caldwell (1967) and Pleiner (1967) argued cogently for crucible smelting of oxidic ores producing small amounts of copper. During the 1966 field season at Tal-i Iblis, R. Pleiner and colleagues conducted the experimental reduction of copper in a crucible modeled on the Iblis type. They smelted fragments of malachite gathered from the site in the crucible placed in a simple fire-pit like that excavated at Iblis. They used a simple bellows and recorded a temperature of 1100°C at the tuyère. After one half-hour some reduced copper was found entrapped in slag. Pleiner (1967:375) suggested that a single crucible could yield sufficient amounts of copper to produce the small-scale artifacts known from Iblis.

Corroborating archaeological evidence comes from a dependable context in Iblis Area G, where a shallow fire-pit was excavated which contained a crucible fragment and fragments of oxidized copper (Evett 1967:252, fig. 19). The excavator of Area G reconstructed this smelting locus as follows: A shallow pit was dug and left unlined, and crucibles charged with malachite and charcoal were placed in the pit. "Contents were then covered with a mass of chaff-tempered clay, presumably for a bellows or to allow for oxygen to enter and hot gases to escape" (Evett 1967:254). This sentence regrettably is not clear. Caldwell (1967:254, n. 1) suggests that a "clay superstructure" would be more likely, by which he may mean a vertical shaft furnace structure. It should be noted that this type of smelting installation would be ideal for co-smelting involving oxide-sulfide interaction, whether accidental or intentional. Three other fire-pits were located on the same level as that just described and were very similar, with the exception that there was no metallurgical debris associated with them other than charcoal. The presumption is that they lay outside, in a courtyard area perhaps, where domestic garbage was being discarded. The excavator of Area G saw this copper smelting as a household-based "cottage industry," one craft among several being practiced.

At Seh Gabi near Kangavar a number of crucible fragments and several "ingot" molds were excavated (Levine and Hamlin 1974:212). However, findings from late fifth millennium Tepe Ghabristan near Qazvin (Majidzadeh 1979, 1989) constitute the earliest strong evidence for smelting in Southwest Asia. Here a workshop was excavated which may comprise just the sort of installation where metallurgical innovations were being pioneered on the Plateau (Moorey 1994:257). The workshop contained 20 kg of broken, nut-sized fragments of malachite (ideally sized for smelting), two hearths, what may be either a mold or a clay tuyère (if the latter, it is possibly the only identified, prehistoric example of this type of artifact from the Plateau and one of the few—if any—in Southwest Asia), a slagged crucible, and four ceramic molds—three for shaft-hole implements and one for bar ingots (Majidzadeh 1979:82–85). Malachite, a rich oxidic copper ore, smelts easily in a crucible and produces little slag as residue (Tylecote 1974). Slagging found in the ingot mold would result from pouring molten copper that had been smelted in the crucible directly into the mold (cf. Muhly 1983:352). Such a crucible would have contained molten copper with slag residues (dross) floating on it. Pouring the crucible could easily result in some dross being transferred to the casting mold. The evidence from Ghabristan and Iblis constitutes very strong support for the idea that the crucible was the vessel of choice for Chalcolithic (and perhaps later) copper production on the Plateau.

During the Chalcolithic and later periods Plateau copper deposits were being mined for their arsenic-rich minerals and ores, and it is to the great multitude of ore deposits that scholars have looked for the evidence of early workings. However, developing a comprehensive understanding of which sources and what types of ores were being exploited at what point in

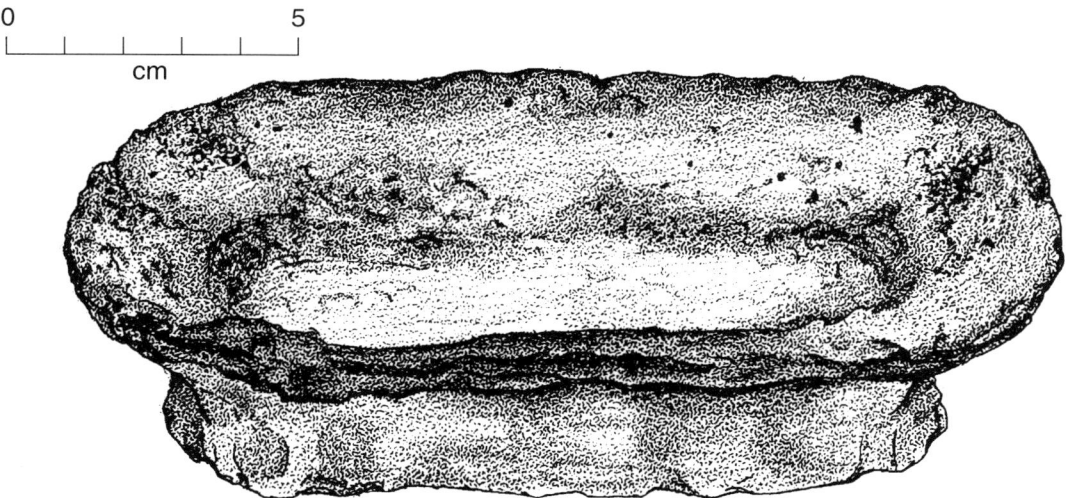

Figure 4.4   *Ceramic crucible (no. 277) from Tal-i Iblis, restored from large fragment. Rest. L. ca. 17 cm. From Dougherty and Caldwell 1967:185, fig. 37:10.*

time has been complicated by the fact that not only are copper deposits with arsenical mineralizations decidedly uncommon on the Plateau but also research in the archaeology of mining on the Plateau has been modest at best. The Wertime expeditions (for the fullest discussions of this research see chapters by Smith et al., Wertime, and Pleiner in Caldwell [ed.] 1967; Wertime 1968; Tylecote 1970) and the French survey (see sources authored by Berthoud with others listed in bibliography below) visited and sampled mining sites on the Plateau, Oman, and Afghanistan. Mining archaeologist Gerd Weisgerber (1990; Weisgerber et al. 1990) of the Deutsches Bergbau-Museum studied ancient mine workings in northwestern Iran.

The most intensively investigated ancient mining complex is that at Veshnoveh, 60 km from Qom and 45 km from Tepe Sialk. The Veshnoveh mines may serve as a model of how early, perhaps even Chalcolithic, shaft and gallery mining was conducted. In addition, though we know little about the mining of the Anarak deposits at Talmessi and Meskani, they clearly were important Plateau deposits, perhaps a major supplier of arsenic-rich minerals.

THE ANCIENT COPPER MINING COMPLEX AT VESHNOVEH

While most ore deposits show evidence of "old workings" (see Bazin and Hübner 1969; also Moorey 1994:246–248), only the copper mines in the Veshnoveh area have yielded some tangible evidence of possible prehistoric exploitation (Holzer and Momenzadeh 1971). This ancient mining complex is the only such site in Iran to have been extensively investigated and published in some detail. Regrettably, in terms of the current discussion, there are no indications that the Veshnoveh deposit was a source of arsenical copper ores.

The mineralization in the four distinct mining loci of Veshnoveh is characterized by the presence of oxides (including malachite in abundance as well as azurite and some native copper) and sulfides (chalcocite, bornite, and chalcopyrite). At the deposit of "Chale Gahr," a single ceramic vessel of the Sialk IV type (ca. 4000–3500 B.C.) was found in one of the many galleries. Other ceramics found scattered on the surface at various locales in the Veshnoveh area suggest much later occupations, including Iron Age, Achaemenid, Sassanian, early Islamic period, and Safavid (Holzer and Momenzadeh 1971:7).

At "Laghe Morad," the long, sinuous mine shafts following presumably rich veins had been made by means of fire-setting (Holzer and Momenzadeh 1971:5). Stone mining mauls, often with an equatorial groove for hafting, were a frequent find. They were used against the weakened host rock. Local sources of stone for the mining mauls were presumably inadequate to the task since only a few tools of these materials have been found. Analysis indicated that the mining mauls of a dark green magmatic rock (spessartite) were imported from an unknown location nowhere near the Veshnoveh area. Most probably they were, as in so many ancient mining sites in the Old World, fluviatile rounded cobbles selected for their weight and hardness from a river bed. The often broken mauls had shattered along pre-existing cleavage and joint planes in the rock. Length ranged from 8 to more than 15 cm and, along with shape, appears to have been "normed." Collected mauls range in weight from 1.2 to 1.9 kg. Stone mauls appear to have been the mining tool of choice, as no pick marks suggesting the use of metal tools were observed.

The combination of ores mined at this site would have been ideally suited to crucible co-smelting. A slag heap of unknown date was located in the southern vicinity of Mazrayeh, about 1.5 km from Veshnoveh. Qualitative spectroscopy of this slag and of copper ore from the Veshnoveh mine area produced almost identical results. Elements with substantial presence in both the slag and ore included Si, Al, Fe, Mg, Ca, Na, K, Ti, Mn, Cr, Sr, V, Ni, and Co. Traces of Ga, Mo, Sc, Zn, and Zr were also noted (Holzer and Momenzadeh 1971:6). These results support the thesis that no substantial arsenical mineralization was present in the four Veshnoveh deposits.

THE ANARAK MINING DISTRICT AND ITS ROLE AS A POTENTIAL BRONZE AGE COPPER SOURCE

The single region on the Plateau with arsenical mineralizations which were readily accessible, and rich and extensive enough to have been exploited over several millennia from the Neolithic into the Bronze Age, is the Anarak mining district, in north-central Dasht-i Lut (Bazin and Hübner 1969:61–63). Within this 10,000 sq km district over twenty polymetallic mineralizations occur, including Cu, Pb, Zn, Ag, Au, Bi, Co, Fe, Mn, Mo, Sb, and U. Two adjacent deposits, namely, Talmessi and Meskani, are considered the primary sources. The richness of Talmessi is emphasized by Maczek et al. (1952:65), who reported that in 1935 native copper was being extracted from a depth of 80 m at a rate of 300 kg per cubic meter. Of particular importance is the fact that not only are Talmessi and Meskani sources of arsenical native copper but they are the only known Southwest Asian deposits with significant quantities of algodonite (Cu_5As) and domeykite (Cu_3As) (Schürenberg 1963; Tylecote 1970:289–290; Heskel 1982:9; Heskel and Lamberg-Karlovsky 1980:86). These two copper arsenides, which often co-occur geologically with native copper, if dropped into molten native copper in a crucible, will simply dissolve (much like sugar in water), releasing their high arsenic content into the melt (W. Rostoker, pers. comm.). Moreover, the high temperature processing of copper-base materials from the Talmessi and Meskani ore bodies, known for their measurable nickel (nickel arsenides) and cobalt (speiskobalt or pyrite of cobalt) content (Schürenberg 1963:213ff.), would result in arsenical copper with nickel and cobalt as impurities. The oft-cited statement that all high-nickel copper came from Oman is, therefore, clearly in dispute and clouds the question of the source of the Cu-As-Ni metal known at sites such as Susa (Moorey 1994:247; and see further discussion below).

The extraction of quantities of native copper at Talmessi may have been accomplished in a manner similar to that used in the Lake Superior District between 4000 and 1000 B.C.:

> Typically, an individual pit was 20–50 feet deep and 10 to 30 feet in diameter, containing about 1000 tons of rock and copper. In some cases, pits were mined adjacent to each other, forming trenches, with ledges of rock left between individual pits for control of water. It is estimated that at least 5000 such pits were dug by ancient peoples in the Lake Superior area with total copper production estimates ranging upward from 5000 tons over a 5000 year period. (Wayman 1985:68)

The technique of fire-setting was used in the Lake Superior mining operation as were stone mining mauls, which littered the mining sites by the thousands. Regrettably, none of those who have visited the Talmessi deposit or other deposits in the Anarak district recorded observations on the ancient workings or commented on the presence of the mining mauls so typical of ancient mining locations.

That these Anarak district deposits could have been supplying an expanded region (Heskel 1982; Berthoud et al. 1982; see also Pigott in press; Pigott et al. in press) is supported by the fact that trade routes were established as early as the Neolithic, when obsidian was extensively traded (Wertime 1973:876; Moorey 1982:84). Analyses of a native copper pin from Sialk (see above, pg. 74) suggest that Talmessi, some 200 km to the east, was the source of the copper (Smith 1968). By the end of the Chalcolithic period, interaction between Plateau settlements and those in lowland Khuzistan and Mesopotamia was well developed (Moorey 1982; Deshayes 1960). It has been argued that analyses of arsenical copper artifacts at fourth millennium Susa, in lowland Khuzistan (see discussion below), show a strong elemental correlation to the Talmessi deposit (Berthoud et al. 1982:43; Malfoy and Menu 1987:364; cf. Seeliger et al. 1985:643; see also Moorey 1994:249).

Scholarly consensus suggests that Chalcolithic metallurgy on the Iranian Plateau was centered on native copper and ores that were arsenical. A workable arsenical copper could have been obtained by one or more processes which involve the simple melting of native copper and copper arsenides in a crucible or through at least two distinct processes of reduction of an arsenical ore (to be discussed later). While we witness a clear elaboration of artifact types throughout the Chalcolithic and Bronze Ages, the technological tradition of metalworking on the Plateau appears basically conservative, as little significant innovation can be identified over the millennia. This is surprising in that from the early Chalcolithic onward, the workshop at Ghabristan and the substantial amounts of metalworking debris found at Hissar, Sialk, and Shahdad suggest full time metalworkers. Their efforts appear to have been directed to elaborating their finished products through casting and hammering rather than evolving different techniques of manipulating basic raw materials.

While it is apparent that the working of copper on a regular basis is evident on the Plateau earlier than in the Mesopotamian lowlands, the first occurrences of tin bronze artifacts appear simultaneously during the fourth millennium on the Plateau and in the lowlands. A bronze needle came from Sialk III-B (Ghirshman 1938:206) and a bronze flat adze was excavated from the Necropolis at Susa I (Berthoud 1979: appendix 13, #974). These initial, random occurrences of bronze may be the result of trade from further east. Afghanistan, with its juxtaposition of abundant copper and tin deposits, constitutes a likely source. But before turning to the the coming of bronze, let us look at fourth and third millennium metallurgy at the site of Susa, the northern urban center of what was to become the Elamite kingdom in the later third millennium. A century of archaeology at this site combined with the excavation and analysis of a large corpus of copper-base artifacts has provided a wealth of information about metallurgical development at Susa, much of it discussed by F. Tallon (1987). The results can be summarized here.

METALLURGICAL DEVELOPMENTS AT SUSA IN LOWLAND KHUZISTAN DURING THE FOURTH AND THIRD MILLENNIA B.C.

While the site of Susa is off the Iranian Plateau and geographically part of the Mesopotamian lowlands, no discussion of Iranian archaeometallurgical developments is complete without consideration of the evidence from this remarkable site. During the fourth millennium B.C., copper-base artifacts, which include axes, awls, needles, and flat adzes from the Necropolis (Susa I), are for the most part very pure copper with arsenic as the notable impurity (in about half, or 37 out of the 68 artifacts analyzed from this period, the average As content was only 0.41%) and with varying concentrations of iron (Malfoy and Menu 1987:365; see also Moorey 1994:256–259 for overviews of Susian metallurgy, Ubaid through Early Dynastic Period).[2] One third, or 23 of the 68 analyzed artifacts, contain not only arsenic (1.6% As on average) but also nickel

(1.1% on average), which is seen as an impurity traveling with the arsenic. Arsenic and nickel in copper would suggest the copper-nickel arsenides known to typify the deposit at Talmessi. In their discussion J. M. Malfoy and M. Menu (1987:364) suggest that the Anarak district (e.g., Talmessi) native copper and oxide and carbonate ores were the source of this metal.

Arsenic content in smelted metal varies primarily according to arsenic content of the ore and particular smelting furnace conditions. Malfoy and Menu (1987:365) suggest that, where present, a higher arsenic content was intentional, as in, e.g., two of ten mirrors analyzed, possibly to enhance their reflective properties (see also Moorey 1994:250). Tallon (1987:313) is of the opinion that this early Susian metallurgy signals a prelude or even the beginnings of true metallurgy in the Mesopotamian-Iranian interaction zone, contrasting the Susa axe, needle, and mirror production with the less sophisticated manufacture of pins, needles, awl, and rings at Tepe Sialk during the same period. However, while casting a mirror is a skillful task, it is doubtful that at this early stage elemental content was being controlled for purposes such as enhanced reflectivity of mirrors (see also Moorey 1994:251). Tallon (1987:314) also suggests that the copper-base axes from this period were done in imitation of stone axes. In shape, the proportions of these two types of implements are quite similar.

By the period spanned by Susa II/IIIA (second half of the fourth millennium B.C.) copper-base artifacts have a more heterogenous composition: they contain quantities of arsenic, silver, antimony, and bismuth, a composition which suggests the exploitation of *fahlerz* (the gray copper ores also known as copper sulfarsenates), possibly the ore sources from the Kashan zone, in the Bardsir or Sheikh Ali Valleys. The first half of the third millennium B.C. at Susa (IVA) is a period characterized by change and contact with the east (Luristan) and the west (Mesopotamia) as well as by the development of maritime trade and the resulting access to copper from Oman. Artifacts from subsequent Period IVA2 contain As, Ni, Ca, Sn, and Bi and appear to be products of smelting a chalcopyrite-type ore, which would account for the substantial iron content in the artifacts. However, it must noted that fluxing the smelting operation with iron oxide to remove siliceous gangue would also contribute iron to the metal.

Tin bronze appears for the first time at Susa at the midpoint of the millennium. The excavation of the "Vase à la Cachette" (actually the cache was within two ceramic vessels) by de Morgan's mission at Susa in 1908 yielded a number of copper-base artifacts including four bronzes with over 5% tin (see Moorey 1994:253–254). The majority of metal artifacts in the cache, however, were either unalloyed or arsenical copper. The arsenical copper contained substantial amounts of cobalt with variable amounts of nickel and iron. Malfoy and Menu (1987:365) state that the previous research of Berthoud and colleagues "demonstrated in a decisive manner that the copper supply came from the Gulf in this era." However, as Moorey (1994:247, 249–250) indicates, Seeliger et al. (1985:643) and Müller-Karpe (1991:108) (see also Hauptmann et al. 1988:34) have strong reservations concerning the Berthoud study and its final conclusions vis-à-vis Oman.

Critical to this argument is the research by Gerd Weisgerber and Andreas Hauptmann from the Deutsches Bergbau-Museum on the early copper mines and smelting sites in Oman (Hauptmann et al. 1988). A focal point of this research as well as of much earlier research (e.g., Peake 1928; Desch 1929) has been the question, "Is Oman the fabled land of Magan which texts record as having supplied copper to Sumer and Babylonia in the third and second millennia B.C.?" Surveys in the mountains of Oman have demonstrated a strong agreement between the "lay of the land" and third millennium B.C. textual descriptions (Bibby 1977 and Weisgerber 1980 in Hauptmann et al. 1988:49). The enormous volume of slag in Oman that dates to the Bronze Age, arguably the largest such concentration from this period known in the Middle East, also provides credible support for levels of production that could have supported a trade in copper to Mesopotamia (including Susa in lowland Khuzistan). However, as Hauptmann et al. (1988:49) report, the argument still lacks the analytical evidence to conclusively identify Oman with ancient Magan (cf. Potts's [1990, i:117–119, also 135, 199–225] cogent arguments to the contrary). Berthoud (1979) attempted to make the analytical link by arguing that in Susa IV (ca. 2600–2400 B.C.), artifacts and Omani copper ores were comparable. However, Seeliger et al. (1985:643) claim that Berthoud's study is flawed both analytically and geologically. Moreover, Hauptmann et al. (1988:34) suggest that "too few objects were analyzed by Berthoud so that the final evidence for determining whether the copper came from Oman or the southern part of Iran is still a problem."

Without going further into this complex discussion of the Omani sources, two final points merit mention. First, a 6 kg hoard of 22 bun-shaped, i.e., planoconvex, ingots and ingot fragments was excavated at Maysar in Oman by the German team, while a second find in Al-Aqir (Bahla) yielded 16 ingots weighing 19 kg (Hauptmann et al. 1988:41, fig. 4.6). Five circular, bun-shaped ingots were found in the "Vase à la Cachette" at Susa and are published by Tallon (1987: nos. 687–692, pls. 262–264). While these are quite similar to the Omani ingots it is possible that their shape was somewhat standard for the period across Southwest and South Asia, since ingots of this shape are known from elsewhere in the Near East as well as the Indus Valley (Moorey 1994:244). Second, it is also important to reemphasize the point made by Hauptmann and colleagues that a potential source of confusion when making comparisons of trace element profiles involving Omani and Iranian sources is the fact that

[a]n overwhelming part of the mineralizations and ore deposits in Oman form veins that occur in gabbros and peridotitic rocks.... The main copper mineral under the surface is chalcopyrite, which is intergrown with small amounts of cobaltite, loellingite, and other Fe-Co-Ni-As minerals.... [Concentrations can be as follows:] Ni up to 0.6%, Co up to 0.12%, As up to 0.2%). (Hauptmann et al. 1988:35)

While the smelting of chalcopyrite would have required a process quite distinct from that necessary for the Anarak sources, one must wonder if the presence of a similar suite of trace elements in the Omani and Anarak sources (i.e., As-Ni-Co) may blur forever attempts to discriminate between these two potential source areas. The analytical program on the Susa metal by Malfoy and Menu does not address this difficulty in discriminating between Iranian and Omani sources.

At Susa, copper-base artifacts from Period IVB (Akkadian period), dating to the closing centuries of the third millennium, were found to be similar in composition to those from the "Vase à la Cachette," and included some unalloyed copper artifacts and others with relatively low arsenic (Malfoy and Menu 1987:370). The researchers report that a single bronze axe inscribed with the name *Ilish-mani* was analyzed. It is of interest that it was a tin bronze artifact that was inscribed, as if to suggest its unique composition merited special attention. The Period IV artifacts that lack precise dates are also similar in composition to the cached metal. Bronzes are present; several daggers have quite high arsenic and nickel contents (in two of the daggers, the Ni is above 4%) (Malfoy and Menu 1987:371). This arsenic-nickel pairing continues to bring to mind the Anarak-Talmessi ore body, as well as Oman. It is possible that multiple sources were being exploited by Susian metalworkers and that this exploitation included the continuing extraction of copper-nickel arsenides from the central Plateau as well as use of imported Omani ores.

Susa VA at the end of the third millennium is represented by human figurines from the Ninhursag Temple at Susa. These were found to be unalloyed copper, some with high lead (one with 1.9% Pb) and some with high trace-element amounts of silver. It is suggested that these foundation figurines may have actually been brought from Sumer by Shulgi, King of Ur, when he constructed the temples of Susa.

Malfoy and Menu (1987:372) suggest the advent of the Isin-Larsa period at the start of the second millennium (Susa VB) was a "decisive point of departure for tin bronze metallurgy." It was a period when complex alloys were being experimented with.[3] They continue to maintain, based on consistent levels of nickel, cobalt, and iron in copper-base artifacts from Susa IVA2 and Susa VB, that even in this period Oman remained a source of copper.

THE BRONZE AGE (CA. 3200–1450/1350 B.C.)

Arsenical copper production continues to be the dominant tradition on the Iranian Plateau during the approximately two millennia of the Bronze Age. Bronze occurs with some frequency toward the end of the period, but it is still not common. It is only with the inception of the Iron Age that tin bronze becomes the dominant copper-base alloy.

While copper deposits occur with reasonable frequency throughout the highland zones of Southwest Asia, tin sources are far less common. The only geologically confirmed traces of tin within the modern political boundaries of Iran are documented in southeastern Iran in the Dasht-i Lut at locations near Sistan (Stocklin et al. 1972:58). It is in Sistan, ancient Drangiana, that Strabo (*Geography* 15.2.10) mentions that tin is present. But significant tin deposits do not occur in Iran. Large deposits, however, have been identified in neighboring Afghanistan during UNESCO-sponsored joint Afghan-Soviet geological surveys (Chymriov et al. 1973; Shareq et al. 1977) and by the French archaeometallurgical survey (Cleuziou and Berthoud 1982; Berthoud et al. 1982). In Afghanistan two primary zones of mineralization are known, the first of which extends from south of Kandahar to Badakhshan province in the northeast and the second from Sistan north to the vicinity of Herat (Shareq et al. 1977; Stech and Pigott 1986:40, 44–45; also Penhallurick 1986:29) (Fig. 4.5). These tin deposits are the most extensively documented in Southwest Asia, and constitute the most likely source of the region's earliest tin.

The ubiquity of copper deposits on the Iranian Plateau has been documented by Ladame (1945) and by the Geological Survey of Iran (Bazin and Hübner 1969) (Fig. 4.6). The Geological Survey recorded at least twelve major copper-rich zones encompassing over 200 deposits. In Afghanistan, a total of 241 copper deposits have been recorded (Shareq et al. 1977:101–135, map 6; see also Berthoud et al. 1978; Heskel 1982:340–381). This abundance of copper on the Iranian Plateau and in Afghanistan, together with the availability of tin, gives rise to the speculation that this region may have been a "heartland" for the early development of bronze metallurgy (Stech and Pigott 1986:48).

The appearance of bronze after 3000 B.C. coincides with that of gold and lapis lazuli (Muhly 1977:76). The earliest simultaneous occurrence of these luxury materials was in Mesopotamia (Early Dynastic Period III, particularly) while their sources lay considerably further east. The northeastern Afghan province of Badakhshan

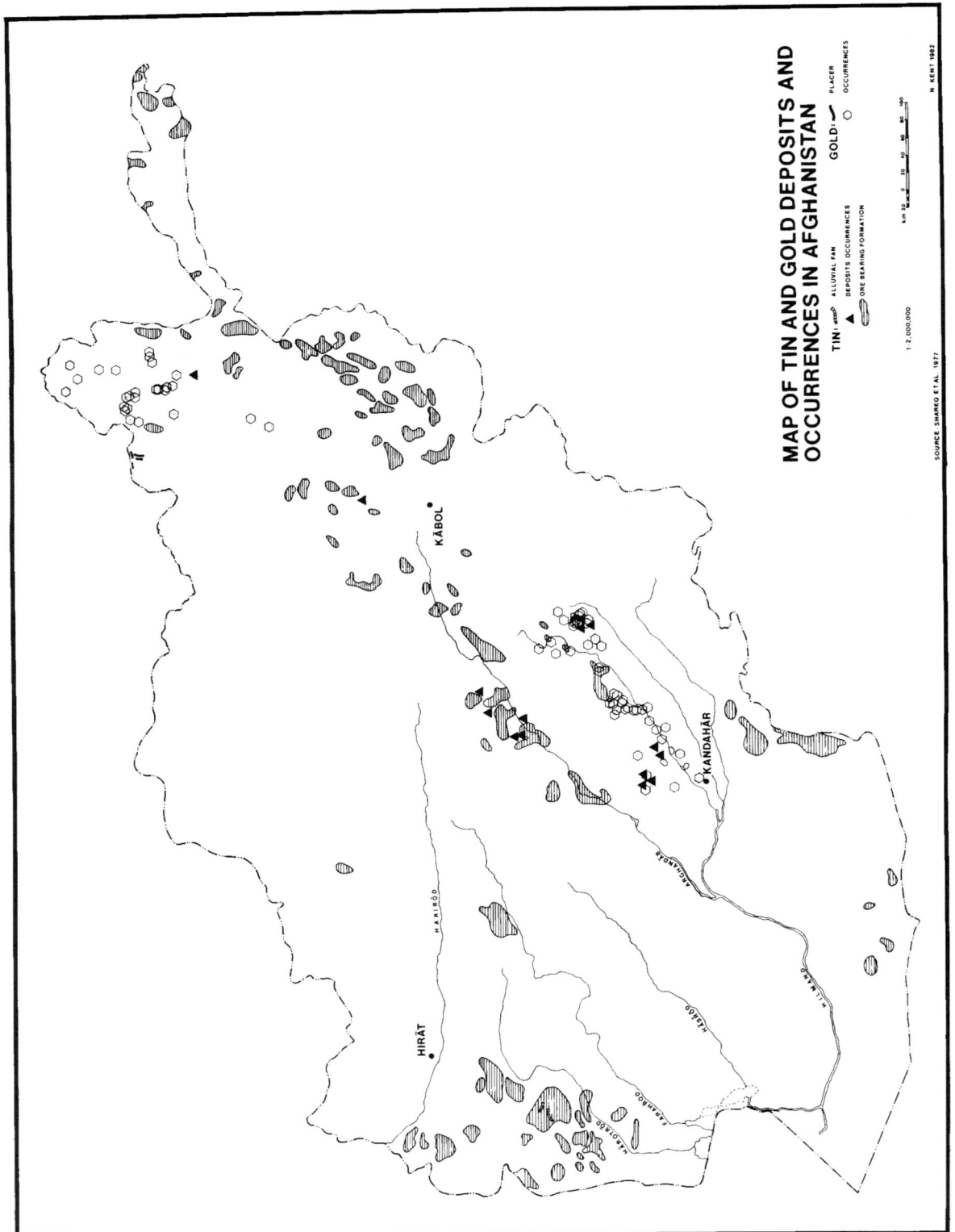

Figure 4.5   Tin and gold deposits and occurrences in Afghanistan. Map by N. Kent after Shareq et al. 1977.

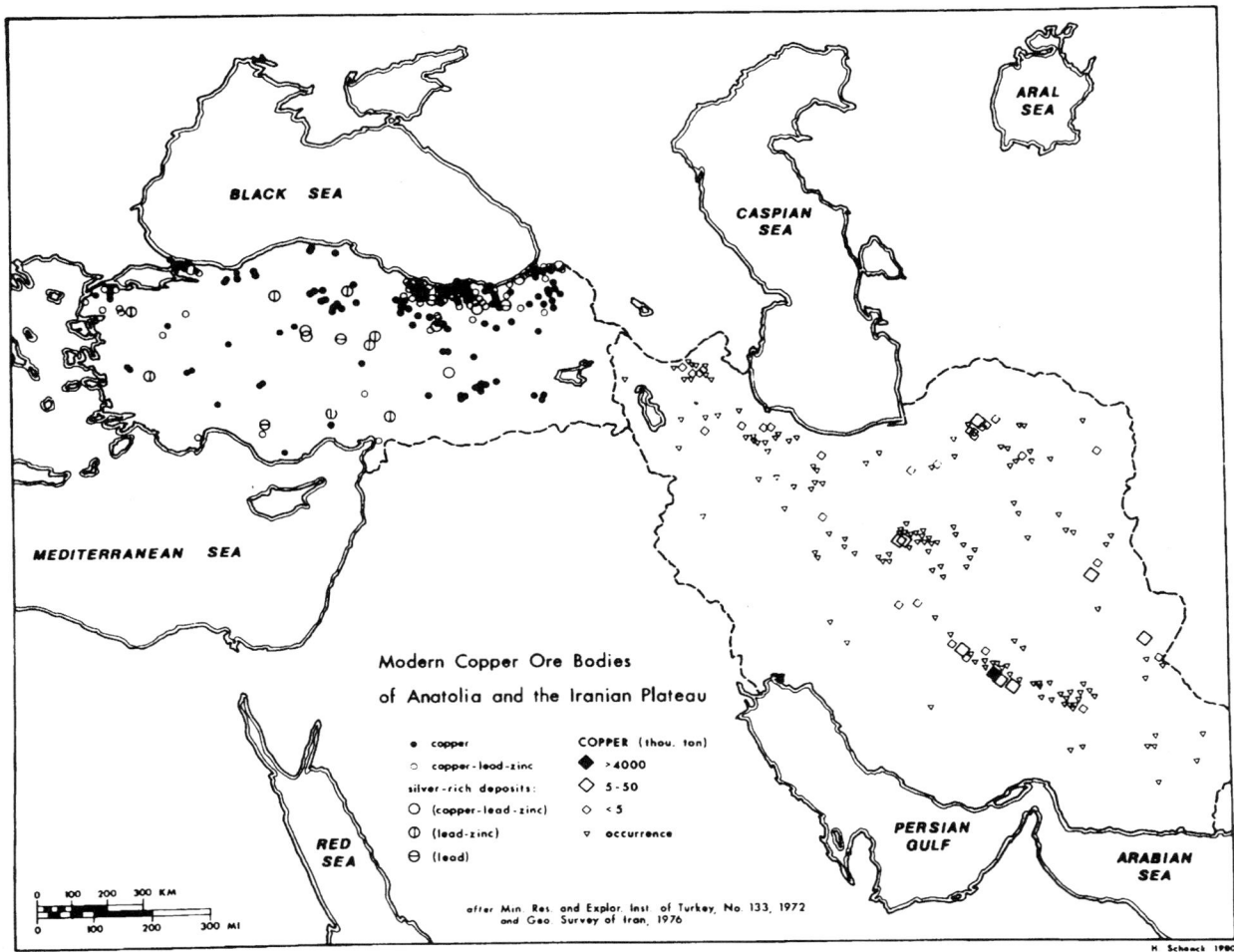

Figure 4.6   Modern copper ore bodies in Anatolia and on the Iranian Plateau. From Pigott 1989b:458, fig. 27.

is a primary lapis lazuli source region (Casanova 1992; Tosi 1974; Herrmann and Moorey 1980–83; Herrmann 1968), and it has been reported that alluvial lapis cobbles occur there (H.-P. Francfort, pers. comm. in Stech and Pigott 1986:46). In addition, small lapis sources are now being reported in the Chagai Hills of northern Pakistani Baluchistan (Jarrige 1988). Gold, and tin for bronze, are also found across much of Afghanistan, concentrated alluvially in the valleys of the major river systems (see Fig. 4.5; Shareq et al. 1977). The alluvial occurrence of these three materials would have made their early exploitation a distinct possibility and their procurement relatively simple. Transport of these raw materials to Mesopotamia meant interaction with the populations on the Iranian Plateau. Finds of lapis lazuli document this contact at sites such as Tepe Hissar and Shahr-i Sokhta, which yielded ample indications of being lapis-working sites (Bulgarelli 1979). Despite substantial interaction with the tin-producing regions, the metalworking traditions of the Plateau sites continued to be dominated by the production of the arsenical copper that characterized the earlier period.

As discussed above, it is only in lowland Khuzistan at Susa that bronze technology is in evidence at this time. The "Vase à la Cachette" filled with a variety of copper-base metal artifacts (Amiet et al. 1980; Berthoud 1979:14) came from a mid-third millennium context. The analyses suggest that tin was being alloyed with arsenical copper (Stech and Pigott 1986:43). Bronze is found with some frequency at Susa by the final centuries of the third millennium and is established by the early second millennium (as it is at Anshan), a period during which Susa shows the strongest cultural ties with Mesopotamia (Amiet 1979:197).

Sumerians were intent on trade and the acquisition of exotic, luxury materials. The rarity of tin may have promoted its status among Sumerians while the peoples of the Iranian Plateau may have remained uninfluenced by such pressures (Stech and Pigott

1986:48). As such, the tin "by-passed" the Plateau on its way to Mesopotamia (Beale 1973:144; Moorey 1982:88). The Plateau metallurgical traditions can be characterized as technologically conservative since, while manufacturing quantities of copper-base artifacts in a variety of forms, they continued to be based primarily on simple melting/smelting of arsenical copper.

ARCHAEOMETALLURGICAL REMAINS FROM BRONZE AGE TEPE HISSAR

While many copper deposits, often with indications of "old workings," are present in northeastern Iran/northern Khorasan, only one arsenical mineralization has been identified by modern geological survey at Taknar (Bazin and Hübner 1969:90). However, the arsenic-bearing gray copper minerals enargite and tetrahedrite are present only in subordinate amounts. It seems improbable that such a source could have supplied arsenical ores to prehistoric sites on the northern Plateau, including Tepe Hissar where Chalcolithic and later levels yielded large numbers of artifacts almost exclusively of arsenical copper (Pigott et al. 1982; Pigott 1989a). Heskel (1982:343ff), who suggests Hissar metalworkers exploited only the gossan zones of nearby copper deposits, as well as those with lead and silver ores, did not address the issue of where the arsenic in the Hissar metal would have come from or the lack of arsenical ores near this site. Thus, the unaswered question remains, "Where did Hissar metalworkers obtain their arsenical copper ores?"

On the one hand, the dearth of identified arsenic-rich copper deposits on the Plateau (at least theoretically, following Heskel's basic premise) would argue in favor of the Anarak sources as a viable option for Hissar and most of the other major Bronze Age copper-producing settlements on the Plateau during the Chalcolithic and perhaps into the Bronze Age. However, in the instance of Hissar, and also Shahr-i Sokhta, the ample amounts of slag and other production evidence suggest mixed ore smelting, not just the melting of copper arsenides in crucibles. Furthermore, the metallography of Hissar and Shahr-i Sokhta copper-base metal artifacts provides no indications of coldworking of native copper/copper arsenides as Heskel has demonstrated for Tepe Yahya. The issue of arsenic and how it ended up in Chalcolithic and Bronze Age artifacts on the Plateau is clearly one that will be the subject of much future study.

Tepe Hissar is one of the major sites whose archaeometallurgical remains have received some focused attention in the laboratory. The site was excavated initially in 1931–32 by Erich F. Schmidt (1937) and much later was the subject of a re-study season in 1976 under the joint direction of Robert H. Dyson and Maurizio Tosi (Dyson and Howard 1989). During the 1976 season Maurizio Tosi and I undertook a surface survey of the ubiquitous production remains (Pigott 1989a) and followed this with a preliminary laboratory study at the Museum Applied Science Center for Archaeology (MASCA) at the University of Pennsylvania Museum (Pigott et al. 1982). The picture which unfolded, based on those analyses, is one of remarkable technological conservatism persisting over the span of Hissar's existence as a settlement from the later fifth to the early second millennium B.C.

The amount of slag and industrial ceramic fragments strewn across the surface of Tepe Hissar (at the time of my visit in early October 1976) was striking (Pigott et al. 1982:215). The same is said to be true of the surface of the south mound at Tepe Sialk (Tylecote 1970:290). But virtually no other mention of the obvious on-site metal production evidence has been made by other archaeologists and scholars visiting or excavating at these two sites. At Hissar scattered slag was evident on some 9.15% or 11,000 sq m of the preserved site. Three areas of dense slag concentration were identified and surveyed. From the surface survey and the study of excavated slag, a rudimentary slag typology was established based on visual observation of collected samples (Pigott 1989a:25–27).[4] The slags are from the furnace smelting of copper and most probably lead. The furnace linings clearly indicate the use of a furnace-based smelting technology at Hissar. The crucible is, in fact, curiously absent from the metallurgical assemblage which we were able to sample from the site.

Furnace lining fragments, a number of which were excavated at Hissar, are not well attested at other Plateau sites, let alone in such volume (Fig. 4.7). For example, only the report on Tal-i Iblis (Caldwell 1967:34) mentions a significant presence of substantial amounts of industrial ceramics, which at that site happen to be crucible fragments. A sample of one of the Hissar furnace linings underwent a refiring experiment to determine the temperature to which the ceramic fabric had been subjected. The temperature range was somewhere between 1220 and 1250°C, quite sufficient to promote the smelting of copper ores, though perhaps difficult to obtain without benefit of a bellows with tuyère (Pigott et al. 1982:225). Prevailing winds on the Damghan Plain, however, are such that they alone may have been sufficient to drive the process of copper smelting (Pigott 1989a:25, n. 3).

The slag analyses performed, though of the most preliminary sort, suggest the smelting of arsenical ores of copper as well as perhaps lead-silver ores. The practice of fluxing not only with iron oxide (to remove high silica content of ores) but also with quartz (to remove high iron content of ores) is suggested by the analyses (Pigott et al. 1982:233). Amounts of iron oxide were found on the site's surface and add support to the contention that it was used in the process of fluxing the smelting operation.

In addition, two ingot fragments were analyzed and contained only trace amounts of arsenic (Pigott et

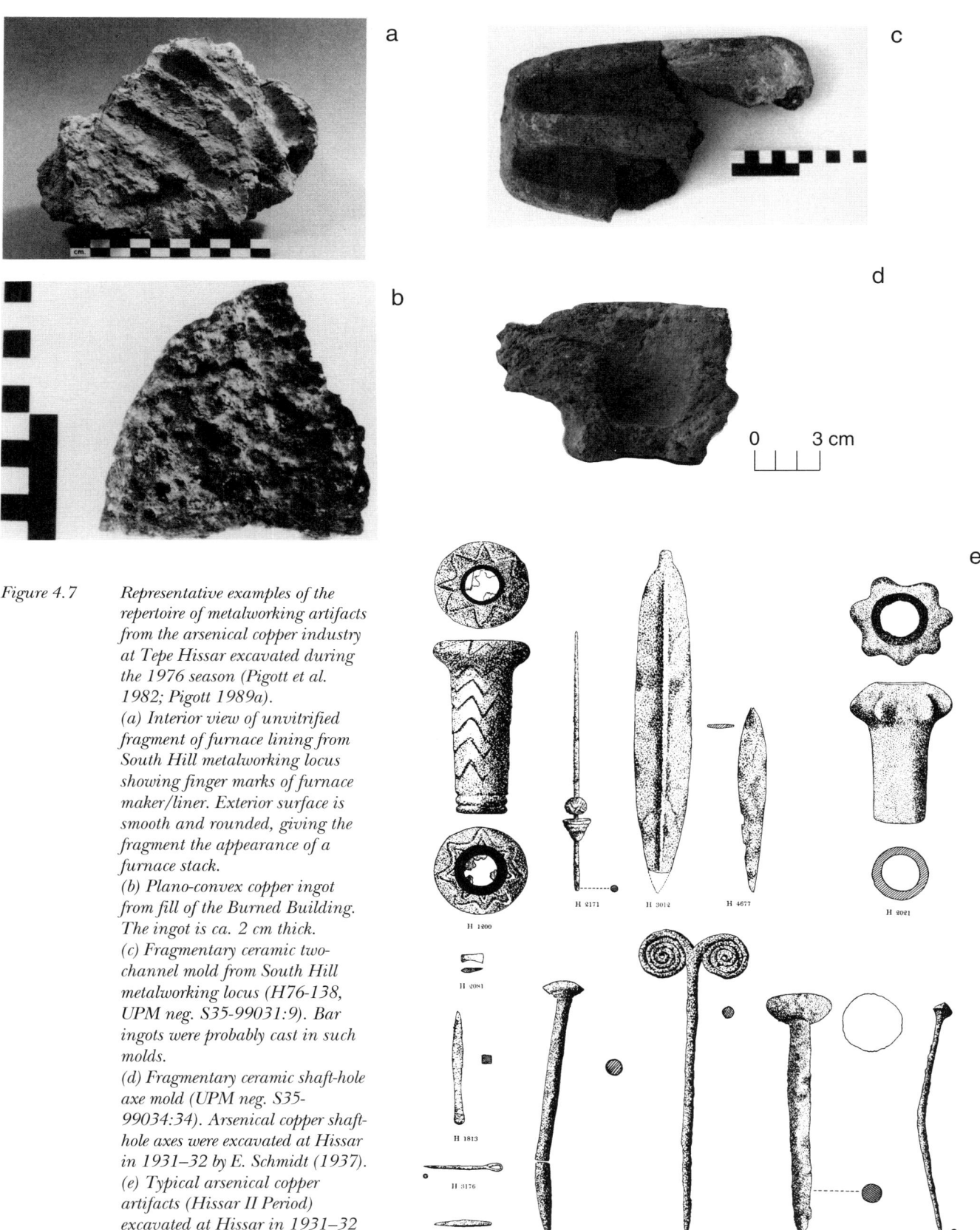

Figure 4.7 Representative examples of the repertoire of metalworking artifacts from the arsenical copper industry at Tepe Hissar excavated during the 1976 season (Pigott et al. 1982; Pigott 1989a).
(a) Interior view of unvitrified fragment of furnace lining from South Hill metalworking locus showing finger marks of furnace maker/liner. Exterior surface is smooth and rounded, giving the fragment the appearance of a furnace stack.
(b) Plano-convex copper ingot from fill of the Burned Building. The ingot is ca. 2 cm thick.
(c) Fragmentary ceramic two-channel mold from South Hill metalworking locus (H76-138, UPM neg. S35-99031:9). Bar ingots were probably cast in such molds.
(d) Fragmentary ceramic shaft-hole axe mold (UPM neg. S35-99034:34). Arsenical copper shaft-hole axes were excavated at Hissar in 1931–32 by E. Schmidt (1937).
(e) Typical arsenical copper artifacts (Hissar II Period) excavated at Hissar in 1931–32 by E. Schmidt (1937).

al. 1982:229, table 2). The Hissar metal finds comprise one of the largest collections of artifacts from an excavated context on the Plateau. An impressive array of artifact types were excavated from the Hissar stratigraphic sequence (Schmidt 1937). Types include (Pigott 1989a:32): (1) personal ornaments (pins, bracelets, earrings, nails, belt hook, diadem), (2) weapons (daggers, spearpoints, mace), (3) tools (axes, chisels, mattock, shaft-hole axe), and (4) other (stamp seals, wands, vessels, mirror, human figurine, animal figurine, trident, rings, tubes, ingots).

The unpublished elemental analyses were conducted in 1962 by Ũfuk Esin as part of the major program of analyses of ancient metal being conducted at the Württemburgisches Landesmuseum in Stuttgart by Junghans and Schroeder. Almost 198 of the 1109 copper-base artifacts excavated at Hissar were analyzed by emission spectroscopy. These analyses, which are summarized graphically in Pigott et al. (1982:230–231) also suggest the exploitation of the gray copper (*fahlerz*) ores found in weathered copper sulfide ore bodies (Pigott et al. 1982:232). Arsenical copper artifacts dominate; however, the mean As-content was 1.32% with a standard deviation of 1.29%. One wonders, given these low values, if the As-content would have had much of an impact on the performance of these artifacts, where strength and durability were useful. Metalworking at Hissar gives every appearance of being a conservative tradition based on the use of arsenical ores of copper and one that perhaps may typify the character of copper production on the Plateau prior to the Iron Age. But we know little about which ore deposits were being exploited to supply the needs of Hissar's metalworkers. Interestingly, tin bronze makes no demonstrable appearance in the Hissar sequence (Reisch and Horton in Schmidt 1937:359; Pigott et al. 1982:230; Berthoud et al. 1982:50, n. 66). This gap lends additional credence to the notion that tin, which may have been carried from Afghanistan and/or Central Asia to Sumer, either was not moving across the Plateau or was "by-passing" or being "by-passed" by metalworkers in major Bronze Age urban centers on the Plateau.

ARCHAEOMETALLURGICAL REMAINS FROM SHAHR-I SOKHTA

Arsenical copper metallurgy characterizes the assemblage at Shahr-i Sokhta, to the southeast of Tepe Hissar (Heskel 1982:97–120; Heskel and Lamberg-Karlovsky 1980, 1986; Hauptmann 1980; see also Tosi's [1983] study of a copper-base statuette from the site). An as yet unpublished, detailed characterization by analytical chemistry of the copper smelting techniques of the metalworkers at this site was conducted by D. Helmig (1986). Ore, matte, and copper in the form of ingots, prills, and artifacts were analyzed, and it was concluded that the excavated ore samples were being processed at the site. However, arsenic and lead content of ores showed "marked discrepancies" with those of the slag and metal (Hauptmann et al. 1988:46).

Helmig demonstrated that arsenical copper with 1–3% As was produced unintentionally and only sporadically using somewhat As-rich ores. He could exclude deliberate addition of (relatively pure) arsenic ores to the metal. With this point he also could confirm Lorenzen's (1965) and Tylecote et al.'s (1977) results, which pointed out that the recovery rate of arsenic in copper is quite high and that arsenical alloys can be obtained through the smelting of suitable ores.

Heskel (1982:97ff) describes the archaeometallurgical remains from this site as being the "largest and best corpus" in Iran, one which contains ores, slags, ingots, and finished and partially finished artifacts. He analyzed 51 artifacts, both excavated and from the site's surface. Copper carbonates, oxides, sulfides, and complex copper-iron sulfide ores were identified visually. What Heskel terms "sulfide metallurgy" was present by the end of the fourth millennium. He suggested that this technology was brought to Shahr-i Sokhta by smiths from the Namazga culture of Turkmenistan. Of particular interest is Heskel's (1982:99) statement that

[i]t is very likely that chalcocite was added to copper oxide and carbonate ore charge in the furnace in proportions that oxidized the sulfur and did not require roasting of the ore. This is known as pyritic smelting and constitutes both economical and sophisticated method [sic] of smelting copper ores (see Rothenberg 1972).

In making this insightful observation, Heskel had keyed on one of at least three distinct processes by which arsenical copper could have been produced in antiquity (see discussion below). He was following the lead of Beno Rothenberg's team's research on the Timna project who in turn must have studied the work of Hofman (1914). William Rostoker (Rostoker et al. 1989; Rostoker and Dvorak 1991) at the end of the 1980s independently encountered Hofman's work and undertook a series of smelting process replication experiments resulting in his model for "co-smelting" (discussed below). Furthermore, Heskel (1982:104) held that, in Iran, "the crucibles would have formed the bowl of the furnace," an observation which is now strongly supported by findings detailed elsewhere in this chapter, and by evidence from Thailand (Rostoker et al. 1989; Pigott et al. 1997) and Anatolia (Yener and Vandiver 1993; Vandiver et al. 1993).

Heskel's recognition of the co-smelting process was an important step in developing an understanding of how copper ores were processed in the earliest stages of metallurgical activity on the Plateau. A final point merits attention in this regard. The act of co-smelting oxides and carbonates may well have proceeded almost unintentionally rather than as a conscious and sophisticated development, as Heskel suggests. In the mining of weathered copper sulfide ore bodies the potential for extracting ore with mixtures of oxide and sulfides is high, particularly at the point of transition from the upper, often surface-exposed, oxide zone into the zone of secondary enrichment. As long as a mixture of oxides and sulfides is charged into the furnace, the likelihood of the smelt yielding copper is quite strong. Ancient smiths need not have been sorting oxides to mix with sulfides to produce their copper— the nature of ore body structure and composition may have solved the problem for them.

Additional evidence indicates that the fourth and third millennia on the Plateau are characterized mainly by the use of arsenical copper with only the rare bronze artifact appearing. This holds true for the assemblages at Shahr-i Sokhta to the southeast (Heskel 1982:97–120; Hauptmann 1980; see also Tosi's [1983] study of copper-base statuette), and at Tepe Yahya to the south (Heskel and Lamberg-Karlovsky 1980, 1986; Heskel 1982:73–97; Tylecote and McKerrell 1971; see also Pigott in press), and probably at Shahdad, also in the south (Hakemi 1992; Moorey 1982:83, 90–91; Salvatori 1978; Salvatori and Vidale 1982; Vatandoost-Haghighi 1978). An arsenical copper shaft-hole axe from a burial at the site of Khurab (Stein 1937:121), in the southeast, has been the subject of several studies (Maxwell-Hyslop 1955; Zeuner 1955; During-Caspers 1972), including a detailed metallurgical analysis of its composition and manufacture (Lamberg-Karlovsky 1969; Lechtman 1970). Further to the west, at Tal-e Malyan in Fars province, in the late fourth and early third millennium B.C. Banesh Period context (Nicholas 1980, 1990), artifacts are exclusively of arsenical copper (Pigott et al. in press). However, by the early second millennium, in the subsequent Kaftari Period, analyses indicate that bronze predominates (Pigott 1980b:107; unpubl. MASCA analyses). Slag was not common at Malyan in either the Banesh or Kaftari Period, which suggests either that smelting was not practiced on an industrial scale at this site or that those areas where production occurred were not excavated. Kaftari Period slags were characterized by analysis as derived from copper/ bronze production (Carriveau 1978:63–66). Metallographic investigation of small metal finds from both periods suggests that many of these finds are remains from working pieces of copper-base sheet and bar stock (Pigott et al. in press). The geographical and, at this time, cultural proximity of Fars province to the lowlands of Khuzistan is of considerable interest in that the Kaftari assemblage at Malyan appears to be the earliest on the Plateau to contain bronze with some frequency.

For the Iranian Plateau the question remains, "Where was the tin coming from, if bronze was being manufactured at Plateau sites and not being imported in ingot form?" It is difficult to argue for the tin source identified in the Kestel–Göl Tepe complex in the Taurus Mountains of Anatolia (Yener and Vandiver 1993) being one with which metalworkers on the Iranian Plateau had direct or even indirect contact. On the other hand, the suggestion that Afghanistan was a "heartland" of bronze metallurgy (Stech and Pigott 1986:48), while attractive theoretically, remains inadequately substantiated archaeologically. The bronze artifacts from Ghar-i Mar, for example, lack firm chronological positioning (Caley 1971, 1972, 1980; Shaffer 1978:89; cf. Moorey 1982:99, n. 62).

The only well-documented artifacts in bronze from the region were excavated at Mundigak in levels dating from the mid-fourth through the third millennium (Shaffer 1978:144; Jarrige 1985:291; see also Lamberg-Karlovsky 1967:146–148). Here, while only a few of the artifacts, principally axes and an adze, were bronze, their wide chronological distribution could indicate regular use of the alloy (Stech and Pigott 1986:47). It is of interest that lapis and bronze co-occur in Mundigak I (Shaffer 1978:172). To the north in Bactria, lapis occurs at the site of Shortugai in the period when Harappan influence is at its peak, although bronze is not similarly documented (Francfort and Pottier 1978:58–59; Francfort 1989). In short, there exists a distinct possibility that Afghanistan constituted a source area for tin and lapis, for Mesopotamia as well as the Indus Valley (Stech and Pigott 1986:47).

Thus, despite regular contact with tin-bearing regions and the appearance of tin bronze in Mesopotamia, the Iranian Plateau maintained its arsenical copper tradition throughout the Bronze Age. Certain technological developments, however, can be postulated. For example, the analyses of copper artifacts from Hissar suggest that miners were penetrating the sulfide zone in copper ore deposits (Pigott et al. 1982). It is likely that in doing so mixed oxidic and sulfidic copper ores were extracted. When charged in a crucible, such a combination of ores will smelt directly to copper in a single step rather than producing *only* the copper-iron-sulfide "matte" that would require further smelting (Fig. 4.8).

The late William Rostoker conducted a series of smelting experiments that demonstrated the efficacy of such oxide/sulfide co-smelting (Rostoker et al. 1989; Rostoker and Dvorak 1991; see also note in the *Journal of Historical Metallurgy* [Anon. 1985] on smelting experiments in the manufacture of Cu-As alloys conducted by a team led by H. Lechtman and R. F. Tylecote; and new article by Lechtman and Klein [1999]). Rostoker found that no reducing atmosphere is necessary, simply high temperatures (ca. 1250°C). Thus, while charcoal would be an excellent fuel, it is not required. With sufficient natural draft available the smelting could be driven by dry wood fuel alone and could be achieved directly and with relative ease in crucibles or bowl furnaces.

Other than the co-smelting of oxides and sulfides, and the melting of native copper and/or arsenides, there is a third method which may have been used to

$$
\begin{aligned}
&1)\quad 3Cu_2O + FeS \longrightarrow FeO + SO_2 + 6Cu & \Delta F_R = -42.63 \text{ kcal}\\
&2)\quad 5CuO + CuFeS_2 \longrightarrow FeO + 2SO_2 + 6Cu & \Delta F_R = -113.5 \text{ kcal}\\
&3)\quad 3CuO + FeS \longrightarrow FeO + SO_2 + 3Cu & \Delta F_R = -65.23 \text{ kcal}\\
&4)\quad 2CuO + S \longrightarrow SO_2 + 2Cu & \Delta F_R = -49.43 \text{ kcal}
\end{aligned}
$$

Figure 4.8   Co-smelting formulae which indicate that the reduction of mixed oxidic and sulfidic ores can yield copper in a direct one-step operation.

produce arsenical copper. Recent analysis of the Banesh Period arsenical copper from Tal-e Malyan suggests this last method may have been practiced there (Pigott et al. in press). Briefly summarized, this process operates at a low temperature and is non-slagging while yielding an arsenical copper of low iron content (Budd et al. 1992:680; Budd 1993). This process can proceed easily in a bonfire or simple bowl furnace. Any of the minerals commonly known as arsenates, i.e., "arsenic- and antimony-bearing secondary minerals from the oxidized zones of cupriferous base metal orebodies" (Budd et al. 1992:681) can be smelted in this manner. Arsenates are not easily distinguished from more common oxidic copper minerals. Reduction can proceed at temperatures as low as 700°C while producing no slag. In the final product the presence of nickel, an element common in ancient Iranian metal artifacts, necessitates temperatures around 1000°C. In arsenate smelting,

> the forming alloy remained solid, limiting the rate of diffusion.... In practice, smelting ... at any temperature less than about 900°C, about the highest temperature achievable in an open fire without a hearth or furnace, always resulted in an alloy with a few percent arsenic, but never more than 5 percent—regardless of the arsenic content of the ore. (Budd 1993:36)

The orebodies at Talmessi-Meskani are once again the focus due to the fact that the oxidation of arsenides "produces an astonishingly wide range of green basic copper arsenates ... many of which ... closely resemble malachite" (Charles 1980:168). J. W. Barnes, cited in Charles (1980:169) "prefers the use of the arsenate-containing materials rather than the primary sulfarsenides as marking the discovery of selected copper minerals to produce alloys of improved properties," i.e., arsenical copper. Ancient prospectors could have detected the arsenical richness of these ores because such ores give off a garlic smell when hammered.

Thus, the most important point with regard to this process is that the arsenic composition of the metal resulting from this low temperature operation (the temperature must remain below that at which the copper arsenic alloy becomes molten) is independent of the ratio of the arsenic to copper in the ore charged in the "furnace"(Budd et al. 1992:680, citing also Charles 1985; Gale et al. 1985; and Rapp 1988). Where analysis of arsenic contents yields a pattern of content below 5%, scholars should look to arsenate smelting as a likely method for the metal's production. One additional point merits mention here. Copper arsenates (and arsenides) are rich in arsenic, yet frequently the arsenic content of ancient artifacts is relatively low, which in turn suggests that significant arsenic was being volitalized as a gas. Working in and around this gas on a consistent basis would, most probably, have proved deleterious to the health of such metalworkers.

It is possible that Bronze Age Plateau metalworkers moved from melting surface-deposited arsenical materials such as native copper mixed with copper arsenides in crucibles to co-smelting oxide/sulfide ores directly in crucibles. Evidence of crucibles and other metalworking remains (discussed above in this paper) and the lack of distinct furnace installations (except at Shahdad and probably Hissar) tend to support this hypothesis. It was not until sulfide ores were smelted exclusively that the need for strong reducing conditions would have required a short shaft furnace. Perhaps the furnaces at Shahdad might shed some light on this issue (see below). Finally, the interesting possibility that arises from the potential exploitation of complex sulfide ores such as chalcopyrite is that metallic iron may result as the end-product of such copper smelting. Rostoker demonstrated this in his crucible smelting experiments (Rostoker et al. 1989:84). With a strong draft and highly reducing conditions iron, rather than copper, can be inadvertently smelted. Through such "accidents" Bronze Age metalworkers could have encountered metallic iron well before its Iron Age florescence.

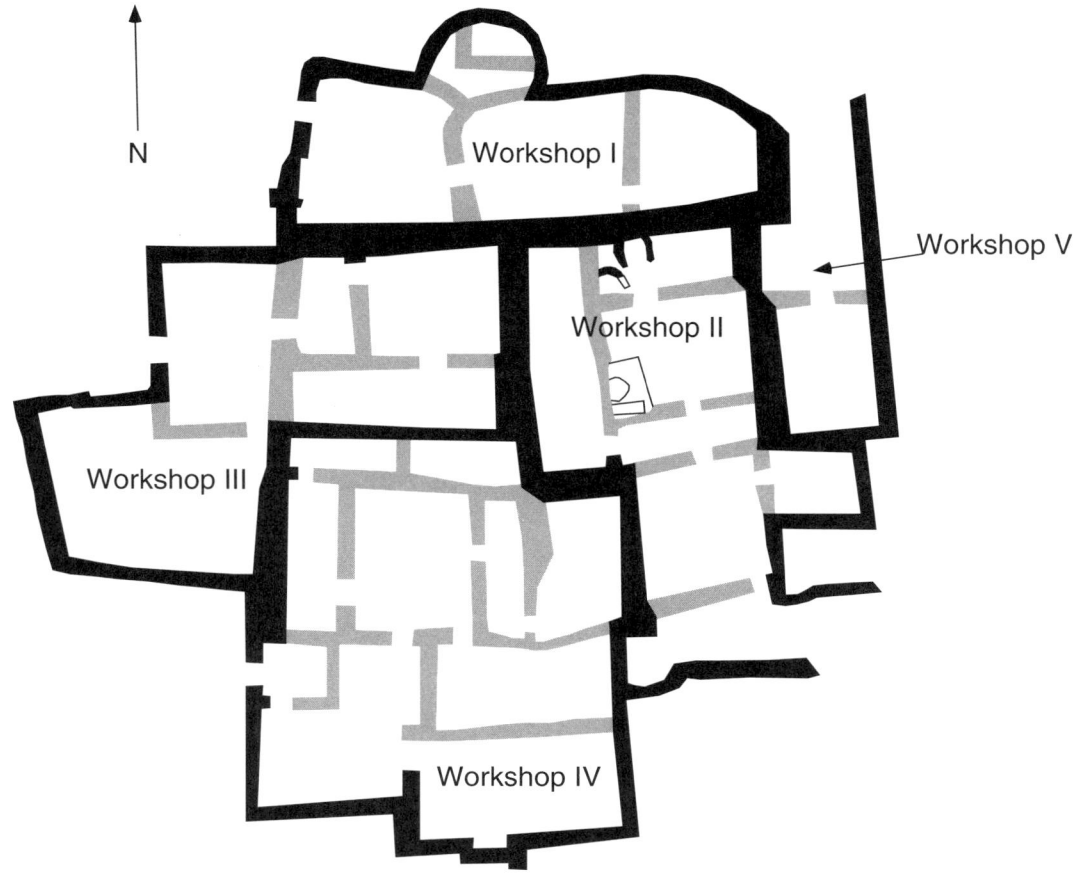

Figure 4.9   The metalworking quarter at Shahdad, designated "Site D," consists of 29 small rooms that make up five workshops. The best-preserved of the furnaces is that mapped here (Fig. 4.10), located in Room 6, Workshop II. By J. P. Edens after Hakemi 1992:fig. 15.4.

THE BRONZE AGE WORKSHOP QUARTER AT SHAHDAD

What is certainly one of the ancient Near East's best preserved metalworking quarters was excavated at the site of Shahdad in the southern Dasht-i Lut of Iran (Hakemi 1992; see also Vatandoost-Haghighi 1978; Moorey 1982:83, 90–91). It is dated to the second half of the third millennium B.C. However, the description of this installation leaves much to be desired as it is often imprecise in its wording. Copper-base metal production and its working were going on here, but what production processes were being used is by no means clear.

The excavated location known as "Site D" at Shahdad is ca. 25 by 25 m, but a larger metalworking locus once existed here which has been heavily eroded. The complex included the remains of twenty-nine small rooms which comprise five quite similar workshops (Fig. 4.9). Entrances opened to the south and west to escape strong prevailing winds. Ores were stored in ceramic circular bins with round flat lids which were always situated next to the rooms that contained the smelting furnaces. These containers appear to have been built in place. Though it is apparently coincidental, the containers look very much like short shaft furnaces. Some of them even taper in a vaguely conical shape.

The actual furnaces are unique in structure and represent the Plateau's only in situ examples; however, equating their structure, as described, with their purported function is difficult (Fig. 4.10). Hakemi (1992:122–124) describes the furnace in Room 6 as follows:

> The furnace, measuring 1.30 x 0.85 x 0.28 m, was raised like a stage and was constructed only with clay. The furnace and the south wall of the chamber had been closed by a short rim, possibly a charcoal hod. The metal smelting firebox, measuring 0.50 x 0.50 x 0.25 m was quadran-

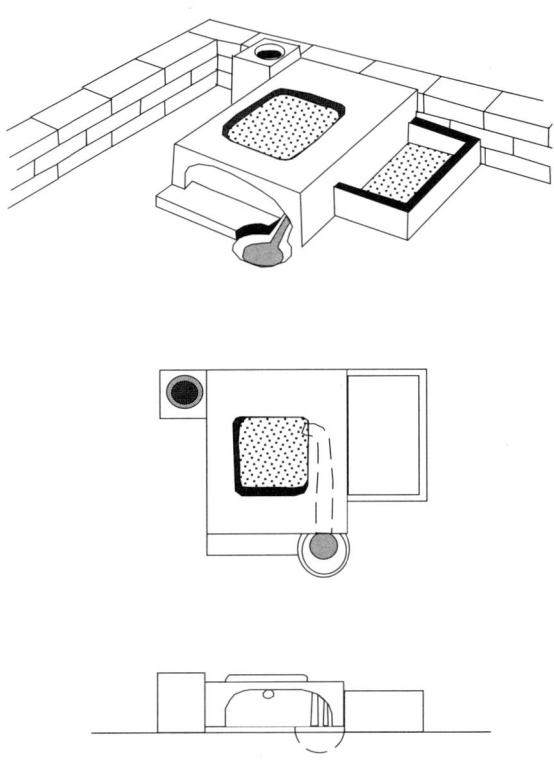

Figure 4.10   Reconstruction, plan, and profile of furnace structure in Room 6, "Site D," Shahdad. By J. P. Edens after Hakemi 1992:fig. 15.11.

The smelting furnaces had the same structure in all five workshops. Large granite stones present in the rooms may well have been anvils for ore crushing. Refractory clay was used to line the fireboxes. Casting of artifacts was accomplished with the aid of what are termed "small molding furnaces." The type of molding furnace was linked to the type of artifact being cast. There were four distinct types of molding furnaces excavated. These appear to be structures associated with melting of copper in crucibles, a number of which were found in the workshops. Crucibles, all of which had different measurements, were found in ten different rooms. Seven out of eight that were depicted are cup-shaped vessels with well-defined pouring spouts. In shape they appear to be best suited to melting and pouring of molten metal rather than smelting.

There were remarkably few small finds in the workshops. The workshops give every impression of having been cleaned out, perhaps prior to their and the site's abandonment. Only a single, open-face, ceramic, socketed axe mold is mentioned and depicted. It was found next to a molding furnace. Two so-called "spigots" are depicted as having an interior pipe along which air or metal might pass. They do not appear to be tuyères and may have been some sort of plug to hold back molten metal as Hakemi suggests. While Hakemi (1992) does not mention the presence of slag at Shahdad, systematic surface survey was conducted at the site by an Italian team headed up by S. Salvatori and M. Vidale (1982). They commented on the "large number" of copper-base artifacts excavated at the site and suggested that production was local. Moreover, they report,

> in fact, large expanses, in an area c. 500 x 500 m . . . are literally covered with slags from the smelting of copper, nor is it rare to see on the ground traces of circular or rectangular structure which denote the presence of furnaces. (Salvatori and Vidale 1982:7)

gular in shape and was located above the furnace. There was a hole behind the righthand corner of the firebox lined to a sloping passageway used for the circulation of molten metals; the sloping passageway continued to the bottom of the chamber and terminated in a fixed conical crucible. The sloping passageway was covered by clay mortar and continued until it reached the right edge of the firebox.

These circular structures are no doubt what Hakemi termed ore storage containers. If they are furnaces one would expect them to be charred, slag-stained, and vitrified in places, but this is not reported.

It is interesting to note that while Bronze Age production remains are present on the Plateau, this cannot as yet be said to hold true for the coming of iron.

THE IRON AGE (CA. 1450/1350–600 B.C.)

In the mid-second millennium, Bronze Age cultures in northwestern Iran were replaced by a new complex of traits, usually said to have been brought by Indo-Iranian tribes (Young 1967:24; Burney and Lang 1972:117; Ghirshman 1979). Increasing evidence shows that Bronze Age traditions persisted into the Iron Age and co-occurred with the new cultural configuration; among these traditions was bronze working, whose practitioners may have had a rudimentary working knowledge of iron, but as yet little need or inclination to make use of this metal. Indigenous metalsmiths may have served recently arrived patrons and worked first in bronze during the Iron I period (ca. 1450/1350–1100 B.C.) and then in both bronze and iron during the Iron II period (ca. 1100–800/750 B.C.; Burney and Lang 1972:117; Moorey 1969:137; Pigott 1977, 1980a).

Analysis of copper-base artifacts from the Iron Age shows tin bronze to be common. It was more predictable to produce artifacts in bronze than in arsenical copper because the proper proportions of copper and tin in bronze could be controlled for products of desired color and mechanical properties, while the amount of arsenic in arsenical copper was determined by the amount in the ore and by smelting conditions. Analyses of excavated early Iron Age finds from Dailaman (Egami et al. 1965, 1966; Sono and Fukai 1968; Fukai and Ikeda 1971) and from Marlik (Negahban 1996) have shown that arsenical copper was still used but tin bronzes with controlled amounts of tin (5 to 12%) were prevalent (Moorey 1982:94–95; Tylecote 1972).

Among the large number of excavated sites with Iron Age contexts, several are of particular archaeometallurgical interest. From the excavations at Haft Tepe in Khuzistan (ca. 13th century B.C.), an unusual pyrotechnological installation was associated with a crafts workroom containing materials that included mosaics of colored stones framed in bronze, a dismembered elephant skeleton used in bone tool manufacture, and a mass of several hundred bronze arrowpoints and small tools.

Situated in a courtyard directly in front of this workroom is a most unusual kiln. This kiln is very large, about 8 m long and 2 and one half m wide, and contains two long compartments with chimneys at each end, separated by a fuel chamber in the middle. Although the roof of the kiln had collapsed, it is evident from the slight inturning of the walls which remain in situ that it was barrel vaulted like the roofs of the tombs. Each of the two long heating chambers is divided into eight sections by partition walls. The southern heating chamber contained metallic slag, and was apparently used for making bronze objects. The northern heating chamber contained pieces of broken pottery and other material, and thus was apparently used for baking clay objects including tablets. (Negahban 1977)

Excavations of a number of graves in the Iron I/II necropolis at Marlik (Negahban 1964) produced abundant evidence of the metalworker's craft: a diverse array of bronze artifacts including human figurines (Negahban 1979a), metal vessels (Negahban 1983), stamp seals (Negahban 1977, 1979b) and weaponry (e.g., maceheads, Negahban 1981; see also Muscarella 1984 for a discussion relating to fibulae and chronology). Iron was not common at Marlik although it is in the Iron II period that iron first appears in quantity in northwestern Iran.

During the tenth/ninth century B.C. iron appears in this region at Iron II sites, including Dinkha, Hasanlu, Haftavan and in central Iran at Sialk.[5] The largest excavated collection dating to this period comes from Hasanlu, where more than 2000 iron artifacts are associated with the ninth century B.C. destruction level of the Citadel (Pigott 1980a, 1989c). Of interest is the fact that almost an equal amount of bronze artifacts also came from this context. Certain artifact types occurred both in bronze and iron. Others, such as hoes, sickles, knives, and saws occurred exclusively in iron, while bronze was most frequently used for equestrian gear, architectural and domestic decoration, personal ornaments, and certain weapons and armor (de Schauensee 1988).

While there is no evidence to identify where iron was worked at Hasanlu, three different excavation loci have produced evidence of bronze working. The first is on the north side of the mound where Aurel Stein's 1936 sounding produced a cache of artifacts, including some possible bronze bar ingots, strips, and at least three stone molds for small artifacts (Stein 1940:390–404, pl. 26). In the Outer Town at Hasanlu, excavation revealed the "Artisan's House" where remains of intense burning, crucible fragments, a possible ingot, a shaft-hole axe mold and a open flat axe mold as well as other mold fragments clearly suggest a workshop. Hematite nodules, which were also among the finds in this locus, are most probably for the working and planishing of sheet bronze (Pigott 1981:48; see also Coghlan 1951:76–77; Woolley 1963:285; Bass 1967:163, n. 78; Shaffer 1984; Hodges 1976:76; and discussions in Kenoyer and Miller and Possehl and Gullapalli, this volume). A third locus is on the Citadel Mound in Burned Building III, where crucible fragments, a pair of bivalve sandstone molds for a ribbed-bladed shaft-hole axe, and fragments of a similar mold in clay were excavated (Pigott 1981:138–139).

At about 800 B.C. or slightly later Urartians are thought to have destroyed Hasanlu (Dyson and Muscarella 1989; Dyson and Voigt 1989). Although we have few analyses of excavated Urartian metal finds, either bronze or iron, the artifacts clearly indicate the Urartians were skilled in metalworking technology (Van Loon 1966:80–84; Kellner 1979:151–156; Seidl 1988).[6] The focus of metalworking activity, however, now shifted to west-central Iran. Here the most extraordinary finds come from Luristan. Ornate bronze lost-wax castings, frequently with zoomorphic motifs, are the best known artifacts from this region. While a great majority of these "classic" Luristan bronzes are unprovenanced (see, for example, Moorey 1971 and Muscarella 1988, 1989b), some have been found in association with more common varieties of bronzes at the sites of Surkh Dum (Muscarella 1981; Schmidt et al. 1980), Baba Jan (II and III) (Goff 1978), and in the gravefields excavated by Louis Vanden Berghe (for the collected references see Vanden Berghe 1979; Vanden Berghe and Haerinck 1981, 1987).[7] These bronzes appear to be the culmination of a long-lived tradition of bronze casting and sheet metal working ranging from the third millennium down to the seventh century B.C. Significantly, some iron artifacts occurred with these bronzes, frequently with bronze cast-on to the iron (e.g., Waldbaum 1971). The presence of this particular technique is a useful indicator of a period of transition from bronze to iron and can be taken, in western Iran, as one of the definitive markers of the Iron II period (Pigott 1981:117; 1989c:72; see also Maxwell-Hyslop and Hodges 1964).

THE IRON DAGGERS FROM LURISTAN

Included in discussions of Luristan bronzes is a unique group of iron daggers bearing both zoomorphic and anthropomorphic images on their pommels (e.g., Moorey 1991; Rehder 1991; Pleiner 1969a, b; France-Lanord 1969:75–126; Smith 1971) (Fig. 4.11). Almost ninety examples are known and show a remarkable degree of similarity (Muscarella 1989a). These very unusual weapons have been the source of much discussion, including whether or not they actually come from Iran. None are known to come from a site, let alone an excavation, and they are thought to have been looted from tombs in the 1920s and 1930s, when masses of bronzes appeared on the antiquities market. Tombs are most certainly involved because the high level of preservation of the swords is comparable to that of iron artifacts excavated from tombs in Luristan by the late L. Vanden Berghe's Belgian expedition. Smith (1971:51) suggests, however, that the low temperature treatment of the iron (air cooling below 750°C) may have conferred some corrosion resistance on the swords.

While these swords have been known for decades and much has been written concerning their art history and technology, the most recent revelation concerns their apparent early date. Accelerator mass spectrometry (AMS) dates on samples taken from the swords housed in the Royal Ontario Museum and at MIT suggest that the iron in the swords was smelted in the period 1094±60 years B.C. (Rehder 1991:14; Moorey 1991). If these swords were actually made in Luristan/Iran, they are among the very earliest iron artifacts known in this area, and rank among the ancient Near East's most important pre-1000 B.C. iron artifacts. Regarding the issue of their origin the weight of the argument swings in favor of an Iranian (Luristan) origin. A number of scholars have linked the swords under discussion with other types of excavated Luristan iron weaponry (Maxwell-Hyslop and Hodges 1966:167ff; Calmeyer 1969:127; Moorey 1971:317; Muscarella 1989a:353; cf. Moorey 1991:8). In contrast, Muscarella (1989a:352) mentions that three scholars, Herzfeld (1941:135ff, 166f), Maryon (1961:174), and Ghirshman (1983:71, 73), have argued for a non-Iranian origin for the swords and have suggested the Pontic coast of the Black Sea as the point of origin. This is an

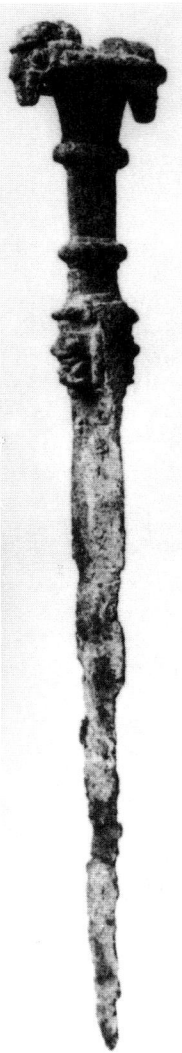

Figure 4.11 Iron dagger from Luristan. UPM 30-38-18, L. 41.9 cm.

intriguing notion in light of the fact that the best-known evidence for iron smelting furnaces in Southwest Asia comes from the Colchis region of the Caucasus at the eastern edge of the Black Sea (Khakhutaishvili 1976; Pigott 1989c).

R. Moorey (1991:6–7) succinctly summed up metallurgist J. Rehder's observations on the sword production technology:

His research swings the chronological pendulum back in favour of a date towards the end of the second millennium B.C. rather than two or three centuries later, whilst emphasizing the fragility rather than the strength of these swords. Significantly, his analysis of the iron-working technology indicates that those who smelted the metal had a knowledge of forge welding, whilst those who made them apparently did not. The argument that these objects were high status display or votive weapons manufactured early in the history of regular iron production by bronzesmiths, who were improvising brilliantly in ignorance of iron's special qualities, has gained significantly in strength.

Regarding Rehder's point about the knowledge of forge welding, it must be said that the process of coalescing a bloom or blooms through a sequence of heatings and hammerings does not necessarily imply a true understanding of forge welding. The presence in two swords of obvious seams which were the result of careless forging in the process of bloom consolidation may be an indication of this (Rehder 1991:15). The coalescing of amorphous masses of iron bloom and the forge welding of shaped parts of a dagger in the making by a bronzesmith may have been two separate processes conceptually in which the first did not automatically lead to an understanding of the second.

It is also interesting to note that the evidence suggests to Rehder that small blooms were being consolidated to form larger pieces of stock which were then forged into these large daggers, which can weigh a kilogram or more. This process suggests that simple bloomery furnaces were producing mini-blooms dispersed throughout the furnace bottom, which probably had to be raked out of a hot furnace or broken out of a slag mass, hand picked, and then reheated for forging into a larger mass.[8]

It is clearly the case that these artifacts reflect the craft of the bronzesmith as opposed to that of the accomplished ironworker, as Rachel Maxwell-Hyslop and Henry Hodges (1966:169) and Cyril Stanley Smith (1971:51) have observed. Moorey (1991:1) makes another important observation regarding the appearance of this iron-working tradition:

> Thus it is that the early decorated iron objects from Luristan provide a rare opportunity to observe the interaction of a traditional metal-working craft, skilled in casting and hammering copper and copper alloys, and a new intrusive one whose products at this stage had to be laboriously forged from many pieces of low-carbon wrought iron; a metal not best suited to the rich ornament expected of metalworkers in their communities.

Production most certainly must have been contemporaneous for all the daggers and they may well reflect the output of a single workshop. This conclusion, as Moorey (1991:2) points out, is based on "homogeneity of style, structure and technique evident in these swords," and has been raised by a number of scholars.

Moorey's iconographic arguments suggest a possible association with imagery linked to a major underworld deity, not unlike Nergal, who, he says, is linked to Assyro-Babylonian, Hittite, and Hurrian cult and who is associated with the sword (Moorey 1991:7–8). It seems quite appropriate for a weapon of war (and death)—one which at the same time may have been designed primarily as a mortuary good—to be associated with a god of the Underworld, where those dispatched by such weapons were sent (cf. Moorey 1991:8). The notion that such weapons might have been funerary in purpose stems, in part, from their very awkward construction which would appear to make them difficult to wield, unless by mounted horsemen at the charge (Maxwell-Hyslop and Hodges 1966:168; also Rehder 1991:16). Maxwell-Hyslop and Hodges make an additional point of interest regarding the sword technology. They argue that the smiths were unaware of the technique of casting a bronze hilt onto an iron blade, which might have simplified the task of producing a decorative hilt (Rehder 1991:172–173). This jibes with a probable Iron I date for the weapons, as casting-on is a marker of the Iron II period and later in Iran,[9] with notable early examples at ninth century Hasanlu and somewhat later in Luristan (Moorey 1971; Pigott 1981, 1989c). It is possible to argue that the early date for these weapons places them in a period when iron was very rare, highly valued, and most probably carrying substantial symbolic connotations with its ownership and use: i.e., the weapons' smiths and the weapons' owners may have wanted such daggers only in this highly prized, remarkable new shiny silvery material.

Rehder (1991:13) states at the outset of his article on the material and manufacture of the decorated iron swords: "In addition, the properties of the iron can provide a reasonable base for opinion on the possible use and significance of the swords." He lists features common to the thirteen swords of which analyses have been published. Two are of particular relevance here: (1) The iron in the swords is a heterogenously carburized, bloomery produced, wrought iron which was softer than work-hardened tin bronze; and (2) the swords were built up by 8 to 15 separate parts which were mechanically fastened by the use of techniques such as crimping and rivets. Certain of these mechanical joins were apparently not capable of withstanding sharp blows such as might be sustained in combat. At the same time Rehder (1991:16) remarks that the choice of the multi-piece construction struck him as "an intentional display of virtuosity." He feels that the lack of hardened iron suggests an early stage in the working of iron as does the fact that the low carbon blades were also usually annealed, leaving them as soft and ductile as possible. Rehder (1991:19) mentions that the forg-

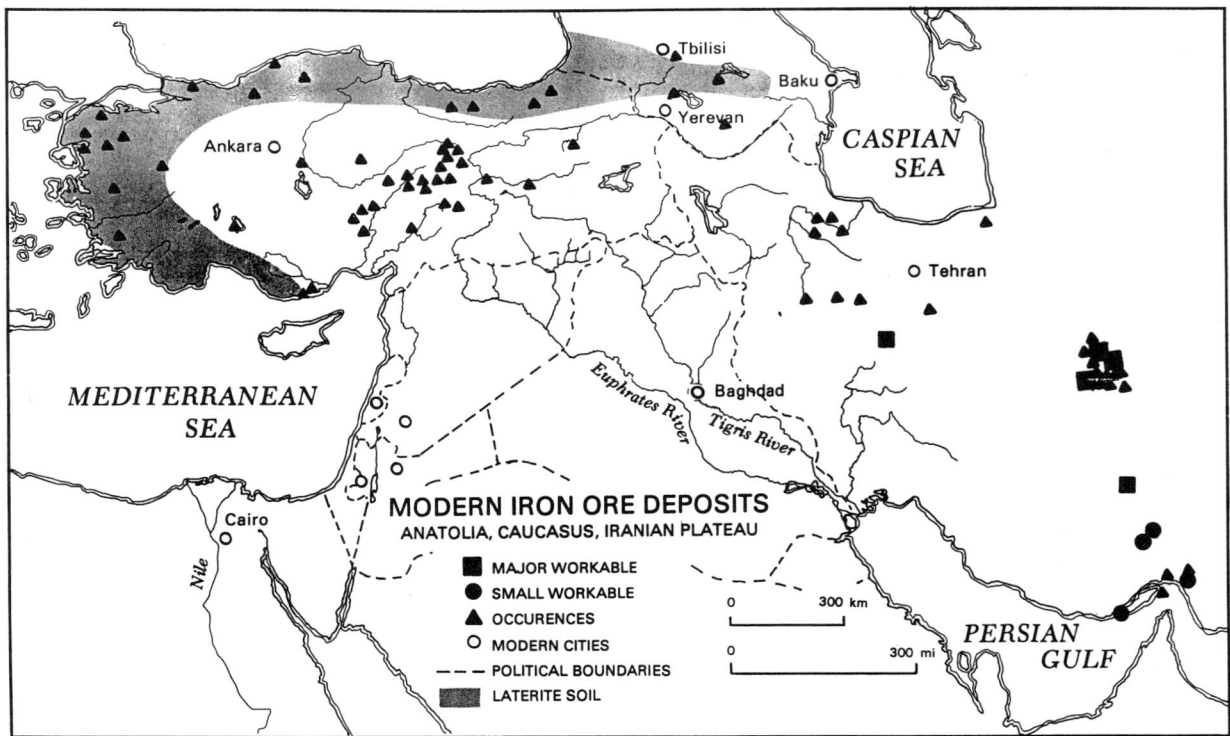

Figure 4.12   Modern iron ore deposits in Iran, Anatolia, and the Caucasus. Map by J. Snyder after Pigott 1981:fig. 2.

ing smiths "used a slow-cool heat treatment to produce maximum softness and ductility." The most likely case is that rather than this being a consciously applied treatment to produce softness it was the fact that iron was left to cool perhaps in the forge or at room temperature and ended up soft as a result. It is hard to envision that at this early point of working iron the smiths had a repertoire of treatments. Most likely they just did it the one way they knew. They observed that soft metal forges easier. Shape was clearly all important in these unique daggers and getting to that shape was a time-consuming undertaking that required soft metal down to the last hammer blow. Rehder (1991:18) sums up the time-honored view of these swords: "It is difficult to escape the conclusion that the swords were made only for ritual, display, or prestige purposes" (see also Muscarella 1989a:351). This is probably true, but does not obviate the possibility that from time to time they were also used to club, cut, or stab as weapons were meant to be used.

THE COMING OF IRON IN WESTERN IRAN

Given the potential for iron-working knowledge to spread among bronzesmiths of western Iran prior to iron's widespread adoption during the Iron II period, the motivation for its adoption is of interest. Clearly iron has certain economic advantages over bronze, since bronze necessitates the importation of tin while iron as an ore is ubiquitous (Fig. 4.12). However, this advantage is offset by the time and manpower needed to forge iron into useful shapes (Smith 1971:51). In the form of well-carburized steel, iron has a clear technical superiority to coldworked, 10% tin bronze, but there is no analytical indication that true steel was intentionally being produced during the Iron Age in Iran. Artifacts that have been metallographically analyzed are generally wrought iron or heterogenously carburized, i.e., low carbon iron with zones of high carbon randomly distributed throughout the metal. Technologically, this iron was, at best, on a par with the bronzes of the period and appears to have been comparable to iron from Assyria. Since iron did not provide obvious economic or technological advantages, the impetus for its adoption may have been provided by the region's iron-using Assyrian neighbors. Iron first occurs *en masse* in Iran at a time when Assyrians were beginning to make significant campaigns into the area. Winter has argued in her work on the "local style" of artistic craftsmanship at Hasanlu for a process of emulation; people from western Iran may even have visited Assyria and upon their return sought to copy what they had seen (Winter 1977). Iron was a high-status material in the Assyrian Empire; texts indicate that it was Assyrian royalty and the military who used it (Pleiner and Bjorkman 1974:286–288).

Whether as a function of this interaction with Assyria or simply concurrent with it, western Iran was in a period of great aesthetic creativity. At Iron II Hasanlu, for example, iron was only one of a variety of decorative materials being experimented with, and it often occurred in unique bimetallic artifacts. Aesthetically sensitive environments, it has been suggested, encouraged experimentation with metals as decorative materials (Smith 1970:501). In western Iran bronzesmiths may have begun working iron as a decorative metal and then moved onto fashioning tools and weapons in it.

Metallographic analysis of artifacts from Hasanlu dating to ca. 800 B.C. shows them to be unsophisticated products of a technology in transition, one in which bronzesmiths are probably working the iron. Iron at Hasanlu is generally a low carbon, heterogeneously carburized "steel" (Pigott 1981:229–267). Bimetallic artifacts are common at the site and certain iron artifacts were worked as if they were bronze, including repoussé iron plaques. By Iron III the two metals, iron and bronze, were increasingly differentiated in use, with bronze being used more decoratively and iron more functionally.

The large quantities of iron which appeared so abruptly in Iron II and III in western Iran suggest that this metal was being produced on some scale. I have argued elsewhere (Pigott 1981, 1989c) that the difficulties inherent in the production process mean that to obtain iron in quantity it has to be mass produced to make its production worthwhile. The technological process involved in its smelting and forging indicates some sort of permanent production organization, a key characteristic in industrialized production (Udy 1964:115). Organization and coordination of the various technological operations were necessary; these included mining, charcoal-making, and smelting, as well as the labor of skilled specialists, the blacksmiths. A review of the archaeological evidence associated with the first significant use of iron in western Iran, at sites such as Hasanlu, suggests that the appropriate sociopolitical conditions for such organization were present (Pigott 1981:202-77). However, just how organized production actually was remains an important issue for discussion, especially in the light of the recent doctoral thesis research by M. McConchie at the University of Melbourne (1998). Her discussions comparing and contrasting iron production in northeast Anatolia in the Urartian and Post-Urartian periods, as well as her presentation of a model for the organization of iron production and smithing in the highland system of the region, demand that the social/technological context in which iron metallurgy developed in western Iran be reviewed with her evidence and arguments in mind.

Hasanlu's strategic location on major east-west trade routes—perhaps as a gateway community—and the complex material culture of wealthy elite in the Citadel suggest the site could have functioned as an important iron distribution, if not an actual iron production center in the region of northwestern Iran (Pigott 1989c:70). What supports the suggestion that iron might have been locally produced at Hasanlu? The evidence is slim but includes boulders of the iron ore magnetite (titanium bearing), clearly broken (not eroded) from a larger ore body matrix (Pigott 1980a:456, n. 40). These boulders were found built into walls and as paving stones. A second supporting body of evidence comes from those iron artifacts that are clearly unique to Hasanlu, including iron shoulder rondels or plaques decorated in repoussé and used as horse gear (Pigott 1989c:71, fig. 6a, b), and iron sidebar cheek pieces with three holes for attachment. These two categories are northwest Iranian in style and are not known on Assyrian reliefs (de Schauensee 1989; Dyson and Muscarella 1989) Another category unique to Hasanlu are the lion pins. These personal ornaments are iron-shanked pins with lions *en couchant* cast-on in bronze (Pigott 1989c:74, fig. 13a–c).[10] In addition, the presence of standardized common utilitarian iron artifact types such as small tanged iron knives with upturned points (Pigott 1989c:74, fig. 12) suggests local manufacture meeting a local demand—sheep shearing, perhaps (Egami et al. 1965). It is also clear that once iron became available, there was also a high level of demand with a strong military orientation, evident in some 2000 iron artifacts excavated in the ninth century B.C. destruction level of the Hasanlu Citadel. It is difficult to envisage a military elite being dependent upon trade access for their armament supply. Local smelting and blacksmithing would seem a more satisfactory solution.

A closing comment on the coming of iron to western Iran sums up current assessments of this much-heralded technological transition:

> At all of these sites [e.g., Hasanlu, Kordlar, Godin, Nush-i Jan, Baba Jan] the indications of wealth and advanced social stratification coincide with the widespread occurrence of iron, as well as with a period of intensified Assyrian presence in each region. Sites such as Godin, Baba Jan, and Tepe Sialk lie directly along the southern route linking Assyria with the Iranian Plateau, and Assyrian stelae and rock sculptures document the importance of the route to Assyrians. . . . Thus Assyrian interest and influence, perhaps stimulated by economic factors such as trade and the acqusition of raw materials, appear to be the critical variables in the process of adoption of iron in western Iran. (Pigott 1981:224)

NOTES

1. This chapter is an updated and expanded treatment of the encyclopedia entries in *Encyclopaedia Iranica* entitled *AHAN* [Iron] (Pigott 1982) and *Bronze* (Pigott 1989b). Chronology is based on Dyson and Voigt 1992.

2. As pointed out by Malfoy and Menu (1987:364) the starting point for research on Susian archaeometallurgy is the extensive fieldwork across Iran, Oman, and Afghanistan and associated laboratory analysis undertaken by Thierry Berthoud and colleagues of the R.C.P. 442 of the C.N.R.S. in France. This was published by Berthoud as his doctoral thesis (1979) and with colleagues (1982; see also Berthoud et al. 1979; Berthoud, Bonnefuss, et al. 1980; Berthoud, Craddock, et al. 1980). A number of other related reports authored by Berthoud with others (see listing of reports with Berthoud as lead author in bibliography) unfortunately are not in the permanent record. Most of the fundamental conclusions of this work are incorporated in the work of Tallon (1987) and Malfoy and Menu (1987) and thus have not been discussed separately (see Moorey's [1994:249] overview of the Berthoud project).

3. Evidence from the southern end of the Elamite kingdom at its other capital, ancient Anshan (modern Tal-e Malyan), suggests a similar scenario. It is in the Kaftari Period (ca. 2200–1800 B.C.) at Malyan that tin bronze becomes the dominant alloy, based on analyses undertaken thus far at MASCA. Interestingly, arsenic and nickel (cobalt was not tested for) continue to be present in more than trace element amounts in the analyzed artifacts from this period at the site.

4. Among the most interesting results was the identification of the reuse of a very glassy type of slag for small blade production. Three cores and one blade were found (Pigott 1989a:29, fig. 3; Pigott et al. 1982:234, figs. 26, 27). Also, a specific type of medium-coarse buff-orangish cooking ware was found to have been tempered with crushed slag and even some copper prills (Pigott et al. 1982:217, figs. 2, 3).

5. Three iron balls were excavated from fifth millennium Period II at Tepe Sialk and were identified as meteoritic (Halm 1939:206). Moorey (1994:279) says that these balls when analyzed showed a Widmanstätten pattern, a crystallographic structure characteristic of meteoritic iron. I suspect that such artifacts were actually nodules of magnetite used as hammerstones and planishers (Pigott 1981:48). The crystal structure may be a misidentification of the normal crystalline structure of the iron ore.

6. For the most recent discussion of Urartian metalworking technology, see Rivka Merhav's 1991 exhibition catalogue entitled *Urartu—A Metalworking Center in the First Millennium B.C.E.* and M. McConchie's doctoral thesis at the University of Melbourne (1998), *Iron Technology and Ironmaking Communities in Northeastern Anatolia: First Millennium B.C.* Of particular importance in Merhav's volume is the chapter "Ore Deposits and Mining in Eastern Anatolia in the Urartian Period: Silver, Copper and Iron" by Oktay Belli. He describes what he believes to be two Urartian iron smelting sites in the vicinity of Van replete with great quantities of iron slag and ceramic tuyères (Belli 1991:37–38). However, no excavation has taken place and a direct Urartian association and the chronology of the sites are not firm (McConchie 1998). Regardless, these sites can be classed among the handful of iron smelting sites (potentially early?) known anywhere in Southwest Asia.

7. Luristan bronzes are not discussed in detail here primarily because they are unprovenanced. For those interested in their composition see, for example, the analyses found in Moorey 1971. For general discussions see Muscarella 1988,1989b.

8. I have discussed the question of mini-blooms in the context of early second millennium B.C. Kültepe in Anatolia in relation to the Akkadian term *amutu;* and, as well, in the context of late second and early first millennium B.C. Assyrian iron production (Pigott 1981:82–87).

9. For an example of casting-on and a short discussion of the technique in the ancient Iranian context, see Pigott 1989c:72. Note also the following observation by Moorey (1994:285): "But there are sufficient surviving artefacts to indicate that this technique, almost certainly long established among the Anatolian industries working more directly with iron, had a role in Assyrian workshops by at least the ninth century BC, when much of the Hasanlu IV metalwork was probably manufactured." Casting-on may be, therefore, yet another direct technological marker of the strong connection between Assyrian culture and that of northwestern Iran.

10. The presence of indigenous metalsmiths at Hasanlu seems a certainty, and such individuals previously schooled in working precious metals, and to a lesser extent copper and bronze, possibly, with the coming of iron, added this new metal to their repertoire of materials to shape. As Smith (1972:119–120) pointed out, "when reasonably pure, [iron is] beautifully responsive to cold working, to raising and repoussé, to punching, tracing and chasing and similar techniques that are seen perhaps at their best in gold and silver."

REFERENCES CITED

Amiet, P.
1979   Archaeological Discontinuity and Ethnic Duality in Elam. *Antiquity* 53:195–204.

Amiet, P.; Berthoud, T.; and Liszak-Hours, J.
1980   La Spectrométrie de masse éclaire et les chemins de la métallurgie. Pp. 85–87 in *La Vie mysterieuse des chefs-d'oeuvres. La Science au service de l'art*, ed. L. Citti. Paris: Grand Palais.

Anon.
1985   The Manufacture of Copper-arsenic Alloys in Prehistory. *Journal of Historical Metallurgy* 19(1):141–142.

Bass, G.
1967 Cape Gelidonya: A Bronze Age Shipwreck. *Transactions of the American Philosophical Society* 57(8).

Bazin, D., and Hübner, H.
1969 Copper Deposits in Iran. *Geological Survey of Iran*, Report 13. Tehran.

Beale, T. W.
1973 Early Trade in Highland Iran: A View from a Source Area. *World Archaeology* 5:133–148.

Belli, O.
1991 Ore Deposits and Mining in Eastern Anatolia in the Urartian Period: Silver, Copper and Iron. Pp. 16–41 in *Urartu: A Metalworking Center in the First Millennium B.C.E.*, ed. R. Merhav. Jerusalem: The Israel Museum.

Berthoud, T.
1979 *Etude par l'analyse de traces et la modelisation de la filiation entre minérai de cuivre et objets archéologiques du Moyen-Orient (IVème et IIIème millénaires avant notre ère)*. Doctoral thesis, Université Pierre et Marie Curie, Paris.

Berthoud, T.; Besenval, R.; Cesbron, F.; Cleuziou, S.; Françaix, J.; and Liszak-Hours, J.
1975 *Etude sur la metallurgie iranienne aux IVème–IIIème millennaires*. Paris: Recherche Coopérative sur Programme No. 442, Commissariat à l'Energie Atomique, Laboratoire de Recherche de Musées de France, Unité de Recherche Archeologique No. 7.
1976 *Les Anciennes mines de cuivre en Iran*. Paris: Recherche Coopérative sur Programme No. 442, Commissariat à l'Energie Atomique, Laboratoire de Recherche de Musées de France, Unité de Recherche Archeologique No. 7.
1977 *Les Anciennes mines d'Afghanistan*. Paris: Recherche Coopérative sur Programme No. 442, Commissariat à l'Energie Atomique, Laboratoire de Recherche de Musées de France, Unité de Recherche Archeologique No. 7.

Berthoud, T.; Besenval, R.; Cesbron, F.; Cleuziou, S.; Pechoux, M.; Françaix, J.; and Liszak-Hours, J.
1979 The Early Iranian Metallurgy: Analytical Study of Copper Ores from Iran. Pp. 68–74 in *Proceedings of the 18th International Symposium on Archaeometry and Archaeological Prospection*. Archaeo-Physika 10. Köln: Rheinland-Verlag.

Berthoud, T.; Besenval, R.; Cleuziou, S.; and Drin, N.
1978 *Les Anciennes mines de cuivre du Sultanat d'Oman*. Rapport Préliminaire Recherche Coopérative sur Programme No. 442. Paris: Université Pierre et Marie Curie.

Berthoud, T.; Bonnefuss, S.; de Choux, M.; and Françaix, J.
1980 Data Analysis: Towards a Model of Chemical Modifications from Ores to Metal. Pp. 87–102 in *Scientific Studies in Early Mining and Extractive Metallurgy*, ed. P. T. Craddock. British Museum Occasional Paper 20. London.

Berthoud, T.; Cleuziou, S.; Hurtel, L. P.; Menu, M.; and Volfovsky, C.
1982 Cuivres et alliages en Iran, Afghanistan, Oman au cours des IVe et IIIe millénaires. *Paléorient* 8(2):39–54.

Berthoud, T.; Craddock, P. T.; Hauptmann, A.; Maddin, R.; Muhly, J. D.; Pigott, V. C.; Stech-Wheeler, T.; and Weisgerber, G.
1980 Production, échange et utilisation des métaux: Bian et perspectives des recherches archéométriques récentes dans le domaine oriental. *Paléorient* 6:99–127.

Berthoud, T., and Françaix, J.
1980 *Contribution a l'étude de la métallurgie de Suse aux IVe and IIIe millénaires. Analyse des elements-traces par spectrometrie d'emission dans l'ultra-violet et spectrometrie de masse à étincelles*. Rapport CEA-R-5033. Gif-sur-Yvette.

Bibby, G.
1977 *Dilmun*. Hamburg: Rowohlt.

Budd, P.
1993 Recasting the Bronze Age. *New Scientist* (Oct. 23):33–37.

Budd, P.; Gale, D.; Pollard, A. M.; Thomas, R. G.; and Williams, P. A.
1992 The Early Development of Metallurgy in the British Isles. *Antiquity* 66:677–686.

Bulgarelli, G. M.
1979 The Lithic Industry of Tepe Hisar at the Light of Recent Excavation. Pp. 39–54 in *South Asian Archaeology 1977*, Vol. 1, ed. M. Taddei. Naples: Istituto Universitario Orientale.

Burney, C., and Lang, D. M.
1972 *The Peoples of the Hills*. New York: Praeger.

Caldwell, J. R.
1967 The Setting and Results of the Kerman Project. Pp. 21–40 in *Investigations at Tal-i-Iblis*, ed. J. R. Caldwell. Illinois State Museum Preliminary Reports 9. Springfield: Illinois State Museum.
1968 Tal-i-Iblis and the Beginning of Copper Metallurgy in the Fifth Millennium. *Archaeologia Viva* 1:145–150.

Caldwell, J. R. (ed.)
1967 *Investigations at Tal-i-Iblis*. Illinois State Museum Preliminary Reports 9. Springfield: Illinois State Museum.

Caldwell, J. R., and Shahmirzadi, S. M.
1966 *Tal-i-Iblis: The Kerman Range and the Beginning of Smelting*. Illinois State Museum Preliminary Reports 7. Springfield: Illinois State Museum.

Caley, E. R.
1971 Analyses of Some Metal Artifacts from Ancient Afghanistan. Pp. 106–113 in *Science and Archaeology*, ed. R. H. Brill. Cambridge, MA: The MIT Press.
1972 Chemical Examination of Metal Artifacts from Afghanistan. Pp. 44–50 in *Prehistoric Research in Afghanistan, 1959–66*, ed. L. Dupree. Transactions of the American Philosophical Society 62. Philadelphia.
1980 Chemical Composition of Some Early Copper Alloys Found in Afghanistan. *Vijnana Parishad Anusandhan Patrika* 23(3):223–233.

Calmeyer, P.
1969 *Datierbare Bronzen aus Luristan und Kirmanshah*. Berlin: De Gruyter.

Carriveau, G. W.
1978 Application of Thermoluminescence Dating Techniques to Prehistoric Metallurgy. Pp. 59–66 in *Applications of Science to the Dating of Works of Art*, ed. W. J. Young. Boston: Museum of Fine Arts.

Casanova, M.
1992 The Sources of Lapis-lazuli Found in Iran. Pp. 49–56 in *South Asian Archaeology 1989*, ed. C. Jarrige. Monographs in World Archaeology 14. Madison, WI: Prehistory Press.

Charles, J. A.
1980 The Coming of Copper and Copper-base Alloys and Iron: A Metallurgical Sequence. Pp. 151–181 in *The Coming of the Age of Iron*, eds. T. A. Wertime and J.D. Muhly. New Haven: Yale University Press.
1985 Determinative Mineralogy and the Origins of Metallurgy. Pp. 21–28 in *Furnaces and Smelting Technology in Antiquity*, eds. P. Craddock and M. J. Hughes. British Museum Occasional Paper 48. London.

Chymriov, V. M.; Stazhilo-Alekseev, K. F.; Mirzad, S. H.; Dronov, V. I.; Kazikhani, A. R.; Salah, A. S.; and Teleshev, G. I.
1973 Mineral Resources of Afghanistan. Pp. 44–85 in *Geology and Mines and Industries of the Republic of Afghanistan*. Kabul.

Cleuziou, S., and Berthoud, T.
1982 Early Tin in the Near East: A Reassessment in the Light of New Evidence from Afghanistan. *Expedition* 24(3):14–19.

Coghlan, H. H.
1942 Some Fresh Aspects of the Prehistoric Metallurgy of Copper. *Antiquaries Journal* 22:22–38.
1951 *Notes on the Prehistoric Metallurgy of Copper and Bronze in the Old World*. Oxford: Pitt Rivers Museum.
1975 *Notes on the Prehistoric Metallurgy of Copper and Bronze in the Old World*, 2nd ed. Oxford: Pitt Rivers Museum.

Desch, C. H.
1929 *Interim Report on Sumerian Copper*. Pp. 437–441. London: British Association for the Advancement of Science.

de Schauensee, M.
1988 Northwest Iran as a Bronzeworking Center: The View from Hasanlu. Pp. 45–62 in *Bronzeworking Centres of Western Asia 1000–539 B.C.*, ed. J. Curtis. London: Kegan Paul International.

1989 Horse Gear from Hasanlu. *Expedition* 31(2–3):37–52.

Deshayes, J.
1960 *Les Outils de bronze l'Indus au Danube IVe au IIe millénaire* I–II. Paris: Paul Geuthner.

Dougherty, R. C., and Caldwell, J. R.
1967 Evidence of Early Pyrometallurgy in the Kerman Range in Iran. Pp. 17–20 in *Investigations at Tal-i-Iblis*, ed. J. R. Caldwell. Illinois State Museum Preliminary Reports 9. Springfield: Illinois State Museum. Reprinted from *Science* 153(3739) (1966):984–985.

During-Caspers, E. C. L.
1972 La Hachette trouee de la sépulture E de Khurab dans le Balouchistan Persan. Examen rétrospectif. *Iranica Antiqua* 9:60–64.

Dyson, R. H., and Howard, S. M. (eds.)
1989 *Tappeh Hesar, Reports of the Restudy Project, 1976.* Florence: Casa Editrice Le Lettere.

Dyson, R. H., and Muscarella, O. W.
1989 Constructing the Chronology and Historical Implications of Hasanlu IV. *Iran* 27:1–27.

Dyson, R. H., and Voigt, M. M.
1989 East of Assyria. The Highland Settlement of Hasanlu. *Expedition* (special issue) 31(2–3).
1992 The Chronology of Iran, ca. 8000–2000 B.C. Pp. 122–178 in *Chronologies in Old World Archaeology*, 3rd ed., ed. R. Ehrich. Vol. I:122–178, Vol. II:125–153. Chicago: University of Chicago Press.

Egami, N.; Fukai, S.; and Masuda, S.
1965 *Dailaman I. The Excavations of Ghalekuti and Lasulkan 1960.* Tokyo: Yamakawa Publishing Co.
1966 *Dailaman II. The Excavations at Noruzmahale and Khoramrud.* Tokyo: Yamakawa Publishing Co.

Evett, D.
1967 Artifacts and Architecture of the Iblis I Period: Areas D, F and G. Pp. 202–255 in *Investigations at Tal-i-Iblis*, ed. J. R. Caldwell. Illinois State Museum Preliminary Reports 9. Springfield: Illinois State Museum.

France-Lanord, A.
1969 La Fer en Iran au premier millénaire avant Jésus Christ. *Révue d'histoire des mines et la métallurgie*, I:75–126.

Francfort, H.-P.
1989 *Fouilles de Shortugaï.* Recherches sur l'Asie Central Protohistorique. 2 vols. Mémoires de la Mission Archéologique Française en Asie Centrale Tome II. Paris: Diffusion de Boccard.

Francfort, H.-P., and Pottier, M.-H.
1978 Sondage préliminaire sur l'établissement protohistorique harappéen et postharappéen de Shortugaï (Afghanistan du N.-E.). *Arts Asiatiques* 34:39–64.

Fukai, S., and Ikeda, J.
1971 *Dailaman IV: The Excavations of Ghalekuti II and I, 1964.* Tokyo: Yamakawa Publishing Co.

Gale, N.; Papastamaki, A.; Stos-Gale, Z. A.; and Leonis, K.
1985 Copper Sources and Copper Metallurgy in the Aegean Bronze Age. Pp. 81–101 in *Furnaces and Smelting Technology in Antiquity*, eds. P. Craddock and M. J. Hughes. British Museum Occasional Paper 48. London.

Ghirshman, R.
1938 *Fouilles de Sialk*, Vol. 1. Paris: Paul Geuthner.
1979 L'Iran. La Migration des Indo-Aryens et des Iraniens. Pp. 63–66 in *Akten des VII. Internationalen Kongresses für Iranische Kunst und Archäologie, München, 7–10 September 1976. Archäologische Mitteilungen aus Iran.* Ergänzungsband 6. Berlin: D. Reimer.
1983 *Les Cimmériens et leurs Amazones.* Paris: Editions Recherche sur les Civilisations, Memoire No. 18.

Glumac, P. D.
1985 Earliest Known Copper Ornaments from Prehistoric Europe. *Ornament* 8(3):15–17.

Goff, C.
1978 Excavations at Baba Jan: The Pottery and Metal from Levels III and II. *Iran* 16:38–65.

Hakemi, A.
1992 The Copper Smelting Furnaces of the Bronze Age at Shahdad. Pp. 119–132 in *South Asian Archaeology 1989*, ed. C. Jarrige. Monographs in World Archaeology 14. Madison, WI: Prehistory Press.

Halm, L.
1939 Analyze chimique et étude micrographique de quelques pièces de métal et de céramique provenant de Sialk. Pp. 205–208 in *Fouilles de Sialk (près de Kashan 1933, 1934, 1937)*, Vol. 2, ed. R. Ghirshman. Paris: Paul Geuthner.

Hauptmann, A.
1980 Zur frühbronzezeitlichen Metallurgie von Shahr-i Sokhta (Iran). *Der Anschnitt* 2–3:55–61.

Hauptmann, A.; Weisgerber, G.; and Bachmann, H.-G.
1988 Early Copper Metallurgy in Oman. Pp. 34–51 in *The Beginning of the Use of Metals and Alloys*, ed. R. Maddin. Cambridge, MA: MIT Press.

Helmig, D.
1986 *Versuche zur analytisch-chemischen Charakterisierung frühbronzezeitlicher Techniken der Kupferverhüttung in Shahr-i Sokhta/Iran*. Diplomarbeit, Faculty of Chemistry, Ruhr-University, Bochum.

Herrmann, G.
1968 Lapis Lazuli: The Early Phases of its Trade. *Iraq* 30:21–57.

Herrmann, G., and Moorey, P. R. S.
1980–83 Lapis Lazuli. *Reallexikon der Assyriologie und Vorderasiatischen Archäologie* 6:489–492.

Herzfeld, E. E.
1941 *Iran in the Ancient East; Archaeological Studies Presented in the Lowell Lectures at Boston*. London: Oxford University Press.

Heskel, D.
1982 *The Development of Pyrotechnology in Iran during the Fourth and Third Millennia B.C.*, Ph.D. dissertation, Department of Anthropology, Harvard University. Ann Arbor, MI: University Microfilms International.
1983 A Model for the Adoption of Metallurgy in the Ancient Middle East. *Current Anthropology* 24(3):362–366.

Heskel, D., and Lamberg-Karlovsky, C. C.
1980 An Alternative Sequence for the Development of Metallurgy: Tepe Yahya, Iran. Pp. 229–265 in *The Coming of the Age of Iron*, eds. T. A. Wertime and J. D. Muhly. New Haven, CT: Yale University Press.
1986 Metallurgical Technology. Pp. 207–214 in *Excavations at Tepe Yahya, Iran 1967–1975*, eds. C. C. Lamberg-Karlovsky and T. W. Beale. Cambridge, MA: Peabody Museum of Archaeology and Ethnology.

Hodges, H.
1976 *Artifacts*. London: John Baker.

Hofman, H. O.
1914 *Metallurgy of Copper*. New York: McGraw-Hill.

Hole, F.
1977 *Studies in the Archaeological History of the Deh Luran Plain: The Excavations of Chogha Sefid*. University of Michigan Museum of Anthropology Memoir 9. Ann Arbor: University of Michigan.

Holzer, H. F., and Momenzadeh, M.
1971 Ancient Copper Mines in the Veshnoveh Area, Kuhestan-E-Qom, West Central Iran. *Archaeologia Austriaca* 49:1–22.

Jarrige, J.-F.
1985 A propos d'un fôret a tige helicoidale en cuivre de Mundigak. Pp. 281–292 in *De l'Indus aux Balkans (Recueil à la mémoire de Jean Deshayes)*, eds. J.-L. Huot, M. Yon, and Y. Calvet. Paris: Recherche sur les Civilisations.
1988 *Les Cités oubliées de l'Indus: Archéologie du Pakistan*. Paris: Association Française d'Action Artistique.

Kellner, H.-J.
1979 Eisen in Urartu. Pp. 151–156 in *Akten des VII. Internationalen Kongresses für Iranische Kunst und Archäologie, München, 7–10 September 1976. Archäologische Mitteilungen aus Iran*. Erganzungsband 6. Berlin: D. Reimer.

Khakhutaishvili, D. A.
1976 A Contribution of the Kartvelian Tribes to the Mastery of Metallurgy

in the Ancient Near East. Pp. 337–348 in *Wirtschaft und Gesellschaft im Alten Vorderasien*, eds. J. Harmatta and G. Komoroczy. Budapest: Akadémiai Kiadó.

Knauth, P.
1974 *The Metalsmiths*. New York: Time-Life Books.

Ladame, G.
1945 Les Ressources métallifères de l'Iran. *Schweizerische Mineralogische und Petrographische Mitteilungen* 25:167–303.

Lamberg-Karlovsky, C. C.
1967 Archaeology and Metallurgical Technology in Prehistoric Afghanistan, India and Pakistan. *American Anthropologist* 69:145–162.
1969 Further Notes on the Shaft-hole Pick-axe from Khurab, Makran. *Iran* 7:163–168.

Lechtman, H.
1970 The Khurab Pick-axe—Corrigenda. *Iran* 8:173.

Lechtman, H., and Klein, S.
1999 The Production of Copper-arsenic Alloys (Arsenic Bronze) by Cosmelting: Modern Experiment, Ancient Practice. *Journal of Archaeological Science* 26:497–526.

Levine, L. D., and Hamlin, C.
1974 The Godin Project: Seh Gabi. *Iran* 12:211–213.

Lorenzen, W.
1965 *Helgoland und das früheste Kupfer des Nordens*. Ottendorf: Niederelber-Verlag.

Maczek, M. von; Preuschen, E.; and Pittioni, R.
1952 Beitrage zum Problem des Ursprunges der Kupfererzverwertung in der Alten Welt. *Archäologia Austriaca* 10:61–70.

Maddin, R.; Wheeler, T. S.; and Muhly, J. D.
1980 Distinguishing Artifacts Made of Native Copper. *Journal of Archaeological Science* 7(3):211–226.

Majidzadeh, Y.
1979 An Early Prehistoric Coppersmith Workshop at Tepe Ghabristan. Pp. 82-92 in *Akten des VII. Internationalen Kongresses für Iranische Kunst und Archäologie, München, 7–10 September 1976. Archäologische Mitteilungen aus Iran*. Erganzungband 6. Berlin: D. Reimer.
1989 An Early Industrial Proto-urban Center on the Central Plateau of Iran: Tepe Ghabristan. Pp. 157-166 in *Essays in Ancient Civilization Presented to Helene J. Kantor*, eds. A. Leonard, Jr. and B. B. Williams. Studies in Ancient Oriental Civilization No. 47. Chicago: The Oriental Institute of the University of Chicago.

Malfoy, J.-M., and Menu, M.
1987 La Métallurgie du cuivre à Suse aux IVe et IIIe millénaires: analyses en laboratoire. Pp. 355–373 in *Métallurgie susienne I*, Vol. 1, ed. F. Tallon. Paris: Editions de la Réunion des Musées Nationaux.

Maréchal, J.
1958 Etude sur les propriétés mécaniques des cuivres à l'arsenic. *Métaux Corrosion Industries* 33:377–383.

Maryon, H.
1961 Early Near Eastern Steel Swords. *American Journal of Archaeology* 65:173–184.

Maxwell-Hyslop, K. R.
1955 Note on a Shaft-hole Axe-pick from Khurab, Makran. *Iraq* 17:161.

Maxwell-Hyslop, K. R., and Hodges, H.
1964 A Note on the Significance of the Technique of 'Casting On' as Applied to a Group of Daggers from North-West Persia. *Iraq* 26:50–53.
1966 Three Iron Swords from Luristan. *Iraq* 28:164–176.

McConchie, M.
1998 *Iron Technology and Ironmaking Communities in Northeastern Anatolia: First Millennium B.C.* Ph.D. thesis, Department of Classical Studies and Archaeology, University of Melbourne, Australia.

Merhav, R.
1991 *Urartu—A Metalworking Center in the First Millennium B.C.E.* Jerusalem: Israel Museum.

Moorey, P. R. S.
1969 Prehistoric Copper and Bronze Metallurgy in Western Iran. *Iran* 7:131–154.

1971 *Catalogue of the Ancient Persian Bronzes in the Ashmolean Museum.* Oxford: Clarendon Press.

1982 Archaeology and Pre-Achaemenid Metalworking in Iran: A Fifteen Year Retrospective. *Iran* 20:81–101.

1991 The Decorated Ironwork of the Early Iron Age Attributed to Luristan in Western Iran. *Iran* 29:1–12.

1994 *Ancient Mesopotamian Materials and Industries.* Oxford: Clarendon Press.

Muhly, J. D.
1977 The Copper Ox-hide Ingots and the Bronze Age Metals Trade. *Iraq* 39:73–82.

1983 Kupfer B. Archäologische. *Reallexikon der Assyriologie und Vorderasiatischen Archäologie* 6(5):348–364.

Müller-Karpe, M.
1991 Aspects of Early Metallurgy in Mesopotamia. Pp. 105–116 in *Archaeometry '90*, eds. E. Pernicka and G. A. Wagner. Boston: Birkhäuser Verlag.

Muscarella, O. W.
1981 Surkh Dum at the Metropolitan Museum of Art: A Mini-Report. *Journal of Field Archaeology* 8:327–359.

1984 Fibulae and Chronology, Marlik and Assur. *Journal of Field Archaeology* 11(4):413–419.

1988 The Background to the Luristan Bronzes. Pp. 33–44 in *Bronzeworking Centres of Western Asia 1000–539 B.C.*, ed. J. Curtis. London: Kegan Paul International.

1989a Multi-piece Iron Swords from Luristan. Pp. 349–366 in *Archaeologia Iranica et Orientalis: Miscellanea in Honore Louis Vanden Berghe*, eds. L. de Meyer and E. Haerinck. Gent, Belgium: Peeter Press.

1989b Bronzes of Luristan. Pp. 478–483 in *Encyclopaedia Iranica*, ed. E. Yarshater, Vol. IV, Fasc. 5. New York: Routledge and Kegan Paul.

Negahban, E. O.
1964 *A Preliminary Report on Marlik Excavation, Gohar Rud Expedition, Rudbar 1961–62.* Tehran: Offset Press.

1977 *A Guide to the Haft Tepe Excavation and Museum.* Tehran: Ministry of Culture and Arts, General Department of Museums.

1979a The Seals of Marlik Tepe. *Journal of Near Eastern Studies* 36:81–102.

1979b Seals of Marlik. Pp. 108–137 in *Akten des VII. Internationalen Kongresses für Iranische Kunst und Archäologie, München, 7–10 September 1976. Archäologische Mitteilungen aus Iran.* Erganzungsband 6. Berlin: D. Reimer.

1981 Maceheads from Marlik. *American Journal of Archaeology* 85:367–378.

1983 *Metal Vessels from Marlik.* Prähistorische Bronzefünde, Abteilung 2, Band 3. Munich: C. H. Beck.

1996 *Marlik: The Complete Excavation Report*, 2 vols. University Museum Monograph 87. Philadelphia: University of Pennsylvania Museum.

Nicholas, I. M.
1980 *A Spatial/Functional Analysis of Late Fourth Millennium Occupation at the TUV Mound, Tal-e Malyan, Iran.* Ph.D. dissertation, Department of Anthropology, University of Pennsylvania, Philadelphia. Ann Arbor, MI: University Microfilms.

1990 *The Proto-Elamite Settlement at TUV.* University Museum Monograph 69, Malyan Excavation Reports 1. Philadelphia: University of Pennsylvania Museum.

Northover, P.
1989 Properties and Use of Arsenic-copper Alloys. Pp. 111–118 in *Old World Archaeometallurgy*, eds. A. Hauptmann, E. Pernicka, and G. A. Wagner. Der Anschnitt, Beiheft 7. Veroffentlichungen aus dem Deutsches Bergbau-Museum 44. Bochum: Deutsches Bergbau-Museum.

Peake, H.
1928 The Copper Mountain of Magan. *Antiquity* 2:452–457.

Penhallurick, R. D.
1986 *Tin in Antiquity: Its Mining and Trade Throughout the Ancient World with Particular Reference to Cornwall.* London: Institute of Metals.

Pigott, V. C.
1977 The Question of the Presence of Iron in the Iron I Period in Western Iran. Pp. 209–234 in *Mountains and Lowlands: Essays in the Archaeology of Greater Mesopotamia*, eds. L. D. Levine and T. C. Young. Bibliotheca Mesopotamica 7. Malibu, CA: Undena.

1980a The Iron Age in Western Iran.

	Pp. 417–462 in *The Coming of the Age of Iron*, eds. T. A. Wertime and J. D. Muhly. New Haven, CT: Yale University Press.
1980b	Research at the University of Pennsylvania on the Development of Ancient Metallurgy: Research at MASCA. *Paléorient* 6:105–110.
1981	*The Adoption of Iron in Western Iran in the Early First Millennium B.C.: An Archaeometallurgical Study*. Ph.D. dissertation, Department of Anthropology, University of Pennsylvania. Ann Arbor, MI: University Microfilms.
1982	AHAN. Pp. 624–633 in *Encyclopaedia Iranica*, Vol. I, Fasc. 6. New York: Routledge and Kegan Paul.
1989a	Archaeo-metallurgical Investigations at Bronze Age Tappeh Hesar, 1976. Pp. 25–33 in *Tappeh Hesar, Reports of the Restudy Project, 1976*, eds. R. H. Dyson, Jr. and S. M. Howard. Florence: Casa Editrice Le Lettere.
1989b	Bronze. Pp. 457–471 in *Encyclopaedia Iranica*, Vol. IV, Fasc. 5. New York: Routledge and Kegan Paul.
1989c	The Emergence of Iron Use at Hasanlu. *Expedition* 31(2–3):67–79.
1996	Near Eastern Archaeometallurgy: Modern Research and Future Directions. Pp. 139–176 in *The Study of the Ancient Near East in the Twenty-first Century*, eds. J. Cooper and G. Schwartz. Winona Lake, IN: Eisenbrauns.
1999	A Heartland of Metallurgy: Neolithic/Chalcolithic Metallurgical Origins on the Iranian Plateau. Pp. 107–120 in *The Beginnings of Metallurgy*, eds. A. Hauptmann, E. Pernicka, T. Rehren, and Ü. Yalcin. Der Anschnitt, Beiheft 9.

Pigott, V. C.; Howard, S. M.; and Epstein, S. M.
1982   Pyrotechnology and Culture Change at Bronze Age Tepe Hissar (Iran). Pp. 215–236 in *Early Pyrotechnology: The Evolution of the First Fire-using Industries*, eds. T. A. Wertime and S. F. Wertime. Washington, D.C.: Smithsonian Institution Press.

Pigott, V. C.; Rogers, H. C.; and Nash, S. K.
in press   Archaeometallurgical Investigations at Tal-e Malyan: The Banesh Period. In *Proto-Elamite Civilization in the Land of Anshan*, ed. W. A. Sumner. Malyan Excavation Reports 3. Philadelphia: University of Pennsylvania Museum and Chicago: The Oriental Institute.

Pigott, V. C.; Weiss, A. D.; and Natapintu, S.
1997   The Archaeology of Copper Production: Excavations in the Khao Wong Prachan Valley, Central Thailand. Pp. 119–157 in *South-East Asian Archaeology, 1992: Proceedings of the Fourth International Conference of the European Association of South East Asian Archaeologists, Rome, 28th September–4th October*, eds. R. Ciarla and F. Rispoli. Rome: L'Istituto Italiano per l'Africa e l'Oriente (Is.AeO-Rome).

Pleiner, R.
1967   Preliminary Evaluation of the 1966 Metallurgical Investigation in Iran. Pp. 340–405 in *Investigations at Tal-i-Iblis*, ed. J. R. Caldwell. Illinois State Museum Preliminary Reports 9. Springfield: Illinois State Museum.
1969a   *The Beginnings of the Iron Age in Ancient Persia*. Annals of the Náprstek Museum 6. Prague: Náprstek Museum.
1969b   Untersuchung eines Kurzschwertes des Luristanischen Typus. *Archäologischer Anzieger* 1:41–47.

Pleiner, R., and Bjorkman, J. K.
1974   The Assyrian Iron Age: The History of Iron in the Assyrian Civilization. *Proceedings of the American Philosophical Society* 118(3):283–313.

Potts, D.
1990   *The Arabian Gulf in Antiquity*. 2 vols. Oxford: Clarendon Press.

Rapp, G.
1988   On the Origins of Copper and Bronze Alloying. Pp. 21–27 in *The Beginning of the Use of Metals and Alloys*, ed. R. Maddin. Cambridge, MA: MIT Press.

Rehder, J. E.
1991   The Decorated Iron Swords from Luristan: Their Material and Manufacture. *Iran* 29:13–20.

Rostoker, W., and Dvorak, J. R.
1991   Some Experiments with Co-smelting to Copper Alloys. *Archeomaterials* 5(1):5–20.

Rostoker, W.; Pigott, V. C.; and Dvorak, J. R.
1989   Direct Reduction to Copper Metal by Oxide/Sulfide Mineral Interaction. *Archeomaterials* 3:69–87.

Rothenberg, B.
1972   *Timna: The Valley of the Biblical Copper*

Salvatori, S.
1978　Problemi di Protohistoria Iranica: Note Ulteriori su di una Ricognizione di Superficie a Shahdad (Kerman, Iran). *Rivista di Archeologia* 2:5–15.

Salvatori, S., and Vidale, M.
1982　A Brief Surface Survey of the Protohistoric Site of Shahdad (Kerman, Iran): Preliminary Report. *Rivista di Archeologia* 6:5–10.

Shaffer, J.
1978　The Later Prehistoric Periods. Pp. 71–86 in *The Archaeology of Afghanistan from Earliest Times to the Timurid Period*, eds. F. R. Allchin and N. Hammond. New York: Academic Press.
1979　Bronze Age Iron from Afghanistan: Its Indications for South Asian Protohistory. Pp. 41–62 in *Archaeology and Palaeoanthropology of South Asia*, eds. K. A. R. Kennedy and G. L. Possehl. Delhi: Oxford & IBH/American Institute of Indian Studies.

Schmidt, E. F.
1937　*Excavations at Tepe Hissar, Iran: 1931–33*. Philadelphia: University of Pennsylvania Press for the University Museum.

Schmidt, E. F.; Van Loon, M.; Curvers, H. H.; and Brinkman, J. A.
1980　*The Holmes Expedition to Luristan.* Oriental Institute Publication 108. Chicago: University of Chicago Press.

Schürenberg, H.
1963　Über iranische Kupfererzvorkommen mit komplexen Kobalt-Nickelerzen. *Neues Jahrbüch für Mineralogie Abhandlungen* 99(2):200–230.

Seeliger, T. C.; Pernicka, E.; Wagner, G. A.; Begemann, F.; Schmitt-Strecker, S.; Eibner, C.; Öztunalı, Ö.; and Baranyi, I.
1985　Archäometallurgische Untersuchungen in Nord- und Ostanatolien, *Jahrbüch der Romisch-Germanischen Zentralmuseums, Mainz* 32:597–659.

Seidl, U.
1988　Urartu as a Bronzeworking Center. Pp. 169–176 in *Bronzeworking Centres of Western Asia 1000–539 B.C.*, ed. J. Curtis. London: Kegan Paul International.

Mines. London: Thames and Hudson.

Shahmirzadi, S. M.
1979　Copper, Bronze, and Their Implementation by Metalsmiths of Sagzabad, Qazvin Plain, Iran. *Archäologische Mitteilungen aus Iran* 12:49–66.

Shareq, A.; Chymriov, V. M.; Stazhilo-Alexseev, K. F.; Dronov, V. I.; Gannon, D. J.; Lubemov, G. K.; Kafarshiy, A. Kh.; and Malyarov, E. P.
1977　*Mineral Resources of Afghanistan*, 2nd ed. Afghan Geological and Mines Survey, United Nations Development Support Project, AFG/74/012. Kabul.

Smith, C. S.
1965　The Interpretation of Microstructures of Metallic Artifacts. Pp. 20–52 in *Applications of Science in the Examination of Works of Art*, ed. W. J. Young. Boston: Museum of Fine Arts.
1968　Metallographic Study of Early Artifacts Made from Native Copper. In *Actes du XIe Congrès International d'Histoire des Sciences Warsaw* 6:237–243.
1969　Analysis of the Copper Bead from Ali Kosh, Appendix 2. Pp. 427–428 in *Prehistory and Human Ecology of the Deh Luran Plain: An Early Village Sequence from Khuzistan, Iran*, eds. F. Hole, K. V. Flannery, and J. A. Neely. Ann Arbor: University of Michigan.
1970　Art Technology and Science: Notes on their Historical Interaction. *Technology and Culture* 11(4):493–549.
1971　Techniques of the Luristan Smith. Pp. 32–52 in *Science and Archaeology*, ed. R. H. Brill. Cambridge, MA: MIT Press.
1972　Metallurgical Footnotes to the History of Art. *Proceedings of the American Philosophical Society* 116(2):97–135.
1976　On Art, Invention and Technology. *Technology Review* 78(7):2–7.

Smith, C. S.; Wertime, T. A.; and Pleiner, R.
1967　Preliminary Reports of the Metallurgical Project. Pp. 318-326 in *Investigations at Tal-i-Iblis*, ed. J. R. Caldwell. Illinois State Museum Preliminary Reports 9. Springfield: Illinois State Museum.

Solecki, R.
1969　A Copper Mineral Pendant from Northern Iraq. *Antiquity* 43:311–314.

Sono, T., and Fukai, S.
1968　*Dailaman III: The Excavations at Hassani Mahale and Ghalekuti 1964*. Tokyo:

Yamakawa Publishing Co.

Stech, T.
1990 Neolithic Copper Metallurgy in Southwest Asia. *Archeomaterials* 4(1):55–61.

Stech, T., and Pigott, V. C.
1986 The Metals Trade in Southwest Asia in the Third Millennium B.C. *Iraq* 48:39–64.

Stein, M. A.
1937 *Archaeological Reconnaissance in Northwestern India and Southeastern Iran.* London: Macmillan.
1940 *Old Routes in Western Iran.* New York: Greenwood Press.

Stocklin, J.; Eftekahr-Nezhad, J.; and Husmand-Zadeh, A.
1972 *Central Lut Reconnaissance, East Iran.* Geological Survey of Iran, Report 22. Tehran.

Tallon, F.
1987 *Métallurgie susienne* I. 2 vols. Paris: Editions de la Réunion des Musées Nationaux.

Tosi, M.
1974 The Lapis Lazuli Trade across the Iranian Plateau in the Third Millennium B.C. Pp. 3–22 in *Gururajamanjarika. Studi in onere di Giuseppe Tucci.* Naples: Istituto Universitario Orientale.
1983 A Bronze Female Statuette from Shahr-i Sokhta. Chronological Problems and Stylistical Connections. Pp. 303–317 in *Prehistoric Seistan* 1, ed. M. Tosi. Rome: IsMEO.

Tylecote, R. F.
1970 Early Metallurgy in the Near East. *Metal and Materials* 4:285–293.
1972 A Metallurgical Examination of Some Objects from Marlik, Iran. *Bulletin of the Historical Metallurgy Group* 6:34–35.
1974 Can Copper Be Smelted in a Crucible? *Journal of the Historical Metallurgy Society* 8(1):54.

Tylecote, R. F.; Ghaznavi, H. A.; and Boydell, P. J.
1977 Partitioning of Trace Elements between the Ores, Fluxes, Slags and Metal during the Smelting of Copper. *Journal of Archaeological Science* 4:27–49.

Tylecote, R. F., and McKerrell, H.
1971 Examination of Copper Alloy Tools from Tal y Yahya, Iran. *Bulletin of the Historical Metallurgy Group* 5(1):37–38.

Udy, S.
1964 Preindustrial Forms of Organized Work. Pp. 115–124 in *Cultural and Social Anthropology*, ed. P. B. Hammond. New York: Macmillan.

Vanden Berghe, L.
1976 *Bibliographie analytique de l'archéologie de l'Iran ancien.* Tehran.

Vanden Berghe, L., and Haerinck, E.
1981 *Bibliographie analytique de l'archéologie de l'Iran ancien. Supplement 1: 1978–80.* Leuven: Peeters.
1987 *Bibliographie analytique de l'archéologie de l'Iran ancien. Supplement 2: 1981–85.* Leuven: Peeters.

Vandiver, P. B.; Yener, K. A.; and May, L.
1993 Third Millennium Tin Processing Debris from Göltepe (Anatolia). Pp. 545–569 in *Materials Issues in Art and Archaeology* III, eds. P. B. Vandiver, J. Druzik, and G. S. Wheeler. Pittsburgh: Materials Research Society.

Van Loon, M.
1966 *Urartian Art.* Istanbul: Nederlands Historisch-Archaeologisch Instituut.

Vatandoost-Haghighi, A. R.
1978 *Aspects of Prehistoric Iranian Copper and Bronze Technology.* Unpubl. Ph.D. thesis, Institute of Archaeology, London.

Voigt, M. M.
1990 Reconstructing Neolithic Societies and Economies in the Middle East: An Essay. *Archeomaterials* 4(1):1–14.

Waldbaum, J. C.
1971 A Bronze and Iron Iranian Axe in the Fogg Art Museum. Pp. 195–210 in *Studies Presented to George M. A. Hanfmann*, eds. D. G. Mitten, J. G. Pedley, and J. A. Scott. Mainz: Phillip von Zabern.

Wayman, M. L.
1985 Native Copper: Humanity's Introduction to Metallurgy? Part 1: Occurrence, Formation and Prehis-

Weisgerber, G.
1980 ". . . Und Kupfer in Oman." *Der Anschnitt* 32(2–3):62–110.
1990 Montanarchäologische Forschungen in Nordwest-Iran 1978. *Archäologische Mitteilungen aus Iran* 23:73–84.

Weisgerber, G.; Kroll, S.; Gropp, G.; and Hauptmann, A.
1990 Das Bergbaurevier von Sungun bei Kighal in Azarbaidjan (Iran). *Archäologische Mitteilungen aus Iran* 23:85–103.

Wertime, T. A.
1964 Man's First Encounters with Metallurgy. *Science* 146(3649):1257–1267.
1967 A Metallurgical Expedition through the Persian Desert. Pp. 327–339 in *Investigations at Tal-i-Iblis*, ed. J. R. Caldwell. Illinois State Museum Preliminary Reports 9. Springfield: Illinois State Museum.
1968 A Metallurgical Expedition through the Persian Desert. *Science* 159(3818):927–935.
1973 The Beginnings of Metallurgy: A New Look. *Science* 182(4115):875–887.

Winter, I.
1977 Perspective on 'Local Style' of Hasanlu IVB: A Study in Receptivity. Pp. 371–386 in *Mountains and Lowlands: Essays in the Archaeology of Greater Mesopotamia,* eds. L. D Levine and T. C. Young. Bibliotheca Mesopotamica 7. Malibu: Undena.

Woolley, C. L.
1963 *The Beginning of Civilization.* New York: The New American Library.

Yener, K. A., and Vandiver, P. B.
1993 Tin Processing at Göltepe, an Early Bronze Age Site in Anatolia, *American Journal of Archaeology* 97:207–238.

Young, T. C., Jr.
1967 The Iranian Migration into the Zagros. *Iran* 5:11–34.

Zeuner, F. E.
1955 The Identity of the Camel on the Khurab Pick. *Iraq* 17:162–163.

5
Metal Technologies of the Indus Valley Tradition in Pakistan and Western India

Jonathan M. Kenoyer and

Heather M.-L. Miller

ABSTRACT   In this paper we summarize the available literature and recent discoveries on the production and use of metals by peoples of the Indus Valley Tradition of Pakistan and western India. Our primary focus is on the Harappan Phase (2600–1900 B.C.), and includes a review of collections and technical analyses of metal artifacts, along with tables of the published analyses from the sites of Mohenjo-daro, Harappa, Lothal, and Rangpur. The potential ore sources for metals are discussed, with particular attention given to copper, arsenical copper, and tin bronzes but also including lead, gold, silver, and iron. We present an overview of evidence for Harappan Phase metal processing techniques, from smelting to finishing, and examine the use of metal in the context of an urban society that still uses stone tools. In conclusion we outline some future directions for archaeological and archaeometallurgical research in the subcontinent. [Final ms. received 10/96.]

INTRODUCTION

The Indus Valley Tradition of Pakistan and western India has been the focus of considerable research over the past two decades and scholars have begun to fill in many of the gaps in our understanding of regional geography, settlement patterns, subsistence, specific technological developments and the chronology of these changes (see Kenoyer 1991; Mughal 1990; Possehl 1990 for summaries). This paper provides an overview of the non-ferrous metal technologies in the northwestern regions of the subcontinent, and of the role of these technologies during the Harappan Phase of the Indus Valley Tradition (2600–1900 B.C.). As the first such overview since Agrawal's seminal work in 1971, we will focus on the presentation of often inaccessible data, summarizing the information available on metal sources, processing, and use.

The Indus Valley Tradition was centered in the greater Indus plain, which was formerly watered by two major river systems, the Indus and the Ghaggar-Hakra (now dry) (Fig. 5.1). Adjacent regions which were culturally integrated at various periods with this vast double river plain include the highlands and plateaus of Baluchistan to the west, and the mountainous regions of northern Pakistan, Afghanistan, and India to the northwest and north. The Thar Desert and the Aravalli Hills formed the eastern periphery. The coastal regions from Makran to Kutch and Gujarat formed the southern boundary and provided access by sea to the resource areas of the Arabian Peninsula (Besenval 1992).

We have chosen to use the chronology defined by Shaffer (1992), which is presented in Table 5.1 along with its correlations to other more widely used but less precisely defined chronologies. As defined by Shaffer

TABLE 5.1
GENERAL DATES AND ARCHAEOLOGICAL PERIODS

INDUS TRADITION	*Early Food Producing Era* Aceramic Neolithic	ca. 6500–5000 B.C.
	Regionalization Era Early Harappan Early Chalcolithic Ceramic Neolithic	ca. 5000–2600 B.C.
	Integration Era Mature Harappan Chalcolithic/Bronze Age	2600–1900 B.C.
	Localization Era Late Harappan	1900–1300 B.C.

Iron Age
  Painted Gray Ware
  Northern Black
   Polished Ware
		+1200–800 B.C.
		
		(?700) 500 to 300 B.C.

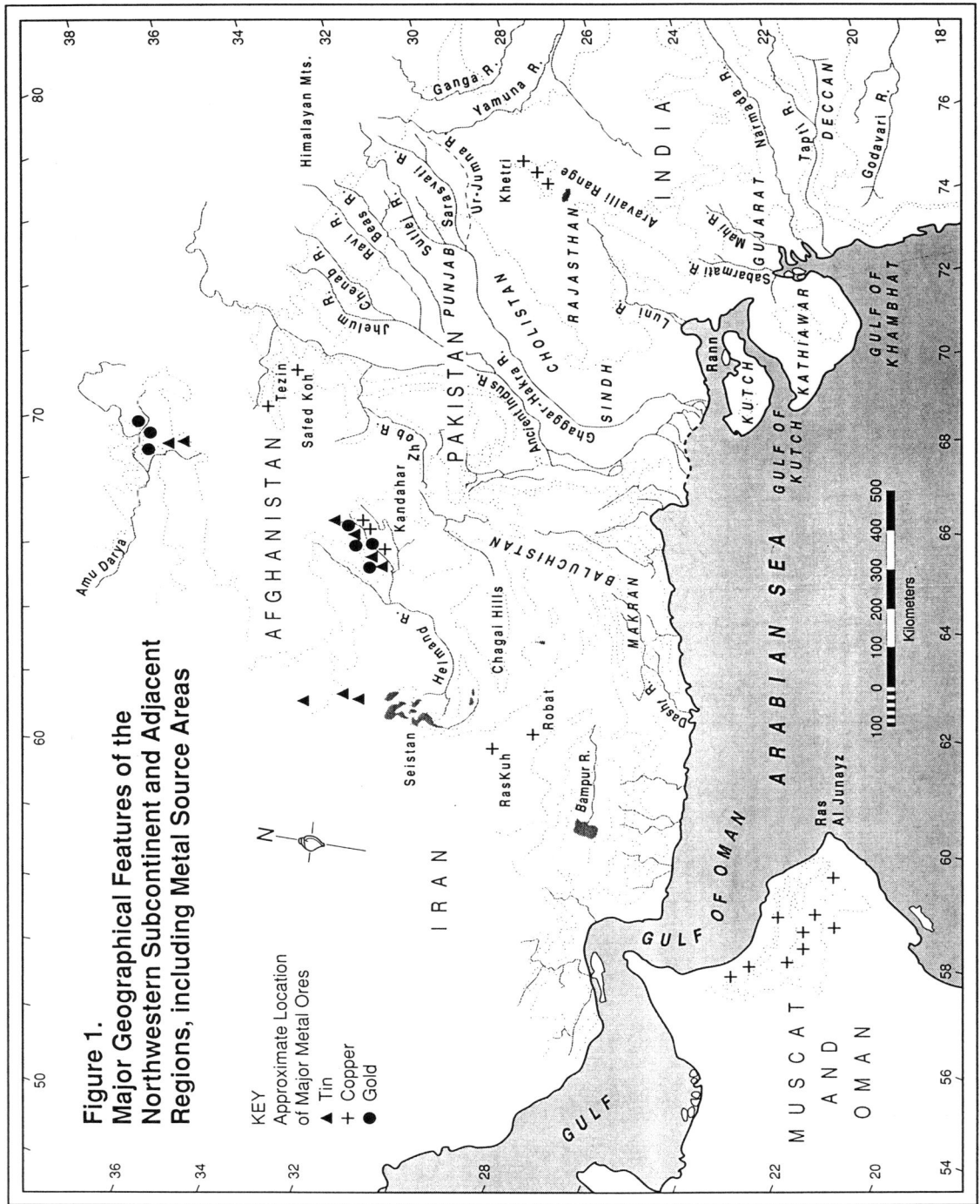

Figure 5.1 Major geographical features of the northwestern subcontinent and adjacent regions, including metal source areas (composed by J. M. Kenoyer from various sources).

TABLE 5.2
ARCHAEOLOGICAL TRADITIONS OF NORTHWESTERN SOUTH ASIA (AFTER SHAFFER 1992)

INDUS VALLEY TRADITION	BALUCHISTAN TRADITION	HELMAND TRADITION
Early Food Producing Era	*Early Food Producing Era*	*Early Food Producing Era*
Mehrgarh Phase	Mehrgarh Phase	Ghar-i-mar Phase*
Regionalization Era	*Regionalization Era*	*Regionalization Era*
Balakot Phase	Kachi Phase	Mundigak Phase
Amri Phase	Kili Gul Muhammad Phase	Helmand Phase
Hakra Phase	Sheri Khan Tarakai Phase*	
Kot Diji Phase	Kechi Beg Phase	
	Damb Sadaat Phase	
	Nal Phase	
Integration Era	*Integration Era*	*Integration Era*
Harappan Phase	Kulli Phase	Shahr-i Sokhta Phase
	Periano Phase	
Localization Era	*Localization Era*	*Localization Era*
Punjab Phase	Bampur Phase	Seistan Phase
Jhukar Phase	Pirak Phase	
Rangpur Phase		

*The Ghar-i-Mar (Dupree 1972) and Sheri Khan Tarakai Phases (Khan et al. 1989) were not identified by Shaffer because the excavations are only recently published or not fully analyzed.

(1992), the Indus Valley Tradition includes all human adaptations in this greater Indus region from around 6500 B.C. until 1500 B.C. and later. This Tradition can be subdivided into four Eras and several Phases (Tables 5.1 and 5.2). The Early Food Producing Era (ca. 6500–5000 B.C.), as defined at the site of Mehrgarh, sees the beginning of domesticated plants and animals, as well as the first find of copper in the form of a bead (Jarrige 1983). The Regionalization Era (ca. 5000–2600 B.C.) follows, with the development of distinct agricultural and pastoral-based cultures associated with various specialized crafts, including the melting and working of copper. During the Integration Era, which is represented by the Harappan Phase (2600–1900 B.C.), we see the cultural, economic, and political integration of the vast region defined above. This paper focuses on the state of metal processing during the Harappan Phase.

The Harappan Phase of the Integration Era represents the first urban civilization in southern Asia and the earliest state-level society in the region (Jacobson 1986; Kenoyer 1991; Meadow 1991). Recent studies suggest that the Indus state was composed of several classes of elites who maintained different levels of control over the vast regions of the Indus and Ghaggar-Hakra Valley. The rulers or dominant members in the various cities would have included merchants, ritual specialists, and individuals who controlled resources such as land, livestock, and raw materials. Although these groups may have had different means of control, they shared a common ideology and economic system as represented by seals, ornaments, ceramics, and other artifacts. This ideology would have been shared by occupational specialists and service communities, who appear to have been organized in loosely stratified groups (Kenoyer 1991). Political and economic integration of the cities may have been achieved through the trade and exchange of important socio-ritual status items, many of which would have been produced by specialized artisans using complex pyrotechnologies to manufacture metal objects, agate beads, steatite seals, stoneware bangles, elaborately painted and specialized ceramics, and faience objects (Kenoyer 1992a).

PROBLEMS IN DEFINING THE ORIGINS AND DIFFUSION OF METAL TECHNOLOGIES

The extensive overlapping exchange networks that connected the greater Indus region to the metal resource areas of West Asia, eastern Iran, and Rajasthan make it difficult to determine the role of diffusion in the origins and dispersal of various metal technologies, especially copper metallurgy. The simplistic yet pervasive model that copper-working technology was developed somewhere in West Asia and diffused to adjacent regions such as the greater Indus region (see this volume and Agrawal 1971) is based on assumptions regarding human cultural interaction and the control of knowledge that are not supported by the archaeologi-

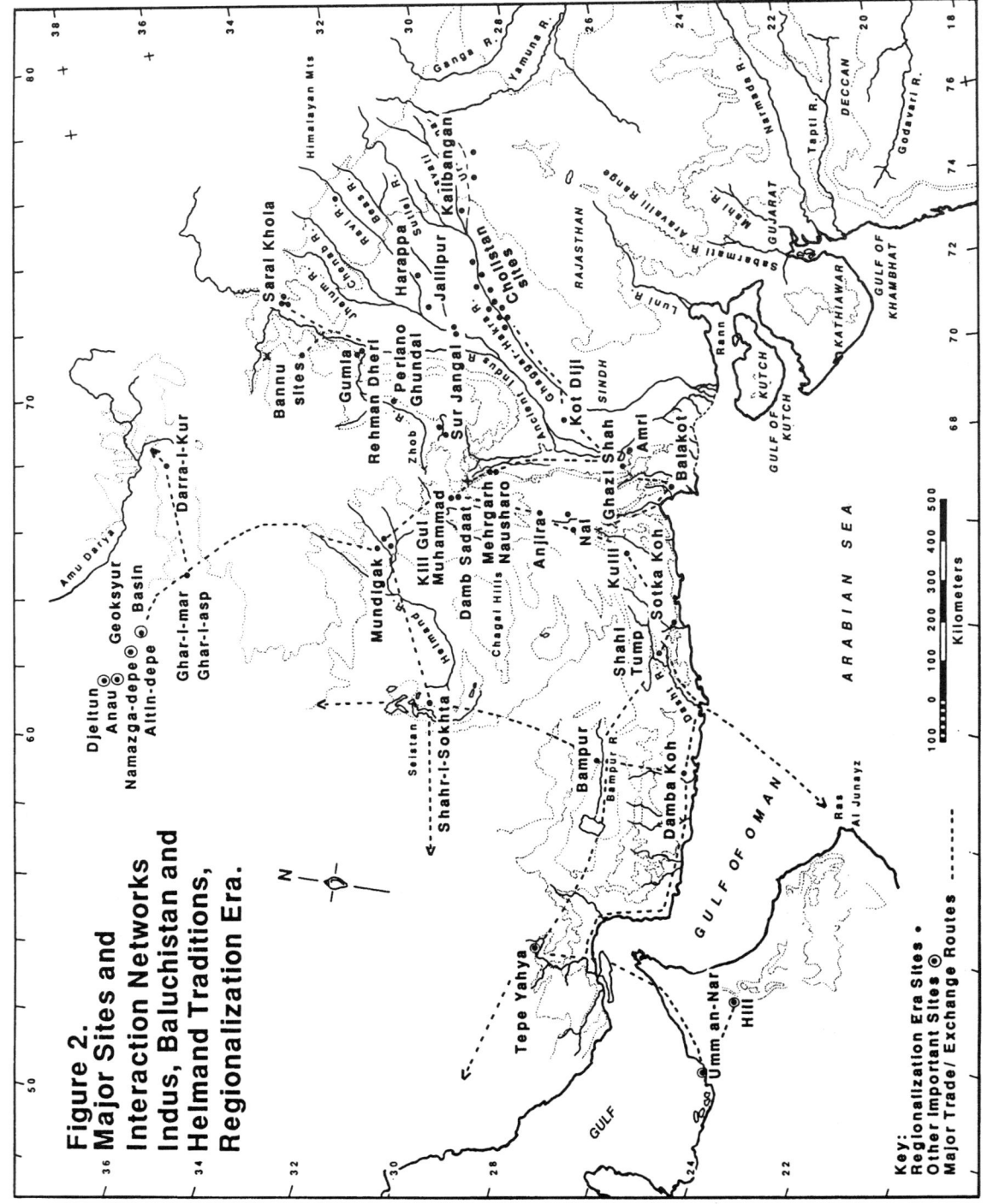

Figure 5.2 Major sites and interaction networks of the Indus Valley, Baluchistan, and Helmand Traditions, Regionalization Era (drawn by J. M. Kenoyer).

cal data currently available for study (Kenoyer 1989). Throughout West and South Asia, beginning in the Palaeolithic and continuing through the Neolithic, we find evidence for a familiarity with fire and its effect on various materials. In the Upper Palaeolithic, iron ores were routinely roasted to make pigments (Schmandt-Besserat 1980) and chert was heated to make it more flakable. During the Neolithic and early Chalcolithic, pyrotechnologies included the firing of different types of clays to make ceramics, and the heating of lithic materials to enhance color, workability, and/or hardness. Although we have no direct evidence for the earliest metal procurement techniques, it is not unlikely that fire setting was being used to extract native copper lumps and granules that could then be further processed by hammering and annealing. The many different pyrotechnologies in practice make it unreasonable to assume that the discovery of metal smelting and melting was simply an accident, and not the result of intentional experimentation and innovation.

For the greater Indus region, the evidence from Mehrgarh and other early sites demonstrates that the pyrotechnological and metallurgical innovations of the Neolithic and Chalcolithic set the technological background for the metallurgical traditions of the Harappan Phase (Jarrige 1985b; Jarrige and Lechevallier 1979). It is clear that the origin and development of copper metal technology occurred *in conjunction with* developments in other technologies. At the site of Mehrgarh during the fifth to fourth millennium B.C., changes were occurring simultaneously in metal production, ceramic production, the drilling of hard stone, production of fired and glazed steatite beads, and shell working. A decrease in the use of certain types of bone and stone tools is also seen at this time (Jarrige 1983). The transitions seen at Mehrgarh between the Neolithic and the Chalcolithic have numerous parallels with similar changes in the highlands of Baluchistan and other regions of the greater Indus region (Fig. 5.2). Sites such as Nausharo (Jarrige 1990), Balakot (Dales 1979), Ghazi Shah (Flam 1993), Rehman Dheri (Durrani 1988), and Kalibangan (Agrawala 1984a; Lal and Thapar 1967) all show evidence for the use of copper in the period prior to the Harappan Phase, along with changes in other technologies.

Throughout southern and northern Baluchistan, Afghanistan, and Rajasthan, the combined resources of metal ores and fuel were available to communities of sedentary agriculturalists and semi-nomadic pastoralists. Such communities were undoubtedly familiar with the properties of ores and how to extract the metal long before it became an important economic process. Furthermore, it is highly unlikely that the process for smelting ores and processing copper was discovered only in one isolated area, since there is increasing evidence that the highland communities of West and South Asia were connected by numerous overlapping networks, both economic and social (Kenoyer 1991).

Since there are many regions of West Asia and South Asia that are rich in both metal ores and fuel, it is quite likely that regional styles of pyrotechnologies evolved according to the physical characteristics of locally available ores. Over time, in adjacent regions such as northern and southern Baluchistan, the regional styles that were less effective and/or practiced by sociopolitically weaker communities would have been eliminated or absorbed through competition. More widely separated regions such as Baluchistan and Rajasthan, which are divided by the Indus Valley flood plains, may have retained their styles and continued to function parallel to each other for a longer period of time. Future studies of regional styles of metal processing and use may provide valuable information for understanding the development of a possible Indus "technological style" or multiple "technological styles" (see Lechtman and Steinberg 1979).

STUDIES OF HARAPPAN PHASE METAL OBJECTS: CATALOGUES AND TECHNICAL ANALYSES

The metal objects have been one of the most neglected of the Indus artifact classes, even though the first technological analyses were carried out in the 1920s and '30s. Although numerous metal objects have been recovered from Harappan Phase sites in Pakistan and western India, relatively few of these have been subjected to metallurgical or compositional analyses. In fact, few of the excavated collections have even been completely published.

CATALOGUES

The most extensive published collections of metal objects are those from the early excavations at the Harappan Phase sites of Mohenjo-daro (Mackay 1931, 1938; Marshall 1931), Harappa (Vats 1940), and Chanhu-daro (Mackay 1943), all in the Indus Valley (Fig. 5.3). The metals from excavations at Rangpur (Rao 1963) and Lothal (Rao 1979, 1985) provide information on the metals of Harappan Phase Gujarat. Information on metal use in the greater Indus region prior to the Harappan Phase comes primarily from the site of Mehrgarh (Jarrige and Lechevallier 1979).

For the Harappan Phase, the best references are the catalogues of metal objects compiled by Yule (Yule 1985a, 1985b), providing descriptions and illustrations of the copper objects from Mohenjo-daro,

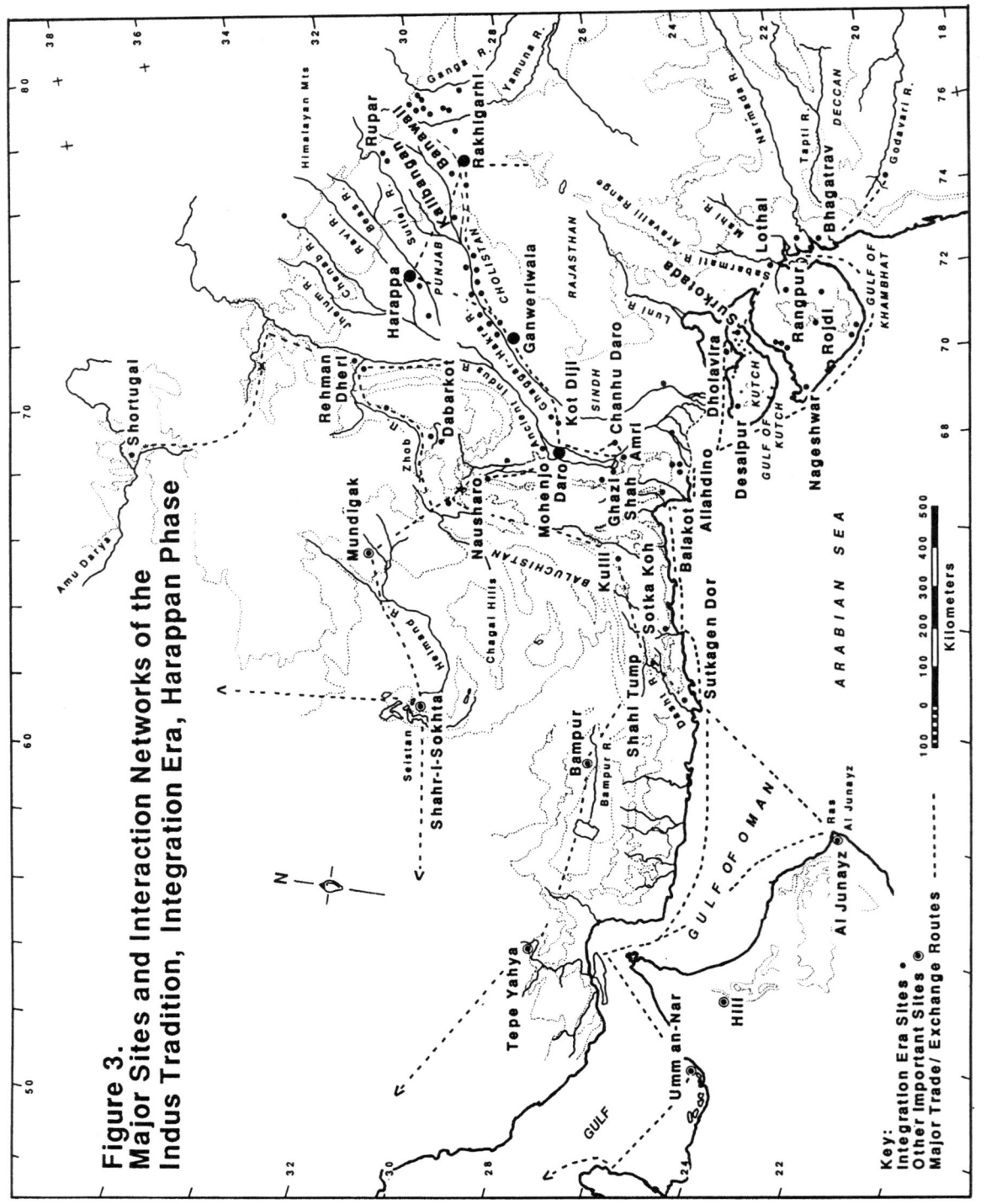

Figure 5.3 Major sites and interaction networks of the Indus Valley Tradition, Integration Era, Harappan Phase (drawn by J. M. Kenoyer).

Harappa, Lothal, and several other sites, including many objects previously unpublished. (Note, however, that these catalogues do not include the objects from Chanhu-daro in the Museum of Fine Arts, Boston.) Herman (1984) has also compiled a catalogue of metal objects from the published Harappan Phase sites, which is particularly useful for its assessment of the stratigraphic relationships of the objects. Haquet (1994) is currently preparing a data base and typology of metal objects from Mehrgarh, Nausharo, and Mundigak, which will be the first catalogue to present metal objects from well-defined stratigraphic contexts ranging from the Regionalization through the Integration Eras.

Full publication of the metal objects from a number of recently excavated sites are still needed, however, before we can confidently discuss changes in Indus Valley Tradition metals over time and in different regions. This includes the metals from recent excavations at Harappa by the Harappa Archaeological Research Project (originally the University of California-Berkeley Expedition), which we hope will be studied within the next year or two. Other important assemblages awaiting study are from the site of Kalibangan (Agrawala 1984a), a site key to our understanding of the northern regions of Rajasthan and Haryana; from Ganeshwar and related sites in Rajasthan, which are very near to the Rajasthani copper mines and extremely rich in copper metal objects (Agrawala 1984a, 1984b; Kumar 1986); and from Ahar and related sites in southeastern Rajastan, where evidence of copper processing has been found (summarized in Hooja 1988; Hooja and Kumar 1995).

We look forward to more detailed publications on the contexts, elemental compositions, and methods of production for the copper objects from Rajasthan. These materials are extremely important for a more complete understanding of the metallurgy of the greater Indus region, particularly its ore sources. They will also provide comparative information about the distinct metallurgical styles and approaches of the Rajasthani cultures that were apparently contemporaneous with the Harappan Phase of the Indus Valley Tradition. (It should be noted that the dating of most of these sites is problematical, as there are very few radiocarbon dates and the relations between the various ceramic types are still highly debated [Kenoyer 1991; Shaffer 1992]. The metal objects themselves have not cleared up the question of chronological affinities because many of the "type" markers are in fact distributed over wide regions and time periods, for example, the double spiral-headed pins, celts, and barbed arrowheads.)

Metal objects from the Localization Era (Late Harappan) are represented by the assemblage of objects from Daimabad (Sali 1986). Unfortunately, until further discoveries are made at the site, the dating and provenience of the metal objects will remain controversial. Consequently, we will not include them in this study. Yule (1985c) has compiled a catalogue of the Copper Hoard objects from India, thought to be roughly contemporaneous with or slightly later than the Harappan Phase. Finally, a summary discussion of metallurgy in the subcontinent has been presented by Kuppuram (1989) which focuses primarily on the historic rather than the prehistoric period.

CHEMICAL AND PHYSICAL ANALYSES

The vast majority of analyzed metal objects come from the major urban sites of Mohenjo-daro, Harappa, and Lothal, and date to the Harappan Phase, between 2600 and 1900 B.C. (Fig. 5.3). It has been impossible to ascertain the total number of chemical and physical analyses conducted to date. Since many of the published tables do not list the field or identification numbers of the object sampled, it has been impossible to determine if an object has been sampled and reported more than once (e.g., Agrawal 1971:tables 18 and 19). Consequently, in our summary tables of chemical analyses (Appendices A and B), we only include analyses reported with field or identification numbers for the object.

Chemical and some physical analyses were done on metal objects from the early excavations at Harappa and Mohenjo-daro (Desch 1931; SanaUllah 1931, 1940; Wraight 1940; Hamid, SanaUllah, Pascoe, and Desch and Carey reported in Mackay 1938). Agrawal's (1971) comprehensive treatise on South Asian metalworking is still of major importance for its critical summary of the earlier published material and many previously unpublished chemical analyses; however, many of his interpretations need to be revised due to the availability of new data from sites such as Mehrgarh (Jarrige and Lechevallier 1979) and Lothal (Rao 1979, 1985), particularly the chemical analyses done by Lal (1985) on material from Lothal.

The most common metal objects were made of copper or copper alloys. SanaUllah (1931:485) defined four categories of copper metal objects at the site of Mohenjo-daro: (1) "lumps" of crude copper directly derived from smelting and rich in sulfur (these are ingots, based on the size and shape description); (2) "refined" copper (i.e., specimens containing few non-copper elements—note that one such specimen is also a "lump" or ingot, however); (3) arsenical copper (SanaUllah's "copper-arsenic alloy"); and (4) tin bronze. At the present no object made from native copper has been reported from an Indus Valley Tradition site.

Other processed metals that have been reported include lead, gold, silver, and electrum. Although there are copper objects with iron components from contemporaneous sites in Baluchistan (Shaffer 1984), no confirmed iron objects have been reported from Harappan Phase sites in the greater Indus region. Finally, no true brass objects (copper-zinc alloy) have been identified from any Harappan sites (but see Other Metals section below).

A serious problem is that most of the published analytical studies of Indus Valley Tradition metals do not outline the specific methods of analysis, so we do not know if the results are really comparable. For example, the very large differences in percent oxygen and acid insoluble materials between metal objects from Harappa and Mohenjo-daro (tested by SanaUllah or Hamid) and from Lothal (tested by Lal) may be due to analytical techniques. Also, we can seldom tell if an element was truly absent from a collection of artifacts, or if no tests were done to determine its presence, such as zinc at Mohenjo-daro (Appendix A). An additional discrepancy factor is introduced by the great disparity in the preservation of metal in different objects and at different sites. For example, many of the objects analyzed from Lothal were less well preserved than those from Harappa and Mohenjo-daro.

Fortunately, additional analytical studies of archaeological materials are currently underway. The compositional and metallographic analysis of recently excavated copper metal objects from Harappa is being conducted at the University of Pennsylvania Museum, MASCA laboratories under the direction of Dr. V. Pigott (Pigott et al. 1989), and a large number of copper metal objects from Chanhu-daro are currently being studied at the Museum of Fine Arts, Boston, under the direction of Dr. Thomas Beale (pers. comm.).

POTENTIAL ORE SOURCES FOR HARAPPAN PHASE METALS

The studies which have been done in the various source areas are discussed by metal type in the sections below. Most of the objects analyzed from sites of the Indus Valley Tradition have been finished copper metal objects, and few analyses have been done of other metals, or of copper ores, slags, metal prills on crucibles, or ingots. The systematic comparison of Indus Valley Tradition copper with copper ores from the variety of sources available has been sorely neglected. This is due in part to the lack of archaeological samples of ores, and in part because many of the ore mineral deposits potentially used in the past are located in border areas or tribal regions that are not easily accessible to modern researchers (e.g., Baluchistan).

Only a small number of actual mineral fragments have been reported from Harappan Phase sites. At the site of Mohenjo-daro, "a quantity of copper ore" was found in a pit in DK area (Mackay 1938:54), and at Harappa, small fragments of chrysocolla and chalcopyrite have been recovered (Dales and Kenoyer 1990; Kenoyer, on-going research). In addition to these copper minerals, a few fragments of hematite, löllingite (arsenic and iron), antimony, cinnabar (sulfide of mercury), cerussite (carbonate of lead), galena, and an unidentified type of lead ore (recently recovered from excavations at Harappa) have been recovered from these two sites as well (Mackay 1938; Marshall 1931; SanaUllah 1931; Vats 1940). It is possible that some of these metallic minerals may have been used in melting and alloying processes, but it is just as likely that they were used for other purposes, e.g., as colorants, cosmetics, medicines, poisons, etc., since the great majority were not found in association with metal processing debris.

COPPER; ARSENICAL COPPER; TIN BRONZE

One of the earliest sourcing studies for Harappan Phase copper was the analysis of material from Mohenjo-daro by Desch (Desch 1931; Desch and Carey reported in Mackay 1938), but Agrawal's (1971) examination of the data and methodology clearly demonstrated the need for new analyses. Agrawal (1971, 1984) suggested that Indus peoples used native copper, oxide ores, and also sulfide ores, at least for the copper objects at Harappa and Mohenjo-daro. This interpretation is based on the percentage of elements found in finished objects, using a method presented by Freidmann et al. in 1966 (Agrawal 1971:tables 14 and 15; 1984). However, given the current debates about sourcing (see this volume and Tylecote 1980), Agrawal's suggestions based on this method may not be valid. As noted above, no native copper fragments or objects have been reported from any Indus Valley Tradition site.

Relatively pure copper objects have been found at all sites in the greater Indus region where copper metal objects have been analyzed, and they comprise the largest percentage of objects (Appendices A and B). (Note that these appendices do not include Desch's work, which only tested for a few elements, nor the analyzed objects without identification information in Agrawal 1971.) Depending on how one defines alloys, tin bronzes are the second largest category and arsenical coppers the third. Out of the 129 copper metal objects that we have tabulated, 36 objects have 1% or more tin, 20 objects have approximately 1% or more arsenic, and 6 of these objects have 1% or more of both tin and arsenic (Appendix B—objects with both tin and arsenic are listed in both tables). It should be emphasized that the analyses of these objects by different scholars are not always comparable, but in general terms the numbers can be useful.

It is important to note that different researchers have used different standards to define alloying (see Stech, this volume, for an excellent discussion of alloying). Agrawal (1971:150, 168) states that more than 1% arsenic or tin constitutes intentional alloying. However, SanaUllah (1931) defined an intentional alloy as containing from 2 to 4.5% arsenic or 4.5 to 13% tin. Some scholars favor the value of 5% tin to qualify as an intentional tin bronze used for functional purposes (Hall and Steadman 1991). This functional criterion ignores the changes in color that occur with the addition of less than 5% tin, and color or ability to resist oxidation may have been more important than hardness or strength for the early metalsmiths and consumers (Lal 1985:653).

Harappan Phase copper alloys are especially difficult to define at present, given the lack of information on copper ore composition and processing technology. In this paper we will follow Agrawal in defining metal objects with 1% or more tin or arsenic as being alloyed. In the lower percentages, however, it is not possible to determine if the tin or arsenic alloy is the result of intentional manufacture or simply a result of the natural ore compositions (e.g., see Tylecote 1980). Some scholars suggest that regardless of the arsenic content, arsenical copper was derived primarily from arsenical copper ores (Pigott 1989).

Morphologically similar objects found at Harappan Phase sites are made from relatively pure copper, arsenical copper, and tin bronze. Possible patterns of alloying are obscured by the lack of a large sample, the absence of any sampling methodology, and the inconsistent manner in which samples from different sites have been studied. It should be noted that most of the objects analyzed were excavated before the introduction of stratigraphic controls, and the variations may have some chronological significance. Another obvious factor contributing to the apparent lack of consistency in alloying is the re-melting of a mixture of metal objects. The recycling of copper/bronze objects is indicated by the numerous caches of broken tools and metal scraps recovered from all of the major sites.

Given these problems with our sample, we prefer to discuss the use of copper and copper alloys as a single group, rather than create artificial divisions based on elemental composition. At this point in our study it appears that Indus metalsmiths did not follow a rigid system of alloying related to specific artifact categories. Furthermore, the lack of patterning seems to be the norm during this period throughout West and South Asia. For example Pigott et al. (1982:231) note that no apparent correlations exist between artifact categories and elemental compositions during any period at Tepe Hissar.

We may be unable to define patterns of alloying because the Indus metalsmiths used alloying for a variety of purposes—functional, aesthetic, ritual, and/or simply expedient. For example, the addition of tin to copper may have been done to increase strength and hardness for some objects, but may have been used to produce particular colors or fulfill ritual requirements in other objects. Or a mixture of alloyed scrap metals may have been the material available for a smith's selection—expediency is difficult to model archaeologically, but too common ethnographically to ignore. (See Lahiri [1993] for an excellent discussion of the variety of reasons for alloying in modern and historic South Asia.) This multiplicity of choice is hinted at by the types of finished objects with high tin contents from Harappan Phase sites. Two categories of objects are high in tin: tools or weapons such as chisels, daggers, and some "celts"; and ornaments such as bangles (Appendix B). When faced with the choice of desired characteristics, including hardness, color, shape, etc., the Indus metalsmiths may have chosen between a number of alternative means of producing a given result. For example, in some instances they may have relied on physical modifications such as forging to harden metal, while in other situations they may have chosen to produce a harder metal by modifying the composition of the metal through alloying. These choices would depend in part on the manufacturing techniques used, and on the stage of metal production (smelting, melting, casting of blanks, etc.) at which the end product was first visualized.

While there is no distinct pattern of alloying relating to specific artifact categories, there does seem to be a pattern in metallurgical traditions on a regional scale. This will be discussed in the Arsenical Copper and Tin Bronze sections below, as these regional patterns are most evident in the varying amount of arsenic in copper metal objects from different parts of the greater Indus region. These compositional differences probably result from the use of more than one copper source by the Indus peoples, rather than from different traditions of alloy use. Therefore, before further discussing copper alloys, the potential source areas for copper are described below, along with any evidence for their exploitation during the Harappan Phase.

COPPER SOURCES

There are three, possibly four, major regions that could have supplied the copper ores or processed metal used by the Indus metalsmiths (Fig. 5.1). The first is the combined area of Baluchistan and Afghanistan, to the west of the Indus Valley, which extends from highland Badakhshan to coastal Makran. This extensive region contains numerous copper deposits and appears to have the earliest evidence for copper processing. A second potential source of copper is the inland mountain range of modern Oman on the other side of the Arabian Sea. A third region, to the east of the Indus and Ghaggar-Hakra Valley, comprises the north-south oriented Aravalli mountain range of Rajasthan. Numerous concentrations of copper ores are found in these ranges along with zinc, lead, and silver ores. A fourth potential source may have been eastern Iran, but so far there is no clear indication that the Indus metalsmiths used Iranian ores or metal, so this source is not discussed here.

BALUCHISTAN-AFGHANISTAN

In the highland plateau west of the Indus Valley flood plains, numerous copper-working areas have been reported over the years, but the most impressive is the region of southern Afghan Seistan, often referred to as Gardan-i-Reg (Dales and Flam 1969; Fairservis 1952, 1961). Here in the windswept wastes of the Helmand Basin there are vast areas of exposed copper slag mixed with pottery and other cultural debris. Dales (1992) mentions that some of this slag was analyzed and contained 14% copper, and that the gold assay was also quite high, but most of the samples have yet to be studied. The copper ores processed at Gardan-i-Reg are assumed to be from nearby deposits, but no detailed report has been published on the mining areas.

The ceramics and other cultural material associated with the copper smelting debris of Gardan-i-Reg correspond to the Helmand Tradition (Shaffer 1992) at the sites of Mundigak, Shahr-i Sokhta (Period III), and Tepe Rud-e Biyaban (Periods II and III). The dating of the ceramics is disputed and while some scholars feel that they fall between approximately 2500–2400 B.C. (M. Vidale, pers. comm.), others suggest that they date to the period prior to 2600 B.C. (J.-F. Jarrige, pers. comm.). The copper smelting activity would be basically contemporaneous with either the late Regionalization Era ("Early Harappan") or the Integration Era, Harappan Phase of the Indus Valley Tradition (Tables 5.1 and 5.2). The occasional discovery of Indus Valley Tradition artifacts at sites in Baluchistan and Afghanistan indicates that there was movement of people and goods between this important mineral resource area and the greater Indus region.

Copper and iron ores that are rich in arsenic are found in limited distributions in Baluchistan (Agrawal 1971; SanaUllah 1940) and the Iranian plateau (Pigott 1989), but it is not clear if these ores were being exploited continuously or only at specific chronological periods. For example, Pigott et al. (1982) note that the arsenic and lead components in copper objects increase in the later periods at Tepe Hissar in Iran (Periods II and III: ca. 3600 to 1700 B.C.; Dyson and Remsen 1989:108–109) and suggest that this increase is due to selection by the metalsmiths. On the other hand this pattern could be the result of changing access to copper ores due to political or trade alliances, and not an intentional act on the part of metalsmiths.

OMAN

Major connections between Oman and the greater Indus region may be inferred from the presence of Harappan Phase artifacts and possible short-term Harappan Phase settlements in Oman (Cleuziou 1984, 1989; Cleuziou and Tosi 1989; Tosi 1982; Potts 1990), combined with the presence of shells from Oman at Indus Valley Tradition sites (Kenoyer 1983). By taking advantage of the monsoon winds, Indus or other maritime traders may have been marketing Arabian copper in the Indus Valley, Baluchistan, and Gujarat.

Much research has been conducted in the important copper mining regions of Oman and Iranian Baluchistan (this volume and Berthoud and Cleuziou 1983; Frifelt 1991; Hauptmann 1985; Hauptmann and Weisgerber 1980a, 1980b; Weisgerber 1981, 1983, 1984; Weisgerber and Yule 1989). Omani copper ores are similar to those of the Aravalli region of Rajasthan (below) in that they have little or no arsenic and have relatively high quantities of nickel, cobalt, and vanadium (Agrawal 1971:152, table 20). They are different from Iranian ores in that they have higher quantities of nickel, cobalt, vanadium, and chromium (Berthoud and Cleuziou 1983). However, in light of the use of arsenic impurities as a sourcing marker by Indus researchers (see below), it is important to note that copper slags and objects containing arsenic *have* been reported from copper processing sites in Oman (Hauptmann and Weisgerber 1980b:135, 137). As is discussed in the section on Arsenical Copper below, the sites in the Indus Valley flood plains may have imported Omani copper, but probably drew on at least one other source as well.

RAJASTHAN

The copper deposits in Rajasthan and the Aravalli mountain ranges have been discussed by SanaUllah (1940), Hegde (1965, 1969), Agrawal (1971, 1984), Asthana (1982), Agrawala (1984a), Hegde and Ericson (1985), Rao (1985), and Hooja (1988:38), but only a few analyses of ore samples have actually been published. Hegde and Ericson (1985:61) also present results from lead isotope analyses of copper ores from eight sites in the Aravallis.

Samples of ores from mines in Rajasthan (Khetri and Alwar), Bihar (Singhbhum), and Afghanistan were examined by SanaUllah, and all contained both nickel and arsenic. SanaUllah (1940:379) proposed that the Rajasthan (Aravalli) mines were the source for most of the metal used in the greater Indus region, because of their relative proximity to Mohenjo-daro and Harappa. SanaUllah did not publish his analyses of Aravalli ores, but Hegde (1969:227) notes that his "sample of Chalco-pyrite obtained from Khetri showed 4.28% of arsenic." In contrast, copper ore impurities from the region of Khetri as reported by the Director, Indian Bureau of Mines, to Rao (1985) are as follows:

Lead	Generally occurs as traces, highest percentage noted is 0.18%
Zinc	Generally occurs in the second decimal, highest percentage noted is 0.18%
Arsenic	Generally occurs in the fourth decimal, highest percentage noted is 0.06%
Cobalt	Around 0.01%
Nickel	Around 0.05%
Iron	15 to 20%

and Agrawal's (1971:table 20, fac. p. 152) analyses of chalcopyrite ores from Khetri (in Rajasthan) and Singhbhum (in Bihar) yielded less than 0.05% arsenic. The question of arsenic in the Aravalli copper deposits is discussed further in the following section.

At this point there is no direct evidence for Harappan Phase mines or smelting sites in the Aravalli copper resource areas, even though these areas have been explored by numerous scholars. The earliest well-dated copper smelting slags are from levels of Ahar dated to the early second millennium B.C. (Sankalia et al. 1969:10; Allchin and Allchin 1982:262; Hegde and Ericson 1985:60). Although Hegde and Ericson (1985) assumed that the smelting furnaces they found in surface surveys in the Aravallis are from the third millennium B.C., these furnaces have not been dated, either by radiocarbon or by associated artifacts. (This is not meant to detract from this very important survey work, but to clarify the dating problems.) If these sources were actually being exploited as early as the third millennium B.C., it is possible that the Indus peoples themselves were not involved in the mining and smelting. These activities may have been undertaken by local communities of the Aravalli region. The Ganeshwar-Jodhpura Culture in northern Rajasthan or the Ahar Culture in southeastern Rajasthan may in fact be some of these groups (Agrawala 1984b; Hooja and Kumar 1995).

However, many Harappan Phase sites *have* been reported in the nearby desert region of modern Cholistan, Pakistan, along the now dry bed of the Ghaggar-Hakra River (Mughal 1980) (Figs. 5.1 and 5.3). This region is close to the copper sources of Rajasthan, and Sir Aurel Stein recovered a copper ingot from Siddhuwala Ther, near Derawar. Many of the sites discovered by Mughal have kilns that were apparently used for "firing pottery, clay objects, bricks and perhaps smelting of copper" (Mughal 1980:96). However, there is no report of ores, slag heaps, or smelting furnaces, which would be required before classifying any of these as copper smelting sites.

With the availability of at least three different major source areas in easy reach, it is not unlikely that the larger urban centers used copper from more than one source over the 700 years of the Harappan Phase. Only future systematic studies will provide the necessary data to elucidate these sources, and the analyses of copper ores, ingots, slags, and metal prills on crucibles are particularly needed. However, the regional patterns of arsenic presence/absence already provide some evidence for the exploitation of more than one source of copper metal.

ARSENICAL COPPER AND SOURCES

Most of the discussion of sourcing of Harappan Phase copper metal has revolved around the presence or absence of arsenic, since it is usually assumed to be an impurity rather than a deliberate alloy, and thus indicative of the source of the copper.

At the Indus Valley urban centers of Mohenjo-daro and Harappa, the great majority of the objects that were analyzed contain at least trace amounts of arsenic, usually less than 1% (Appendices A and B; as noted, these do not include work by Desch or Agrawal). The overall composition of the copper items from the smaller Gujarati sites of Lothal and Rangpur is very different (Appendices A and B). While all four sites contain artifacts with variable amounts of nickel and iron, arsenic is noticeably absent at Lothal and Rangpur (Lal 1985; Rao 1963).

Thus, it appears that the two major cities situated in the actual Indus Valley flood plains were using copper derived from sources containing significant amounts of arsenic, or possibly alloying to produce arsenical copper. These sites were part of the major trade and exchange networks connecting the western highlands, the central plains, the eastern riverine areas, and the coasts of the Indian Ocean. The most probable source areas for arsenical copper ores are the mines of Baluchistan and Afghanistan, or possibly even eastern Iran. The Indus Valley cities may also have used copper from the Aravalli or Oman sources, and mixed these with arsenical copper objects through remelting or recycling, as these sources are usually represented as containing little or no arsenic (but see discussion of source areas, above).

The absence of arsenic from the finished objects at Lothal and Rangpur could be taken as circumstantial evidence for the exploitation of the Aravalli copper ores by the Indus peoples in Gujarat. However, Rao (1985:524) insists that the traces of arsenic in the Aravalli ores show that the arsenic-free Rangpur and Lothal copper was not coming from Rajasthan, but rather from Oman. Such a fine distinction is hard to support, given the values of less than 0.06% arsenic in modern Aravalli ores that Rao himself quotes, as well as the evidence for traces of arsenic in some Omani ores (above, Hauptmann and Weisgerber 1980b).

Data from sites in Rajasthan itself complicate rather than clarify the issue. The site of Ganeshwar is close to the Aravalli ore sources, within 10 to 15 km of copper mining areas at Ahirwala and only 75 km from the Khetri mines, and we might assume that copper objects from the site were being made from local ores. A single copper celt and a number of copper arrowheads (exact quantity unknown) from the site of Ganeshwar have been analyzed by the Geological Survey of India, Jaipur, with the following compositions reported: (1) for the celt: Cu 97%; Ag 0.2; Pb 1.0; As 0.3; Sn 0.01; Ni 0.6; and (2) for the arrowhead(s) (it is unclear whether this is an average of several samples, or the result from a single arrowhead): Cu 96.5%; Ag 0.3; Pb 0.03; As 1.0; Sn 0.2; Ni 0.04; Zn 0.25; Fe 0.2 (Agrawala 1984a; Agrawala and Kumar 1982). At least in these objects, tin and arsenic are present. These figures match the higher arsenic levels in the copper used at Harappa and Mohenjo-daro, rather than the arsenic-free copper used at Lothal and Rangpur (Appendices A and B).

We considered the possibility that arsenic was deliberately added, although this seems unlikely. Actual arsenical ores may have been traded to the smelting areas or even to the major cities of the greater Indus region for use in copper metallurgy. However, the only evidence for such arsenical ores are a few fragments of löllingite or leucopyrite found at both Mohenjo-daro and Harappa. SanaUllah (1931:690) notes that the fragments from Mohenjo-daro were heated, a necessary step to release arsenic, although he suggests the arsenic was used for medicines or poisons rather than copper alloying. In addition, the presence/absence pattern of arsenic in arsenical copper objects holds for the tin bronze objects from Harappan Phase sites as well. That is, the tin bronzes from Harappa and Mohenjo-daro often have high percentages of arsenic, while those from Lothal and Rangpur do not (Appendix B). This further supports the idea that the arsenic is an impurity in the copper from one of several sources, and was *not* intentionally added as an alloy in some parts of the Indus region.

We have no definite conclusions about the source(s) for the arsenical copper at this point; the possibilities are varied, since the data base is small. However, we cannot rule out the possibility that arsenical copper deposits within the Aravalli copper ore beds have been mined out, and previously contained ores that could have supplied the arsenical copper used at Mohenjo-daro, Harappa, and Ganeshwar.

In contrast, it is clear that Lothal, Rangpur, and probably most of the other sites in Gujarat were using copper derived from sources with little or no arsenic. Perhaps after all Rao was correct (although not for the reasons he cites), and the Gujarati sites imported copper from Oman rather than Rajasthan. It is significant that the sites in Gujarat were not consumers of the arsenical copper used by the major cities in the Indus Valley itself, and it will be interesting to see what the copper objects from major Saurashtran urban sites such as Dholavira are like. Probably the Gujarati sites were participating in different trade networks, dealing with peoples in Rajasthan and/or Oman, but not Baluchistan. The Indus Valley Tradition peoples in Gujarat would thus have used other techniques for making hard tools or decorative ornaments without resorting to arsenical copper. One such alternative would be the use of tin bronzes, discussed below.

TIN BRONZE AND SOURCES

Unlike lead and even arsenic, there are no known tin objects or tin minerals from Harappan Phase sites. Tin bronzes were definitely used by the Indus peoples, however, as is seen in Appendices A and B. If tin was being added as a separate metal to form copper alloys, it was carefully conserved and has not yet been discovered in the archaeological record. However, it is also possible that previously alloyed tin bronze ingots and scrap were traded to Indus peoples, rather than tin being traded as a separate metal. Future analyses of ingots and slags at Harappan Phase sites may help answer this question; for example, one of the copper "lumps" (ingots) from Mohenjo-daro analyzed by SanaUllah (1931:485) contained 12.13% tin. This could, of course, be a secondary ingot; the other four "lumps" analyzed all contained little if any tin, and considerably more sulfur (Appendix A).

From the large site of Mohenjo-daro only 24 analyses of copper metal objects have been published, and of these, 12 objects have more than 1% tin (Appendices A and B). At the present time 9 out of 29 copper metal objects analyzed from Harappa contain more than 1% tin. At Lothal, 71 out of the total of 1500 (metal?) objects recovered were analyzed (Lal 1985); this is a relatively large sample for an Indus site. Of the 64 copper metal objects published, few are alloyed with tin, and only 8 have more than 1% tin. Twelve Harappan Phase copper metal objects from Rangpur were analyzed, out of a total of less than 25 recovered. However, all of these analyzed samples from Rangpur have some trace of tin, and 7 out of the 12 objects contain more than 1% tin (Agrawal 1971; Rao 1963).

As discussed above, the fact that the tin bronzes from Lothal and Rangpur contain little or no arsenic indicates that these tin bronzes were being made locally, or imported from sources that were different from those supplying Harappa and Mohenjo-daro (Appendices A and B).

The major sources of tin used during the Harappan Phase probably derive from what is now modern Afghanistan. Some alluvial deposits are reported in western Afghanistan in the Sarkar Valley south of Herat (Berthoud and Cleuziou 1983) and major deposits occur in the central regions north of Kandahar (Pigott 1989; Stech and Pigott 1986; see Pigott, this volume, for more discussion). Other tin deposits occur in northern Afghanistan near the ancient lapis lazuli mines. It is unclear who was controlling the access to the tin resources during the third millennium B.C., but the largest settlements of the Helmand Tradition, Mundigak and Shahr-i Sokhta, are located at strategic points along the trade routes that would have connected these resource areas to the consumers in Mesopotamia. The Harappan Phase site of Shortugai is located at a northern source and may reflect a competitive situation where the Indus peoples chose to develop their own mining and distribution rather than rely on alliances with the sites of Mundigak or Shahr-i Sokhta. However, it is not clear if Shortugai was indeed a trading settlement for *all* of the different available minerals (Francfort 1989). Lapis lazuli and gold working is evidenced from the excavated materials, but there is no clear evidence for the processing of either copper or tin.

It is not unlikely that there was some overland trade of tin to Mesopotamia from northern Afghanistan through northern Iran and from Seistan through southeastern Iran (Moorey 1994:298–299). However, Mesopotamian texts sometimes refer to Meluhha as being a supplier of tin (Berthoud and Cleuziou 1983; Sollberger 1970) and this may indicate that some of the trade was conducted via the Indus Valley or along the Makran coast.

To interject a final note of caution, although the differences in copper alloy compositions between the Indus Valley sites and the Gujarat region sites appear quite striking, more conclusive interpretations must await a larger archaeological sample, further analyses of a wider range of elements, and comparative studies of ores, ingots, and slags.

LEAD

Numerous lead objects have been found at Harappan Phase sites and it is clear that lead was used as a separate metal. Small masses of metallic lead were found in the excavations at Chanhu-daro and a number of lead objects have been reported from Mohenjo-daro and Harappa (Mackay 1938, 1943; Marshall 1931; Vats 1940). One object from Mohenjo-daro described as a "net-sinker" (Marshall 1931:464) has a rough convex surface that appears to have been cast in sand. This object has recently been examined by Kenoyer and appears to be a plano-convex disc-shaped lead ingot. There is a perforation in the center and a lateral perforation that extends across part of the flat surface. Other forms of lead objects include vessels, such as a lead dish (Mackay 1938:pl. CXXVIII, 21), lead cones, and so-called "plumb-bobs" (Marshall 1931). Another use for lead is seen in the form of a rivet used to fill a hole in the bottom of a shell ladle (Dales and Kenoyer 1990). Lead may have been deliberately added to a few copper objects (see Appendices A and B) and may have been important for casting, as the addition of lead causes molten copper to flow more easily.

One lump of lead from Mohenjo-daro analyzed by Desch (reported in Mackay 1938:600) was composed of 99.7% lead and 0.15% copper and had traces of silver. The ores used to make Indus Valley Tradition lead could have been cerussite (lead carbonate) or galena, which is found in many regions of Baluchistan and Rajasthan (e.g., Ajmer) (Pascoe 1931). Craddock et al. (1989) describe in detail the lead, silver, and zinc ores of the Aravallis, in Rajasthan. (See Silver section below.) Cerussite was found at Mohenjo-daro, and a footnote mentions that powdered cerussite was found in a faience vessel at Harappa (SanaUllah 1931:691). Several fragments of what appear to be galena have been recovered from recent excavations at Harappa, along with an unidentified variety of lead ore (possibly lead and arsenic combined) (Griffin and Fenn in Meadow and Kenoyer 1992). Finally, cerussite and lead slag have been reported from Area D at the site of Nal in Baluchistan (Hargreaves 1929; Agrawal 1971:15). This area also has evidence for burned structures and it will be important to determine if the slags represent intentional production or accidental burning of lead minerals used for cosmetic or other purposes.

However, neither of the lead objects analyzed by Lal (1985:656) from Lothal contained silver, copper, iron, tin, or zinc: (1) lead piece 4280 contained 91.42% Pb, traces of Ni, 2.2% acid insoluble residues, and 6.38% oxygen (by difference); and (2) object 10092 contained 99.54% Pb, and 0.46% oxygen (by difference).

SILVER

Silver objects are not uncommon at Harappan Phase sites and practically every major excavated site has objects made of this metal. Silver was used primarily to make vessels that were similar to copper metal or ceramic forms. Silver ornaments are also quite common and include beads, bangles, and rings, as well as fillets and perforated discs. Marshall (1931) claims that silver objects were more common than gold, in contrast to Mesopotamia or Egypt, where silver was rarer. Asthana (1982:276) also notes that silver was much more common at Mohenjo-daro and Harappa than at Lothal and Kalibangan.

Nevertheless, only five samples of silver have been analyzed, two from Mohenjo-daro and three from Lothal (Table 5.3). They all contain significant traces of copper, and three contain lead. The sources of Indus Valley Tradition silver are not known, but on the basis of copper and lead traces in their samples, SanaUllah (in Mackay 1938:599) suggested that most of the silver from Mohenjo-daro was extracted from argentiferous galena. Pascoe (1931) notes the presence of silver mines in Baluchistan and Afghanistan, but to date no Harappan Phase extraction sites have been reported. Silver deposits in the Aravallis are described in Craddock et al. (1989), but again there is no evidence for exploitation until after the Harappan Phase; the earliest dated mines are from the second millennium B.C.

GOLD AND ELECTRUM

Gold ornaments or flakes of gold leaf have been recovered from most excavated Harappan Phase sites. All of the relatively complete pieces of gold ornaments have been recovered from hoards where objects have been stored in copper or ceramic vessels and buried within a house. Fragments of gold leaf or tiny beads are not uncommon in the excavations of Harappan Phase sites; the gold leaf may be derived from beads or other objects that were covered with decorative gold, and the tiny beads undoubtedly derive from broken necklaces. Only a few small gold beads have been recovered from Harappan Phase burials (Dales and Kenoyer 1990).

Very little of the gold recovered from Indus Valley Tradition sites has been subjected to chemical analysis. The earlier excavators used visual criteria to discrimi-

TABLE 5.3

ANALYSES OF SILVER METAL SAMPLES (HAMID AND SANAULLAH IN MACKAY 1938:480, 599; LAL 1985:656, 658)

	MOHENJO-DARO			LOTHAL†		
	DK 5774*	DK 11337,o**		5034	4176	15114
% Silver	94.52	95.52	% Silver	54.65	86.53	71.20
% Lead	0.42	1.40	% Lead	1.64	–	–
% Copper	3.68	3.08	% Copper	2.67	7.87	4.13
			% Iron	3.29	–	trace
			% Nickel	trace	–	–
% Insolubles	0.85	–	% Insolubles	3.06	2.32	16.29
			% Oxygen (by diff.)	34.69	3.28	8.38
TOTAL %	99.47	100.00	TOTAL %	100.00	100.00	100.00

*Also listed as DK 6129 (cf. Mackay 1938:480, 599); not clear which sample actually tested.
**Note that this is a derived estimate (see Mackay 1938:599).
†Tin and zinc were also tested for in the Lothal samples, but no traces were found.

nate between pure gold and a gold/silver alloy. The gold/silver alloy was thought to be either a natural electrum or an artificial alloy made by the Indus gold/silversmiths.

In the course of recent analyses of materials from Harappa by Kenoyer, one gold/silver object and four gold objects have been subjected to initial microprobe elemental analysis. One gold object from Allahdino has also been analyzed. All of the samples were examined with an electron microprobe (Kenoyer assisted by E. Glover, Department of Geology, University of Wisconsin-Madison) to determine the overall ratio of gold to silver, and one object was analyzed for copper as well. The proportion of gold to silver was between 91% and 94% in the five gold objects, but further analysis is necessary to determine if other elements are present. The gold/silver object from Harappa was a lump of partly melted and hammered metal visibly composed of gold and silver. The gold-colored portions had a high ratio of gold to silver and the silver-colored portions had a high ratio of silver to gold. This object was obviously in the process of manufacture, and may reflect a stage in the production of artificial gold/silver alloy.

Two "gold" objects from Lothal have been analyzed by Lal (1985:664–665) and contain 33.45% and 41.48% silver, but no copper, nickel, lead, or zinc. He concludes that the high percentage of silver and the absence of lead indicate that these items were made from electrum rather than an artificial mixture of silver (derived from galena) and gold. If the Indus peoples did use natural electrum, which has a relatively limited distribution in South Asia, it will be much easier to source this material. In contrast, gold has a wide distribution in alluvial deposits throughout South Asia, although mine deposits are more restricted.

There has been much discussion on the possible origin of Harappan Phase gold and, as with copper, there are several potential resource areas. While some earlier scholars considered the South Indian gold mines as a major source area, there is no conclusive evidence for trade between the Indus Valley Tradition cities and the Kollur gold-producing area of South India. The most obvious source of alluvial gold is the upper reaches of the Indus Valley itself and the streams of northern Afghanistan (Pascoe 1931; Stech and Pigott 1986). Significant quantities of gold are found in the tributaries of the Amu Darya, and the Kokcha river itself cuts through deposits that have gold ores. The most convincing indications to date of gold working at a Harappan Phase site have been found at Shortugai in Afghanistan, where the excavators found a fine globule (*gouttelette*) of gold imbedded in the cuprous vitrified internal surface of a crucible fragment (Francfort 1989:136).

IRON

No specific iron objects have been reported from Harappan Phase sites, but there are a few objects with iron components from sites in Afghanistan that were roughly contemporaneous with the Harappan Phase. While these few occurrences indicate that iron was known and used, Shaffer (1984:48–49) may be overstating the case when he concludes that in the late third millennium in Afghanistan, iron was being used to make luxury items and "iron ore was a culturally recognized and valued item, selected for its hardness and functional utility."

Three items incorporating manufactured iron were found at the site of Mundigak, an important urban center that probably controlled the trade from central Afghanistan of tin, copper, and possibly gold (Casal 1961). At Mundigak the iron was always combined with copper/bronze objects, and it appears to have served an ornamental or symbolic function. These objects include a small copper/bronze bell with an iron clapper, a copper/bronze rod with two iron decorative buttons, and a copper/bronze mirror handle with a decorative iron button (Shaffer 1984).

Two other sites in Afghanistan demonstrate the use of unprocessed iron. At Deh Morasi Ghundai a single utilized magnetite nodule was recovered in association with what Dupree (1963) referred to as a shrine complex. At the site of Said Qala Tepe, 28 specular hematite nodules were found that appear to have been used as hammerstones, but Shaffer (1984) suggests that they may also have had a socio-ritual function. (See Fabrication section below for the possible association of such objects with metalworking, at least in Iran.)

In the Indus region itself, ferruginous lumps or possible iron objects have been reported, but where analyses have been conducted (Lal 1985) there is no evidence for actual manufactured iron objects (see Slags section below). On the other hand, the Indus artisans were quite familiar with the properties of iron minerals (limonite, hematite, magnetite, etc.), using them in pigments and slips for ceramics and steatite, and perhaps for coloring faience glazes as well.

OTHER METALS

At this time there is only a little evidence for the use of other metals. Antimony is found in "appreciable proportions" in some copper metal objects at Mohenjo-daro and Harappa, almost always in copper objects containing greater than 1% tin (Appendix A) (SanaUllah 1931:485). Several pieces have also been found at Harappa as an unworked mineral, but mostly from surface contexts (and so are perhaps from the modern fair held at the site). Zinc is also found in traces within a few copper objects at Harappa, and is even present in greater than 1% in two cases (SanaUllah 1940, Pigott et al. 1989). Note that where modern techniques of analysis were used (Pigott et al. 1989), zinc was found in amounts greater than 0.5% in every artifact tested (Appendix A). It seems that the zinc was an impurity in the original copper ores at Harappa, in contrast to Lothal and Rangpur (Appendix A). Finally, cinnabar was found at Mohenjo-daro (SanaUllah 1931:691), but the context is not given. Cinnabar may be the deep red coloring that appears in some shell and steatite inlay (Kenoyer, on-going research).

HARAPPAN PHASE NON-FERROUS METAL PROCESSING

Overall, the Indus metalsmiths appear to have been familiar with the techniques used to process the major metals and alloys, except iron and brass, and we will briefly summarize the important metallurgical processes, techniques, and artifact types in this section.

Given the problems with identifying the uses of firing structures at Indus Valley Tradition sites, as described below, as well as the fact that metal processing is only one of the pyrotechnologies we investigate, we prefer the more general term "kiln" rather than using "furnace," a term specific to metal processing. A detailed discussion of the terms referring to "slags" is presented in Miller (1994b). Other specific terms used in this section will be defined as they are introduced.

Except for the site of Shortugai, where there is evidence for gold processing, most of the indicators for metal processing at Harappan Phase sites are associated with copper processing. These indicators are assessed below, grouped according to the various stages of metal processing (Fig. 5.4; see Miller 1994b).

The major indicators for metal processing at a site include: (1) fragments of ores; (2) kilns, or fragments of kilns, attributed to metal processing; (3) metallurgical slag, from the reduction of ore to metal; (4) tools used for metal processing, such as crucible fragments with metal prills, molds, anvils, stakes, hammers, chisels, etc.; and (5) metal objects, including smelting and melting ingots, semi-finished and finished objects.

Significant amounts of ores and/or metallurgical slag fragments are the most convincing evidence for smelting at a site. A variety of clay-based non-metallurgical slags, including fragments of kilns and crucibles, can represent either the smelting of ores or the melting of metal, depending on the exact nature of these indicators (Bayley 1989; Craddock 1989; Miller 1994b). Metalworking tools (other than crucibles with prills or molds) are usually difficult to attribute directly to metal processing, except when found in well-documented contexts in association with a number of other metalworking indicators. Metal objects, including metal ingots, must also be from such contexts to show their production at a site, as these objects may have been imported from other sites.

METAL PRODUCTION: SMELTING

No evidence for the processing of copper has been found during the Neolithic period in the greater Indus region. The initial stages of copper processing (extraction of native copper or smelting) were probably being carried out closer to the source areas (probably Baluchistan and Afghanistan). The question of how this technology evolved is still unclear and will not be fully understood until new surveys and excavations are carried out. Nevertheless, by the fourth and third millennia it is clear that mining and smelting of ores was being carried out in many localities throughout

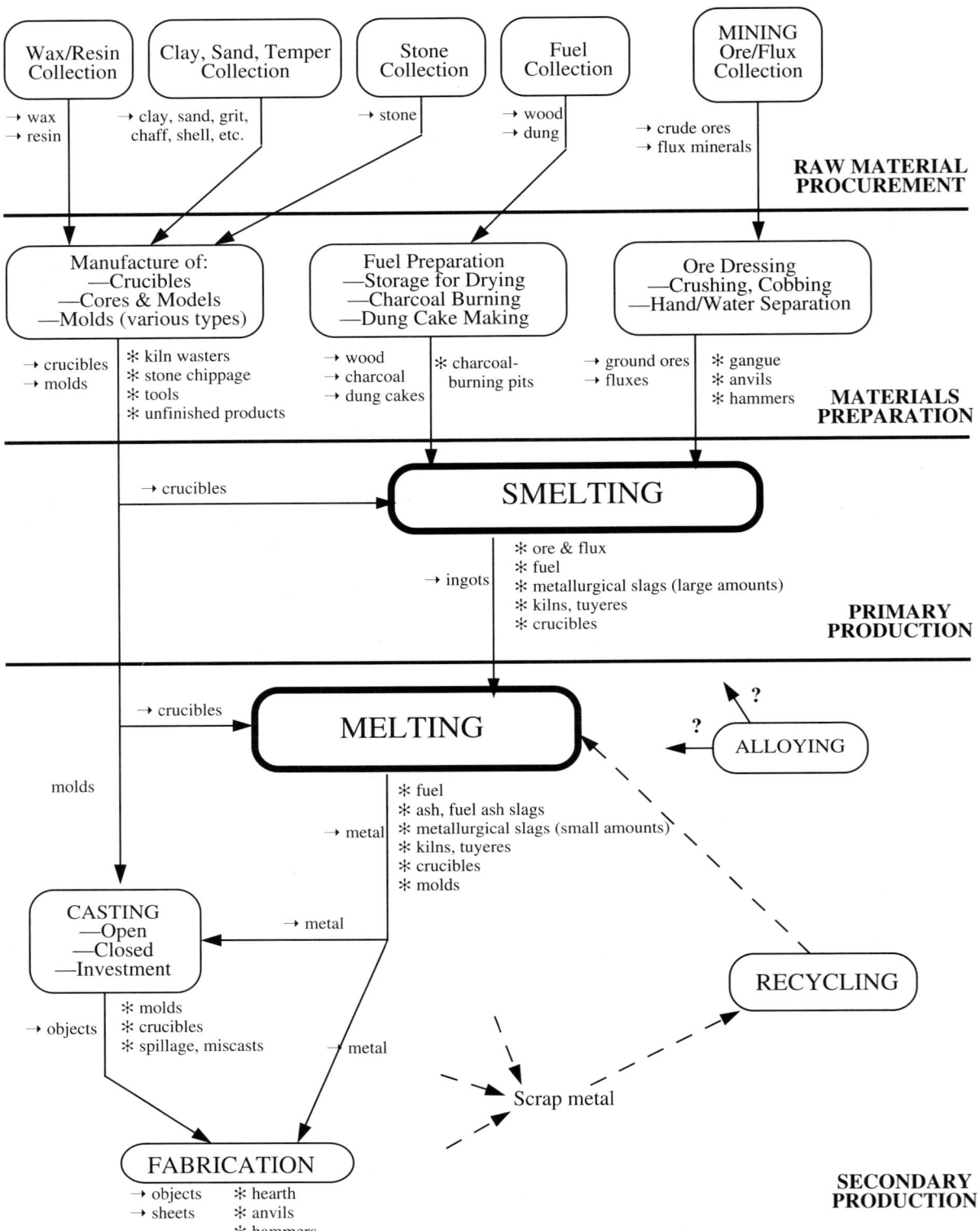

Figure 5.4 Generalized reconstruction of non-ferrous metal processing, showing end-products (→) and by-products (✶) (from Miller 1994b).

Baluchistan and Afghanistan (Dales and Flam 1969; Fairservis 1952, 1961; Jarrige and Tosi 1981; Stech and Pigott 1986) (Figs. 5.1 and 5.2).

The paucity of metal ores from Indus Valley Tradition sites was noted above; there is a similar lack of metallurgical slags. Smelting of metal ore usually results in fairly conspicuous accumulations of manufacturing debris and broken furnaces. In particular, the smelting of most ores results in the production of weather-resistant metallurgical slags, vitrified masses of silica, and other fused minerals, which generally accumulate in conspicuous mounds near the smelting furnaces. On the basis of quantity and type of slag, the small amounts of copper metal slag found at Indus Valley Tradition sites seem to be more representative of melting rather than smelting (Miller 1994b). It is not clear if the gold processing at the site of Shortugai (above) involved gold smelting or simply gold working. The often cited copper smelting slags from Ahar are from levels dated to the second millennium B.C., after the end of the Harappan Phase (Sankalia et al. 1969:10; Allchin and Allchin 1982:262; Hegde and Ericson 1985:60).

The absence of slag heaps and smelting debris may be due to the fact that most excavated sites, particularly of the Harappan Phase, are located some distance from the primary sources of metal ores. Further collections and analyses are needed in the Aravallis, as well as in the regions of Gardan-i-Reg and Cholistan, to determine if there are Indus Valley Tradition or contemporaneous smelting sites in these regions. Archaeologists working in Rajasthan have made a laudable effort to look for copper slags, but specialist surveys are needed. The surveys done by Lowe in South India to locate and analyze iron-working sites, and the work done by Craddock and others on the silver and zinc mines of Zawar, Rajasthan, provide good models for future studies of third millennium copper smelting (Lowe 1989a, 1989b; Craddock 1989, 1991; Craddock et al. 1989).

In view of the present evidence, we must concur with SanaUllah's (1931:485) and Mackay's (1938:451) earlier interpretations that most of the copper used at the Harappan Phase sites was imported in ingot form, and little or no smelting was done at the major sites of the Indus Valley Tradition. Plano-convex disc-shaped ingots (almost all of copper) have been recovered from Chanhu-daro (Mackay 1943:188), Lothal (Rao 1979), Harappa (not reported but found in the reserve collections), and Mohenjo-daro (SanaUllah 1931:485; Mackay 1938:451). Although relatively few ingots were reported in these publications, recent examination by Kenoyer of the reserve collections for Harappa, Mohenjo-daro, and Chanhu-daro (Museum of Fine Arts, Boston) have revealed the presence of additional examples. Many of these ingots in the reserve collections at Harappa and Mohenjo-daro appear to have been smashed and broken in half or into smaller wedge-shaped fragments, apparently to facilitate further processing. Mackay (1938:450–453) also noted that several of the ingots and "castings" (secondary ingots) from Mohenjo-daro had chisel or saw marks partway into the ingot, which was then broken.

MELTING OF METAL

Melting of metal is a necessary stage in the production of cast objects, both semi-finished and finished. The remelting of original smelting ingots to produce secondary or refined ingots is also a common intermediary stage between the production of the original smelting ingot and the final fabricated or cast object. This secondary ingot production is undertaken for one or more reasons: to remove slaggy impurities left in the original smelting ingots; to break up large smelting ingots into more workable or transportable ingots; to form metal alloys; and/or to melt down metal scrap, which is usually varied in composition.

For example, SanaUllah (1931:484–485) remarked in the very first published analysis of Harappan Phase ingots (his "copper lumps"), that three of the four ingots were the "crude product of the smelting furnace" which were too rich in sulfur to be forged. He suggested that such ingots would be remelted for refining; otherwise this metal could only be used for casting "heavy or plain objects."

As noted above, much of the archaeological evidence for melting (and casting) resembles the evidence for smelting, i.e., the presence of crucibles, kilns, slag, and metal ingots. However, careful analysis of these objects, particularly identification of the types of slags present, usually allows differentiation of these processes (see Miller 1994b for a full explanation and references).

CRUCIBLES

Evidence for melting of copper in crucibles is first found during Period III at Mehrgarh, dated to 4000–3500 B.C. Crucibles with traces of melted copper have been found in rubble associated with a Period III firing structure made of brick (Jarrige 1983, 1985a). Jarrige (1985a:289) notes that the crucibles are comparable in interior size and shape to the ingot recovered from Lothal.

Only a few crucible fragments with copper prills have been reported from Harappan Phase sites, and although they could represent small-scale smelting of high-quality copper, copper melting seems more likely; further archaeometric tests may allow definite discrimination between the two processes (see Miller 1994b).

A crucible rim fragment was found during excavations at Mohenjo-daro, and SanaUllah (1931:485 and pl. CXLII, 9) gives the opinion that it represents the re-melting of crude metal for refining. No analyses were ever published, but the photograph of this rim shows slagging, and looks quite similar to fragments found by the Aachen/IsMEO surveys at Mohenjo-daro (below). A single crucible is also reported from the excavations at Harappa; Vats (1940:471) mentions in a footnote that "a fragmentary earthenware crucible whose contents show that it was used for melting bronze" was found near furnace "Fa" on Mound F at Harappa. Again, no analyses were ever published. Finally, the so-called "crucibles" reported from Lothal must be some other kind of container, as they had no traces of slagging and upon analysis contained no trace of copper (Lal 1985:661; see also section on Molds below).

The evidence from surface surveys and recent excavations at Harappa and Mohenjo-daro is much more encouraging. At Harappa, half a dozen crucible rim fragments with slagging and copper prills have been found in surface surveys of the southern slope of E Mound, and a number of both rim and base fragments of similarly slagged crucibles have been found in nearby excavations (Miller 1994a, 1994b). At least two crucible fragments with slagging and copper prills were also found in surface surveys at Mohenjo-daro, as well as an entire small ceramic cup (Pracchia et al. 1985; Tosi et al. 1984). This cup is similar to the crucibles used by modern goldsmiths in South Asia, but there are no visible traces of metal or slagging, and microprobe analyses found no traces of metal (M. Vidale, pers. comm.). All of the crucible fragments from Harappa, and one of the two from Mohenjo-daro, show a quite small diameter; the second fragment from Mohenjo-daro is much larger. Based on the associated types of slags, these seem to represent melting crucibles, not smelting crucibles, but archaeometric research still has to be done.

KILNS

Four kilns at Harappan Phase sites have been attributed to copper-processing: one at Harappa (Vats 1940:470–473, fig. d, pl. XVIIa and b), one at Lothal (Rao 1979:522), and two at Mohenjo-daro (Mackay 1938:49–50, 452). In all these kilns the interior is heavily vitrified, and the blackened melted surface indicates a reducing atmosphere. This evidence of high temperatures seems to have been the primary reason for attributing these kilns to metal processing. However, evidence of high heat alone is not sufficient to show that a kiln was used for metal processing. Many of the excavated Indus Valley Tradition pottery kilns show similar vitrification (e.g., Dales and Kenoyer 1990), as did experimental mudbrick pottery kilns fired at Harappa during the past seven seasons. Temperature data taken during the experimental firings show that the vitrification of the ceramics and the walls of the kiln occurred somewhere in the range of 1000 to 1100°C, in what was usually a heavily reducing atmosphere (Meadow and Kenoyer 1992; Kenoyer 1994).

Kiln "Fa" at Harappa was one of 16 pyrotechnological structures excavated by Vats in a relatively small area of Mound F. Contrary to much of the literature, this kiln was the only one of the 16 that Vats ever genuinely suggested might be related to metal processing. When first excavated, half of the original structure was present (split vertically), and the interior was well preserved due to heavy vitrification. The kiln is described by Vats as a pit 3 ft. 4 in. (1 m) in diameter and 3 ft. 8 in. (1.1 m) to 5 ft. 3 in. (1.6 m) deep. The interior was covered with a sand-tempered mud plaster (Vats 1940), and recent observations reveal that it was at least partly constructed with mudbrick (Miller in Meadow and Kenoyer 1992). Vats found evidence for a vaulted roof with four "flues" remaining, and a fifth "flue" entering at a slightly lower level. Using a modern Punjabi metal melting furnace as a parallel, Vats (1940:471) proposed that the latter flue was "used as an air channel worked by bellows from above," and further suggested that the roof flues were intended as outlets for smoke and/or inlets for fuel that could be closed during operation. No copper prills or copper slags were found in Vats' original excavations or in subsequent surface surveys by the present authors (Dales and Kenoyer 1990 and Miller in Meadow and Kenoyer 1992). Consequently, upon closer examination there is no evidence to support the implication that kiln "Fa" was used for metal processing.

The circular kiln at Lothal was 6 ft. (1.8 m) in diameter, 2 ft. 3 in. (0.7 m) deep, and made of mudbrick with mud plaster. The use of this kiln for melting copper ingots was attributed from its location near the so-called "coppersmith's workshop" (Rao 1979:522). However, the kiln itself showed no signs of copper melting. Indeed, most of the evidence for use of this area for copper melting has been refuted: the analyzed terra-cotta crucibles/molds show no signs of copper or other metals, the "slag" turned out to be corroded copper metal, and the "stone molds" for casting are more likely to be for grinding of stone beads or other materials (see discussions elsewhere in this section). At present there is no published evidence for copper melting at Lothal.

The two kilns at Mohenjo-daro were approximately the same depth and diameter at the top, but varied in size at the base: 4 ft. 3 in. (1.3 m) deep, 3 ft. 3 in. (1.0 m) in diameter at the top, and 2 ft. 10 in. and 3 ft. 2 in. (0.9 m and 1.0 m) in diameter at the base (Mackay 1938:49). They were paved with wedge-shaped brick and covered with mud plaster, with a 4 inch (10 cm) wide ledge on the inside (at different heights), perhaps to support a crucible or grating (Mackay 1938). The interior walls of these kilns were highly vitrified, but again there is no mention of metal slag, prills, or crucibles associated with these kilns. Mackay (1938:50) himself expresses the opinion that these kilns from Mohenjo-daro were used for "something more fusible than copper, owing to the lack of a draught or vent in their lower portions." Although in another section he contradicts himself, saying that the furnaces *were* for copper working, based on the find of three discarded copper/bronze castings in a nearby area (Mackay 1938:452), in later writings Mackay (1948:94) states firmly that there are no known copper smelting kilns from Mohenjo-daro.

Finally, we should mention the "circular furnace pit, 1.5 m in diameter and about 0.6 m deep, full of white ashes and metal-worker's slag (pl. IX, fig 3)" from Ahar, according to Misra (1969:300). Misra reports that this pit was found in the lowermost level of the site, just above virgin soil. However, we were unable to find any mention of such a pit in the Ahar site report (Sankalia et al. 1969), and the only slags mentioned in that report are from Phase Ib and Ic levels (second millennium B.C. levels)—no slags are reported from the Phase Ia levels, tentatively thought to be contemporaneous with the end of the Harappan Phase (see below, Slags section). In any case, such a pit is very large for a furnace, and as no signs of burning are noted, probably represents a dump.

In summary, based on the published data, there are no known metal processing kilns from any Harappan Phase sites, contrary to secondary sources claiming copper smelting furnaces at Harappa. However, new research is beginning to produce the types of data needed to define the presence of copper processing kilns. For example, many kiln wall fragments with embedded copper prills have been found on the southern slope of Mound E at Harappa (Miller 1994a, 1994b), and M. Vidale (pers. comm.) reports the discovery of a small circular kiln on the surface of Mohenjo-daro between VS and Moneer Areas. This kiln was approximately 0.8 to 1 m in diameter and was surrounded by hundreds of copper prills and overfired pieces of clay with copper traces.

SLAG

There is no mention of metal processing slag from any of the early excavations at Harappa or Mohenjo-daro. At the site of Lothal, several objects originally identified as copper slag have been analyzed and were found to be lumps of corroded copper metal (Lal 1985:659–661). On the other hand, two possible candidates for copper or iron processing by-products are listed in Lal's (1985:658) tables of analyses; namely, two unshaped objects composed primarily of iron and copper:

	15211 (Lump)	15212 (Cu frag.)
% Cu	43.1	9.3
% Fe	39.1	66.1
% Sn	–	–
% Pb	–	–
% Ni	1.5	tr
% Zn	–	–
% Ag	–	–
% Acid Insol.	11.5	17.2
% O by diff.	4.8	7.04
TOTAL %	100	100

No information is given about the archaeological contexts or phases of these objects, however, so it is not clear if they are even from the Harappan Phase. Similarly, as noted above, the only well-dated copper slags from Rajasthan were found at Ahar in levels dating to the second millennium B.C., after the Harappan Phase (Sankalia et al. 1969:10; Allchin and Allchin 1982:262; Hegde and Ericson 1985:60).

A few tantalizing statements have been made in preliminary reports on the site of Banawali, in Haryana (Bisht 1982, 1984). Bisht (1984:90–91) mentions "a few contiguous rooms" with reddened floors, "several hearths, ovens and fire-pits," and "pieces of copper and copper slags," and concluded that this area was probably used by metalsmiths. If subsequent analyses show that these are indeed copper slags, this information from Banawali will make a welcome addition to the scanty information on metal processing, especially as these rooms are attributed to the Sothi or pre-Harappan (Kalibangan I) levels of Banawali (Bisht 1984:90).

Also, as with the crucible fragments, several discrete scatters of metal processing slags have been found during surveys at Harappa (Miller 1994a, 1994b) and Mohenjo-daro (Pracchia et al. 1985; Tosi et al. 1984). These slags are primarily vitrified clay-based materials with copper prills; metallurgical slags are quite limited in number and do not appear to constitute the type or quantity of slag associated with smelting (Miller 1994a, 1994b). Based on both the characteristics of these slags and the relatively small amount of copper slag found at the excavated Harappan Phase sites, they are more likely to be by-products from the re-processing of copper ingots or scraps than from the original smelting of ores.

It is highly unlikely that copper processing was not being carried out in these large cities and in smaller settlements, but future research is needed to recover the types of indicators needed to define different stages of metal processing.

OBJECT PRODUCTION: CASTING AND FABRICATION

The production of metal objects can be divided into two categories, casting and fabrication, depending on the state of the metal during the actual working. Casting is done when the metal is molten and fabrication is undertaken when the metal is not molten. These two categories divide the methodologies of metalworking artisans as well as the states of the metal itself. Fabrication involves the direct shaping of metal, while casting begins with the shaping of other materials into which the molten metal is poured.

The tools and techniques of the two categories overlap to some degree, and ancient metalworking ateliers may have been involved in both fabrication and casting. Some objects, however, may have been cast by one group of artisans and finished or fabricated by another group in a separate workshop. The possible division of manufacturing stages into discrete and often exclusive activities practiced by different artisans is an important part of metalworking that has not been investigated for the Indus Valley Tradition primarily because metal production areas have not yet been conclusively identified. The implications of such separation of activities are touched on in the final section of this paper.

CASTING

By defining casting as the shaping of molten metal we include a wide range of metalworking activities, including the production of secondary ingots as well as the casting of semi-finished or finished objects. Evidence for casting thus includes all of the indicators discussed above in the section on Melting, as well as several types of indicators directly related to casting, such as molds, semi-finished objects, and finished objects. (Lahiri [1993] notes the importance of recognizing the presence of secondary ingot casting in South Asian metalworking.)

The best evidence for casting activities at a site is the presence of molds. Ancient mold types include open stone, terra-cotta, or sand molds; bivalve stone, terra-cotta, or sand molds; and terra-cotta-based "lost model" molds. Horne (1990) suggests the term "lost model" rather than "lost wax" since the technique employs other materials besides wax, such as tallow, resin, and tar. At present, no convincing examples of any type of mold for casting metal have been reported from Harappan Phase sites. (Note: we have been unable to find any reference to "open stone molds" reported from Chanhu-daro, contrary to Agrawal's [1984] statement.)

The only published stone "molds" from a Harappan Phase site are from Lothal, where Rao identifies two grooved stones as open molds for casting pins/rods (Rao 1979:557, 568, fig. 121, nos. 3 and 4, pl. CCLIIb). The first problem with Rao's identification is that the grooves are much larger than the metal pins found at Harappan Phase sites. Second, the grooves are not straight along their length or sides and taper off in depth at each end, which seems highly impractical and unlike known casting molds from other regions and periods. Third, there is no report of discoloring or spalling, which would occur if molten metal were repeatedly poured onto a stone surface. Similar stones (made of sandstone and quartzite) with long parallel grooves have been found at Harappa and Mohenjo-daro, and at Chanhu-daro they are clearly associated with agate bead manufacturing areas (Mackay 1943:213–214, pl. XCIII). Our opinion is that the grooved stones from Lothal probably do not represent open stone molds for metal casting.

The only published terra-cotta molds are also from Lothal, and might be fragments of open molds, although Rao calls them "crucibles" (Rao 1979:pl. CCXIIIa, b, c). Scrapings collected from the interiors of these "crucibles" gave negative tests for tin, lead, nickel, arsenic, and copper, even upon repetition of the copper test (Lal 1985:661). Future test results showing the presence of some metallic traces would be necessary to define these objects as molds for casting metal (but see Bayley 1989:298–299). Finally, there are no reports of sand molds of any kind nor of fragments of lost model molds.

Paradoxically, the few metallographic analyses that have been done on Harappan Phase copper/bronze objects indicate that some tools were definitely made by open or bivalve casting (Agrawal 1971; Pigott et al. 1989; SanaUllah 1940; Wraight 1940). In addition, the modeled forms of the famous metal figurines of animals and humans can be attributed to lost model casting. For example, two exquisite objects reported from Chanhu-daro that may have been made by lost model casting have often been overlooked: a covered cart with the wheels missing, and a complete cart with a driver holding a goad (Mackay 1943:39, 41, pl. LVII, 1 and 2). A similar model cart was also recovered in excavations at Harappa (Vats 1940:27). Some of the small vessels may also have been made by lost model casting (Mackay 1938:446; 1943:176).

Due to the low percentages of lead found in his analyses, SanaUllah (1940) did not think that the Indus peoples used lost model casting techniques. While unalloyed copper is difficult to cast, it should be pointed out that lead is not a prerequisite for lost model casting. Other alloys may be used (e.g., copper-tin), as most alloys are usually fluid enough to cast well. Furthermore, the types of objects tested by SanaUllah (see Appendix A) were unlikely to have been made by lost model casting in any case; no figurines or vessels were tested.

The presence of cast objects and no molds could be taken as an indicator of importation of finished metal objects, but this seems unlikely. Many of the cast Harappan Phase metal objects, particularly the figurines, are definitely Harappan in style, corresponding to the morphology and subject matter of objects in other materials such as faience, stone, and ceramic. There is no evidence for Harappan-style metal objects or molds from sites outside of the greater Indus region; indeed, Yule (1985b) notes that Harappan Phase metalware is very different in style from contemporaneous products in other regions.

It is more likely that this paradox is due to the lack of identified metal processing areas at Indus Valley Tradition sites and also perhaps a problem of the archaeological *identification* of molds. The Egyptians and Mesopotamians obligingly left molds of stone, clay, and even metal (Moorey 1985; Scheel 1989); the Indus peoples were not so helpful. Given the scarcity of stone in the Indus Valley flood plains, it is possible that stone molds were reused for other purposes. However, this seems extremely unlikely, as the detailed examination of every fragment of stone and fired ceramic recovered from five seasons of excavations at Harappa, as well as examination of the Mohenjo-daro reserve collections, has not revealed any fragments of stone or ceramic molds (Kenoyer, on-going research). Another explanation is that the Indus peoples used mold materials, such as sand or sandy clays, that leave little trace in the archaeological record.

Two techniques using sand or sandy clay molds are sand casting and lost model casting, both of which leave almost no archaeological traces. Sand-based molds are used for casting in modern South Asia and many other areas of the world (Untracht 1975), and employ a finely powdered sand, sometimes mixed with water and organics such as dissolved sugars. This mixture can be used to make an open mold or packed into a hinged wooden box to make a bivalve mold. A form made of wood or some other material is impressed into the sand mixture, which is cohesive enough to create a mold into which the molten metal is poured. It is well suited for flat objects, such as celts, axes, adzes, knives, or spears. It may even leave characteristic bivalve lines on the objects, often taken to be indicative of the use of stone or terra-cotta bivalve molds.

Since forms are used to impress the sand and create the mold, some degree of duplication of objects is possible, and creation of the sand molds is obviously quite rapid. Although these molds have a great resistance to heat, making them an excellent casting material, they break down quickly into sandy deposits when exposed to weathering from water and wind. In addition, modern sand molds are usually ground and reused, and ancient molds would probably have been similarly recycled.

The materials used to make lost model molds are also quite ephemeral. These materials form a continuum with the fine sticky sand used for sand casting, but employ a more cohesive sandy clay so as to better retain the complex three-dimensional features of the object to be cast. Lost model casting often employs several grades of material. First, the model of wax, resin, tar, etc., is coated with a fine sandy clay. This inner coat will form the details of the object to be cast, so the finer the detail desired, the finer the texture of the coat. Increasingly coarser sandy clay is used to form the bulk of the mold. The crucible containing the metal can be built into the mold, as is done in India and Africa, or metal can be poured into the mold from a separate crucible, as was done in the Americas and Egypt (see Emmerich 1965; Fox 1988; Fröhlich 1979; Horne 1990; Reeves 1962; Scheel 1989 for ethnographic and historic descriptions of lost model casting).

An essential component of the lost model process is the use of a sandy clay that will not sinter under high temperatures. Thus, in addition to the fact that the molds are broken to remove the cast object, such molds also break down very quickly when exposed to weathering. As with sand casting, the broken pieces of the mold are also often recycled, increasing their archaeological invisibility.

The ability to produce molds rapidly and to recycle the mold materials is one of the great advantages of sand and lost model casting over stone-mold casting. While this is beneficial for the artisan, it is a nightmare for the archaeologist. Perhaps with increasing awareness of these methods and their ephemeral remains, archaeologists will begin to look more closely at patches of sand for the tiny fragments that may still retain the contours of cast objects.

FABRICATION

Fabrication of metal objects includes all of the various types of modification of non-molten metal: shaping, via forging, turning, and drawing; cutting; cold and hot joining; and finishing via planishing, filing, polishing, coloring, engraving, and so forth. The metal can be worked while cold or hot, but at a heat below the molten state. Ingots (either secondary or smelting ingots) can be worked either into sheet form or directly into a finished object. Cast semi-finished objects, such as those found by Mackay (1943:175) at Chanhu-daro, also would be subsequently worked to form the finished object.

SHAPING

In its broad sense, shaping is the controlled mechanical stretching of metal. This includes stretching by forging, including sinking and raising; by spinning or turning; and by drawing.

Forging. The most common form of shaping is forging, "the controlled shaping of metal by the force of a hammer," usually on an anvil or stake (McCreight 1982:36). Whereas the term "hammering" is sometimes used for non-ferrous metals, and "forging" reserved for ferrous metals, forging is the term most often used by coppersmiths themselves, and will be used here.

The hammer and the anvil or stake can be made of a variety of materials, such as metal, stone, wood, bone or horn, or even leather. Finds of such tools, or of the marks left on objects by such tools, comprise one type of archaeological evidence for forging. The other main source of evidence for forging comes from metallographic examination of artifacts. Forging can be done while the metal is hot or cold. Forging not only shapes the object, it also hardens it, and so forging is an important step in the manufacture of edged tools. Annealing is the re-heating of an object after working, and most metalworking involves cycles of annealing and hot or cold "hammering" (forging). Metallographic studies help to establish the use of such methods.

There is some evidence for forging during the Harappan Phase, both from finds of tool marks and from metallography (Agrawal 1971; Mackay 1938; Pigott et al. 1989; SanaUllah 1940; Wraight 1940). The Indus peoples forged objects both from castings, including bar or block ingots and semi-finished shapes, and from sheet metal. For example, chisels, which are one of the most numerous tool types, appear to have been forged from cast bars (e.g., Vats 1940:87–88). Metallographic analyses of other edged tools confirm that at least some of these tools were cast and then forged (Agrawal 1971; Pigott et al. 1989; SanaUllah 1940; Wraight 1940). Thin razors are thought to have been cut from copper sheets, then forged to a sharp edge (Mackay 1938, 1943).

Sheet manufacture is a type of forging, and fragments of copper, silver, and/or gold sheets have been found at Chanhu-daro (Mackay 1943), Harappa (Vats 1940), Lothal (Rao 1979, 1985), and Mohenjo-daro (Mackay 1931, 1938; Marshall 1931). The Indus method of sheet manufacture is unknown, and there are no published anvils or hammers. However, many of the noncubical "weights" from Harappan Phase sites are very smooth cylindrical or semi-hemispherical stones with highly polished surfaces. These types of objects might well have been used for metalworking, and the highly polished surface would be necessary to avoid marring the metal. The reports of hematite and magnetite "hammerstones" from Afghanistan (Dupree 1963; Shaffer 1984) are particularly interesting, given the use of these materials in copper metal sheet manufacture in western Iran (Pigott, this volume). Use-wear studies of these stone objects may help to clarify their function. It is also possible that wood or other perishable materials were used. At this time, the most hopeful method for elucidating the methods and materials used by the Indus peoples for forging is through the analysis of objects for tool marks, particularly on the better-preserved gold objects. Experimental studies aimed at discriminating marks left by various tool materials (e.g., stone, wood, metal) will also be useful.

Sinking and *raising* to form vessels from metal are also types of forging. As the names imply, sinking is the forming of metal by hammering from the interior of an object into a depression in an anvil, while raising employs hammering from the exterior of the object over a shaped stake or form. There are a number of ways to raise objects: both from sheets and from ingots directly, while the metal is cold and while it is hot, by an individual artisan or by a group working together. The Indus peoples are known to have raised vessels (Mackay 1938:chap. XIII; Vats 1940), and appear to have made some jewelry by similar forming techniques, particularly the gold "cones." Unfortunately, no studies have been done to show what techniques of raising or sinking were employed.

Hollow metal tubes of copper or copper alloys were also made. The method of production is not exactly known, but the tubes are seamed and were produced from sheets, and so were forged in some fashion (Mackay 1943). The edges do not overlap, but rather abut, and there is no mention of the use of solder in the published accounts. The handled pans (Mackay 1931, 1938, 1943) also show the production of tubes by raising, in that their handles are tubes made by lapping over the opposite sides of the metal. Two pieces of a long, one-inch diameter copper tube were also found at Mohenjo-daro, but no details of its construction are given. Marshall (1931:198) refers to this tube as a "coppersmith's blow pipe," but gives absolutely no support for such a conjecture.

Decorative uses of forging include the use of stamps or punches, hammering of thin metal sheet (often gold) into or over patterns, and chasing. Hammering into or over patterns is not noted so far, but simple circular punch marks were used in the decoration of gold objects and fillets (e.g., Mackay 1938:526, no. 4; Marshall 1931:527, pl. CXVIII, 14), and sheet metal was beaten out from one side to make designs in at least one piece (Vats 1940:64, [d] no. 8). There are no published reports of true chasing (the working of metal from both sides).

Spinning and turning. Spinning and turning are methods of mechanical stretching with results similar to sinking and raising but using a lathe rather than a hammer. While Mackay originally suggested that some of the vessels from Mohenjo-daro may have been made by turning (1931), no lathe marks or evidence for a lathe have yet been discovered to support this conjecture, and Mackay seems to discard it in later writings.

Drawing. Wire production has not been studied at all, although there are abundant examples of copper "wire" and silver "wire." This is probably due to the poor preservation of the metal, so that details of manufacture are not discernible. There is no evidence at present to indicate whether Harappan Phase wire was drawn or forged, although the latter is far more likely for the third millennium B.C. Wire was used to make rings, possibly for adorning the fingers, toes, and ears, as well as for other purposes. Some stone beads have pieces of copper wire inserted through the hole, possibly in imitation of the method seen with gold wire, so that the bead could be hung as a pendant from a larger composite necklace.

When corroded wire remains inside a bead it will often split the bead into two fragments, making the bead look as if it was broken in manufacture. Occasional reports of copper "drill bits" found stuck in beads may be such fragments of corroded wire. For example, Jarrige (1985b) mentions that a long barrel-shaped lapis lazuli bead from Mehrgarh, Period VII, with a "copper bit" still inserted in the hole, was broken in two during the drilling process. This may instead be a fragment of the copper wire used for stringing the bead. However, further investigation of these reports are needed, as the possible use of copper for drill bits is an important unanswered question for the Indus Valley Tradition (Kenoyer and Vidale 1992). Jarrige's (1985a:290) discussion of the helicoidal bronze rods from Lothal, Mundigak, and perhaps Mehrgarh are of great interest for this topic. These rods are exactly like modern auger drills, and the rods from

Lothal and Mundigak still retain their pointed tip. Jarrige (1985a:290) also notes that the helicoidal rod from Mundigak corresponds in diameter to the hafting holes in wooden handles found at Shahr-i Sokhta.

CUTTING

The Indus peoples are thought to have cut many of their thin objects out of sheet metal. Cutting was probably done for the most part with chisels (Mackay 1938:chap. XIII, and many other cases); the V-shaped or double-edged chisels, which are the most common type, can be used to cut metal. Most of the chisels analyzed to date are of tin bronze or arsenical copper, and therefore relatively hard. In addition, there are some saw marks in metal objects (Mackay 1938:368, 452, 475, 583), although Mackay was unable to distinguish if these were from an ordinary blade saw or a wire saw. From the evidence of both chisel and saw marks, the usual procedure seems to have been to cut a groove in the metal mass on one or more sides, then snap the piece in two. Similar methods are seen in Indus Valley Tradition stone working.

JOINING

The number of identified Harappan Phase methods of joining metal is limited. There are very few examples of cold joining, comprising primarily the use of rivets to attach metal handles to metal vessels (Marshall 1931). Hot joining is represented by evidence for soldering of gold and silver (SanaUllah 1940; Rao 1979), and by one example of pouring molten copper metal over a join to attach a copper handle securely to a copper vessel (Mackay 1931:489). (Hot joining is considered a fabrication technique even though the joining material is molten, because the body of the metal is not molten.) There are as yet no published examples of soldering of copper materials, but this is probably due to the generally poor state of preservation of the copper metal and the lack of analysis. For example, a large cooking vessel found at Chanhu-daro appears to have been joined at the carination. However, it is not yet clear if this join was cold or hot joined, due to the poor preservation of the object (Kenoyer, on-going research at the Museum of Fine Arts, Boston).

One possible example of hot joining without solder is seen in a gold tubular bead from Harappa, measuring 33.95 mm in length and between 2.14 and 2.41 mm in diameter. The thickness of the gold wall of the tube is between 0.34 and 0.39 mm and there is no visible longitudinal seam. Further studies of the bead are needed to confirm the precise techniques, but one possible method of manufacture was reproduced in silver by a skilled jeweler in Karachi (Kenoyer, on-going research). A long silver wire was wrapped around a solid rod, and the wire was fused into a tube by vigorous burnishing and annealing. If this were the manufacturing method of the ancient tubular bead, casts of the interior of the gold tube should reveal whorls along the circumference.

Another find from Harappa perhaps related to joining is the discovery of a tiny snippet of gold that is a swollen pillow shape (4.35 mm length, 1.39 width, 1.58 mm thickness). This type of object is often seen in the process of manufacturing granules of gold for use in granulated decoration. However, such granules are also a standard by-product of soldering, for the intentional or unintentional heating of a tiny gold, silver, or copper fragment often results in such a shape. The essential step defining granulation is not the *making* of the granules, but the *attachment* of the granules. At present no ornaments with gold granulation have been found from Indus Valley Tradition sites.

FINISHING

There have been few discussions of Harappan Phase metal finishing techniques (Mackay 1938). While this is understandable for the copper-based objects, which are generally heavily corroded, the silver and especially the gold objects should provide evidence for Indus methods of finishing. Finishing techniques which need to be investigated include the planishing of forged (especially raised) objects; filing and polishing to smooth surfaces; engraving; inlay and stone-setting; and surface coloration via plating or enrichment.

Mackay (1938) notes the presence of hammer marks on many of the vessels, but does not indicate if this represents *planishing*. (Planishing refers to fine, even hammering with a highly polished hammer, used particularly to create a smoother surface on forged objects.) Evidence for *polishing* comes from the gold objects from recent excavations at Harappa, such as the copper and gold beads described below, which show minute striae oriented in groups. Polishing materials could have included ground and polished stone objects, perhaps even magnetite nodules (Pigott, this volume), sand- or silt-sized powders, plants, or even leather. *Engraving* to produce designs or accentuate details is known from the figurines and the numerous inscriptions on copper tablets and other metal objects (Yule 1985a, 1985b). Stone burins or metal engravers may have been used for this technique, and could probably be identified through use-wear studies. Finally, a number of *inlaid* pieces are found among the Indus jewelry, although the techniques of setting have not yet been studied.

At present there is no evidence for *surface coloration* via plating or chemical enrichment. However, the mechanical wrapping of gold sheet over copper and paste beads is known from Mohenjo-daro (Mackay 1938) and Harappa (Dales and Kenoyer 1990), as well as from Allahdino (Fairservis, pers. comm.). At Harappa, the gold sheet used for wrapping the core ranges in thickness from 0.07 to 0.1 mm in thickness; it appears to have been only wrapped and hammered around the core without having been actually fused to it.

More precise information on finishing, and manufacturing techniques in general, would benefit greatly from the restudy of all objects in a standardized fashion, using xeroradiography of the objects and selected metallography. Methods of manufacture would give us more information on the diffusion and independent invention of metal processing techniques in the Indus Valley Tradition. It would also allow investigation of possible regional styles of production both within the Indus Valley Tradition (e.g., Indus Valley vs. Gujarati sites), and in contrast with other Traditions (e.g., Indus Valley vs. the Rajasthani cultures, or even Mesopotamian groups).

THE ROLE OF METAL DURING THE HARAPPAN PHASE

In order to better understand the role of metal objects in Indus society we need to have detailed information on the contexts in which metal objects have been found. The vast majority of metal objects available for study come from the earlier excavations and we have little information on the stratigraphic or chronological context of the artifacts within each site. Nevertheless, some generalizations can be made on the basis of the current evidence. Based on the published reports from sites in the greater Indus region (including parts of Afghanistan, Baluchistan, northwestern India, and Gujarat) and firsthand observations of some unpublished material, we have made a preliminary tabulation of the use of metal and non-metal materials to manufacture specific types of objects (Table 5.4). This table is preliminary and by no means comprehensive, but some of the patterns seen in these rough data are discussed below.

ORNAMENTS AND MIRRORS

The use of copper as a form of ornament has a long history in the greater Indus region and can be traced back to the early levels at the site of Mehrgarh, where there is evidence for a single copper bead from the Neolithic levels (Period IB), circa 6000 B.C. (Jarrige 1983, 1985b). Several copper ornaments have been reported from the subsequent layers (Jarrige 1983), but these objects have not yet been analyzed, so details of composition and manufacture are still unknown.

Deposits of Period III at Mehrgarh (4000–3500 B.C.) also contained corroded fragments of rods and pins. Two double spiral-headed pins show that the discovery of such pins in the later Harappan Phase sites no longer indicates connection with the much later Namazga III sites in the western and northern highlands. A fragmentary copper object from one grave of Mehrgarh Period III may be a compartment seal, again indicating the presence of these objects in a context that pre-dates the Quetta-Namazga III complexes (Jarrige 1985b).

The excavations at Mehrgarh and other sites of the Regionalization Era provide clear evidence for the gradual increase in importance of metal between the Early Food Producing and the Regionalization Eras. During the Harappan Phase of the Integration Era, it is evident that metals supplement rather than replace the earlier materials used to manufacture ornaments (Table 5.4). In fact, many metal ornaments are copies of beads, pendants, or bangles made in non-metal materials. There are some unique new types of metal ornaments, however, specifically those that incorporate gold, silver, and electrum. Metal also comes to be used in composite ornaments with other valued or symbolic materials, such as faience, various colored stones, and shell.

At present, the best clue we have to the role of metal ornaments in Indus society is their archaeological context. Unlike contemporaneous sites in Mesopotamia, almost all of the complete ornaments (e.g., necklaces and belts) found at Harappan Phase sites have been recovered from hoards rather than from burials. Fentress (1977) tabulated the metal objects from Mohenjo-daro and Harappa that were found in hoards vs. those found in non-hoard contexts, including burials (Table 5.5).

Although Fentress' table needs to be updated and to include other sites, it presents some noteworthy patterns. Copper/bronze ornaments as well as copper/bronze tools have been recovered primarily from non-hoard contexts. In contrast, gold and silver ornaments and silver vessels have been found almost exclusively in hoards. It is interesting that copper/bronze *vessels* have been found almost equally in hoard and non-hoard contexts.

When considering the distribution of ornaments, it is thus necessary to discriminate between copper/bronze and metals such as gold and silver (Table 5.5). Gold and silver ornaments have been found stored in ceramic, copper, or silver vessels that appear to have been deliberately hidden away. Some of these hoards include broken ornaments and melted lumps of gold or silver that would undoubtedly have been remelted and made into new ornaments. The hoards often contain numerous stone beads made from agate, carnelian, jasper, turquoise, and other varieties of colored stones.

Copper beads and spacers are also included with some of the hoards (e.g., Allahdino; Fairservis, pers. comm.), but copper ornaments have primarily been recovered in non-hoard contexts, such as in the debris accumulating in the streets or habitation areas, or in some of the burials. Out of 168 total copper/bronze ornaments reported, 130 were found in non-hoard contexts and only 38 were found in hoards, generally in association with gold and silver ornaments. The

TABLE 5.4
METAL AND NON-METAL OBJECTS OF THE INDUS VALLEY TRADITION (COMPILED BY J. M. KENOYER)

OBJECT CATEGORIES	EARLY FOOD PRODUCING ERA NEOLITHIC		REGIONALIZATION ERA* EARLY CHALCOLITHIC		INTEGRATION ERA HARAPPAN PHASE	
	NON-METAL OBJECTS	METAL OBJECTS	NON-METAL OBJECTS	METAL OBJECTS	NON-METAL OBJECTS	METAL OBJECTS
VESSELS	ceramic, stone	– none –	ceramic, stone	– none –	ceramic, stone	copper/bronze/silver/lead
MIRRORS	– none –	– none –	– none –	– none –	– none –	copper/bronze
ORNAMENTS						
beads, pendants	stone, shell, bone, ceramic	copper/gold	stone, shell, ceramic	copper/bronze/silver/gold	stone, shell, ceramic	copper/bronze/silver/gold
bangles	ceramic, shell	– none –	ceramic, shell	copper/bronze	ceramic, shell	copper/bronze/silver/gold
pins/awls	bone, antler	– none –	bone, antler	copper/bronze	bone	copper/bronze
finger/toe rings	– none –	– none –	? shell	copper/bronze	stone, shell, ceramic	copper/bronze/silver/gold
fillets/head bands	? beaded, fabric, leather	– none –	? beaded, fabric, leather	– none –	? beaded, fabric, leather	gold
composite/inlay	stone	– none –	shell, stone, faience	– none –	shell, stone, faience	gold, silver
TOOLS						
arrowheads	stone (microliths)	– none –	stone (microliths)	copper/bronze	– none –	copper/bronze
points	bone, antler	– none –	bone, antler	copper/bronze	bone, antler	copper/bronze
axes	stone	– none –	stone	copper/bronze	– none –	copper/bronze
adzes	stone	– none –	stone	copper/bronze	– none –	copper/bronze
chisels	stone	– none –	stone	copper/bronze	– none –	copper/bronze
drills	stone	– none –	stone	– none –	stone	copper/bronze
blades	stone	– none –	stone	copper/bronze	– none –	copper bronze
spear points	stone	– none –	stone	copper/bronze/silver	– none –	copper/bronze
small knives/razors	stone	– none –	stone	copper/bronze	stone	copper/bronze
fishhooks	– none –	– none –	– none –	– none –	– none –	copper/bronze
sickles	stone blades	– none –	stone blades	– none –	stone blades	– ? none –
saws	denticulated stone blades	– none –	denticulated stone blades	– none –	denticulated stone blades	copper/bronze
INSCRIBED/DESIGNS						
seals, tokens, plaques	stone, bone, ivory, ceramic	– none –	stone, shell, bone, ivory, ceramic	copper/bronze	stone, shell, bone, ivory, ceramic	copper/bronze
FIGURINES						
animal	ceramic, stone	– none –	ceramic, stone	– none –	ceramic, stone	copper/bronze
human	ceramic, stone	– none –	ceramic, stone	– none –	ceramic, stone	copper/bronze

*Sites of the Nal Phase have copper and silver tools and weapons similar to those found in the Harappan Phase of the Integration Era. There are problems with the actual dating of Nal burials and settlements.

TABLE 5.5

MOHENJO-DARO AND HARAPPA: METAL OBJECTS FOUND IN HOARD AND NON-HOARD CONTEXTS (BASED ON FENTRESS 1977:243, TABLE 28)

Vessels	Hoard	Non-Hoard
copper/bronze	39	28
silver	3	0
lead	0	1
Ornaments		
copper/bronze	38	130
gold	2,133	5
silver	47	4
electrum	0	2
Tools		
copper/bronze	72	314

gold and silver ornaments found in non-hoard contexts are usually tiny beads or gold foil fragments that were probably lost in the muddy streets or courtyards.

Although very little metal was buried with the dead, burials, like hoards, provide a context in which metal ornaments were intentionally placed by the Indus peoples. Metal objects found in burials are almost all of copper/bronze, and include mirrors, finger rings, bangles, and occasional beads. In one instance three gold beads were found, strung together with three stone beads. While the mirrors are invariably placed with female burials, the other metal ornaments have been found with both male and female individuals (Dales and Kenoyer 1990). It should be noted that no utilitarian copper/bronze tools have been found in the burials.

The occasional pilfering of burials (see below) cannot be used to explain the low incidence of metal in undisturbed burials, especially the absence of elaborate gold and silver ornaments. Other ornaments that are noticeably absent are those made from exotic materials or involving complex production techniques, e.g., carnelian, faience, or stoneware. This pattern suggests that ornaments that represented wealth or status were passed on from generation to generation and recycled, much as is done today in the subcontinent (Kenoyer 1992b).

The presence of mirrors in the burials is intriguing, as they were made of a considerable amount of metal that could have been recycled. Metal mirrors are a new object during the Harappan Phase, as mirrors were not previously made in any material, either polished stone or metal (Table 5.4). Perhaps mirrors were needed by women in the afterlife, or the personal use of a mirror made it an object that could not be passed on to another individual at death. However, mirrors are not found in *all* female burials. One possibility is that such beliefs may have been important to certain groups within Harappan Phase communities, while others held different beliefs. However, another possibility is that this difference may reflect looting by grave diggers. Many of the burials recently excavated at Harappa were disturbed by Harappan grave diggers in the course of digging pits for later burials, and shell bangles and copper mirrors were generally missing from the disturbed female burials (Dales and Kenoyer 1990).

The fact that precious metal vessels and ornaments have been found primarily in hoards suggests that, unlike copper/bronze metal objects, they were used more overtly to define wealth, status, and power. The types of ornaments depicted on figurines may represent gold, silver, and stone ornaments that had symbolic as well as ornamental value. However it is curious that, unlike Mesopotamia or Egypt, the Indus terra-cotta or stone figures are never depicted holding metal tools or weapons as symbols of status or power. The only pictorial or glyptic representation of the use of weapons or tools is seen in the pictographic signs of the undeciphered Indus script, or in narrative scenes on seals or molded tablets.

TOOLS AND VESSELS

As noted above, the earliest use of metals by Indus peoples is for ornaments or amulets and not for functional tools. During the Regionalization Era at sites such as Mehrgarh, Nausharo, Balakot, Jalilpur, Rehman Dheri, etc. (Fig. 5.2), we see an increase in the use of copper to make tools as well as ornaments. These copper/bronze tools are used in conjunction with the stone or bone tools and supplement rather than replace them (Table 5.4).

During the final Phases of the Regionalization Era in both the Indus Valley and Baluchistan, there is evidence for the introduction of new forms of copper/bronze tools that anticipate the major surge in metal tool production during the subsequent Harappan Phase of the Integration Era (see Yule 1985a, 1985b; Herman 1984 for catalogues of Harappan Phase metal objects; Haquet 1994 for work in progress on a catalogue for Baluchi sites).

During the Harappan Phase, some types of tools that had previously been made of stone do become totally replaced by metal tools (Cleland 1977). The most important include the barbed arrowheads (probably originally made with geometric microliths—Agrawala 1984a), spearheads, axes, adzes, hoes, chisels, and large blade tools. Copper fishhooks may be a new type of tool, as so far no hooks have been identified in bone or shell, or they may replace earlier hooks made of bamboo or some other perishable material.

Interestingly enough, however, there are still some tool categories where non-metal materials are supplemented rather than replaced by metal versions (Table 5.4). These tool categories include stone drills used for perforating a variety of materials, stone blade engraving tools (burins and engravers), various denticulated blades used as saws or scrapers, unmodified stone blades used as knives or scrapers, and a range of stone and bone points or awls.

The first metal vessels, primarily in copper, also appear during this time, supplementing but not replacing previous materials. Many of the metal vessel forms are similar to terra-cotta prototypes, with the "cooking pot" being a predominant form. Yule (1985b:25) notes that of the examples available for his study, "about one half of the types/forms exist in pottery, but many are peculiar to metal." It is not clear how many of the unique shapes in metal are due to the variations that result from the techniques of manufacturing metal, rather than the desire to make new forms. However, the distinctive long-handled pan is a definite example of a new form, as even short handles are almost unknown for Indus Valley Tradition ceramics.

Finally, some uses of metal which might be expected do not occur. The primary example is the continuing use of chert blades to make sickles during the Harappan Phase; only one possible metal sickle (or dagger) has been found, a curved copper blade from Mohenjo-daro (Mackay 1938:471). Even if other more convincing examples of copper sickles do turn up, it is clear that this type of tool was made predominantly by hafting stone blades. A distinctive type of denticulated sickle blade continues to be used in the later Localization Era, e.g., at Pirak (Jarrige and Santoni 1979).

Many scholars specifically focusing on the replacement of some stone tools by metal have concluded that the role of metal (i.e., copper/bronze) during the Harappan Phase was primarily utilitarian (Cleland 1977, Shaffer 1982). However, this conclusion is based on the distribution of *all* metal objects at Harappan Phase sites; ornaments and tools are not separated.

For example, at the small rural site of Allahdino the apparently utilitarian nature of metal objects is further supported by their recovery in all areas of the excavations and not in just one area of the site (Shaffer 1982). In a comparison of artifact distributions at Mohenjo-daro and Harappa, Fentress (1977) notes that the highest frequency of metal objects at Mohenjo-daro was in the so-called "Lower Town" area and not in the so-called "citadel" mound. In contrast, she finds that metal objects are distributed evenly throughout the site of Harappa (Fentress 1977). Recent excavations at Harappa suggest that this type of approach to the presence or absence of metals in different sectors of the large urban centers is not appropriate for understanding the role of metal in the different areas. The main problem lies with site formation processes at Harappa and Mohenjo-daro, where garbage was being dumped in empty structures and platforms were raised by filling them with trash from other areas of the site. Eventually, this process could result in a fairly even distribution of metal objects regardless of where they were originally used or discarded.

Nevertheless, these general observations on the use and distribution of metal objects has led Shaffer to conclude that "metal artifacts were manufactured for use in daily activities and were available to a broad segment of Harappan Phase society, urban or rural" (Shaffer 1982:47). The widespread use of metal tools is thought to reflect a pattern that is distinct from Mesopotamia and Egypt, where metal was generally associated with elites (Hoffman and Cleland 1977). At this point no quantitative data have been given to support this conclusion, and given the current investigations of non-elite contexts in Mesopotamia, it is not unlikely that our understanding of metal use in West Asia will change.

The fact that metal tools are found throughout most sites does not necessarily mean that all segments of the population had access to metal tools. It is possible that the use of metal tools was limited to specific groups or individuals who themselves may have been distributed throughout the settlements (Kenoyer 1989). Given the lack of more precise distributional or chronological data, we can only say that some types of copper/bronze tools and other utilitarian objects appear to have been generally available both at the large urban sites and at smaller rural settlements.

It is clear that metal objects were not simply utilitarian or symbolic, but that they played a variety of roles in the economy, technology, and socio-ritual/ornamental aspects of the Harappan Phase. As we continue to study the role of metal in early urban societies, it is important to remember that a piece of metal in itself does not indicate status or power. It is the context in which metal is used that is most important for understanding its role in a specific society. One clue to such contextual uses may be gleaned from the inscriptions on metal objects.

HARAPPAN PHASE INSCRIPTIONS ON METAL OBJECTS

Although it has not been possible to make an exhaustive study of the types of metal objects on which inscriptions occur, the recent publications of inscribed objects from the Indus Valley Tradition sites (Shah and Parpola 1991; Joshi and Parpola 1987; Parpola 1994a, 1994b; Yule 1985a, 1985b) reveal some interesting patterns. Large axes, adzes, spears, chisels, and sheets of copper often have one or more signs chiseled into one or both sides. The inscriptions are usually in a vertical line down the center but in the case of some celts, they are located at the butt edge. In most instances the script would be obscured by hafting or damaged during use, and this suggests that these metal tools may have had some specific ritual or symbolic function.

One category of metal object that was used almost exclusively for inscriptions is flat, square to rectangular copper tablets. At Mohenjo-daro hundreds of these inscribed tablets have been recovered (Marshall 1931; Mackay 1938; Yule 1985a). The inscriptions consist of Indus script and animal motifs, usually on both faces of the tablet. The engraving may have been done with either a stone burin or a bronze graver. So far this type of engraved tablet is unique to Mohenjo-daro, but copper tablets with raised script were found at Harappa by Vats (Vats 1940) and also by the Harappa Archaeological Research Project (Meadow and Kenoyer 1994). Due to heavy corrosion, the techniques of manufacture have not yet been determined. The script on both types of tablets was not written in reverse and therefore these tablets were not intended to be used as seals, but represent some sort of ritual or economic token (Parpola 1992).

Usually, Harappan Phase seals are made from fired steatite, but there are two examples from Mohenjo-daro of silver seals made in the standard square shape with a boss (Mackay 1938:348, pls. XC, 1, XCVI, 520). Mackay suggests that they were first cast with the animal motif, script, and boss, and later touched up using a graver. The edges have been scraped and pared, but the rest of the surfaces are too corroded to determine the exact nature of manufacture (Kenoyer, on-going research).

Careful examination of some gold ornaments from a jewelry hoard in DK-E area at Mohenjo-daro by Kenoyer has revealed the almost invisible traces of inscriptions that have been overlooked by previous scholars. Of particular interest are gold pendants that have often been referred to as "needles" (Marshall 1931:251–253, 521, pl. CLI, B 3, 4, 5). Two of three pendants are inscribed. The third pendant is currently on display at the National Museum in Delhi and it has not been examined, but it too may be inscribed. On one pendant the five signs encircle the entire object, while on the second example a sequence of five different signs is inscribed along the length of the pendant. From this same hoard came two pairs of gold caps or terminals that would have been affixed to the ends of beads. One of each pair of gold caps has an identical inscription consisting of a single sign.

All of the inscriptions appear to have been made by the same sharp and very pointed tool, and give the impression of being written in the same "handwriting." These inscriptions are extremely important because they are clearly different from the types of inscriptions found on the large copper celts and chisels. These inscribed gold objects were found inside a copper cooking pot that had been covered by a copper plate. Other items in this hoard include a massive belt made of carnelian and bronze beads, gold ear studs, stone beads, and some copper vessels. It is possible that before this wealth was hidden away, the names of specific owners were incised on some of the gold jewelry (Kenoyer, on-going research).

CONCLUSION

We will conclude this long summary of data relating to Indus Valley Tradition metal processing with a brief outline of selected future challenges for work on Indus Valley Tradition metals. High archaeological and analytical priorities include the search for metal processing areas, both at Indus Valley Tradition sites and in the resource areas, and the investigation of production techniques through systematic analyses of objects, slags, and ores. Among the highest theoretical and comparative priorities are understanding the organization and control of metal craft production, and the relationships between Harappan Phase peoples and the copper-using, apparently contemporaneous cultures in Rajasthan.

ARCHAEOLOGICAL AND ANALYTICAL CHALLENGES

As noted above, recent surface surveys have located a few metal processing areas at Harappa and Mohenjo-daro (Miller 1994a, 1994b; Pracchia et al. 1985). But given the fumes, smoke, and fire associated with metal processing, the most likely location of melting and casting areas would be on the periphery of settlements. Very few metalworking areas have been found on the high, mounded portions of the Harappan Phase settlements, and therefore the majority may be buried beneath the alluvium or at the edges of the site. It is also possible that the infrastructure of Harappan Phase craft production encouraged the establishment of "villages" of specialized producers, such as the site of Chanhu-daro, which seems to be a specialized lapidary production site. The currently unexcavated sites in Cholistan and on the Pakistan-Indian border may well supply evidence for a great deal of metalworking, as hinted at by large-scale surface surveys (Mughal 1982). The implications of these sorts of spatial distributions form part of on-going dissertation research by Miller (1994a) on the organization of high-temperature manufacturing areas in relation to both the physical structure and the socioeconomic structure of the Harappan Phase cities.

Systematic compositional and physical analyses of objects, slags, and ores are needed for the determination of Indus Valley Tradition metal sources, and of production techniques. The study of Indus Valley Tradition ore procurement would be greatly enhanced if further detailed investigations of copper ore deposits could be conducted in the Aravalli and Baluchistan regions, as has been done in Oman and for non-copper ores in the Aravallis (see section on Ore Sources above). Some work is in progress on the investigation of production techniques, via examination of finished objects (Pigott et al. 1989 and on-going research) and examination of by-products and manufacturing sites (Miller 1994b and on-going research). However, a great deal remains to be done on all stages of processing.

THEORETICAL AND COMPARATIVE CHALLENGES

The Indus Valley Tradition peoples were talented craftspeople and technological innovators, and the study of craft production is producing important data on many aspects of their economy, society, beliefs, and political structures. Metal production has been one of the least studied industries, and further work on metals will be a welcome addition to on-going discussions about craft production, particularly examinations of control of various crafts.

It is difficult to discuss control for the Indus Valley Tradition, due to the still elusive nature both of Indus elites and of their power base(s). The majority of Indus craft production studies have so far focused on segregation/aggregation of craft production stages, and their use in defining control of craft industries (Bondioli and Vidale 1986; Kenoyer 1989, 1992a; Pracchia et al. 1985; Tosi 1984; Vidale et al. 1992; Wright 1991; summarized in Miller 1994a). These discussions suggest that some Harappan Phase crafts were apparently structured on the basis of social networks and were decentralized in terms of state control. Other crafts may have involved long-distance social networks and alliances that could be decentralized in terms of direct political control, but required some centralized support to maintain long-distance trade relations. It has also been suggested that crafts that were difficult to control directly may have been less important for state economy, while easily controlled crafts could have been important for state economy (Kenoyer 1991, 1992a). The location of the various metal processing stages will form an important new data set for these segregation/aggregation-based discussions.

Other aspects of craft production used to indicate control of production have received less attention for the Indus Valley Tradition (summarized in Miller 1994a). This is in part due to inherent difficulties with the data, including the lack of written texts as well as past methods of excavation and record-keeping, and geomorphological transformations of deposits. Nevertheless, some information about degrees of control can be gleaned from considering such aspects of craft production as the distribution and association of various craft production areas, and the degree of restriction of access to such areas. Standardization of products, although not simply indicative of "attached" vs. "independent" production (Costin 1991), also provides information about the producers and the structure of production. Costin's discussion of different types and degrees of control will be helpful in the continuing process of developing more appropriate interpretive models for the complex Indus Valley Tradition situation.

Finally, some of the more intriguing results of Indus Valley Tradition metal studies will be the opportunities for comparison of metal use and production both within the Indus Valley Tradition (regional variants) and with contemporaneous groups to the west and to the east.

The comparison of production techniques and uses of metals within the Indus Valley Tradition itself will be extremely important. Through such approaches we can begin more meaningful discussions of independent and derived invention, shared and restricted knowledge, the adoption of technological innovations, and technological style, including the deliberate choice of methodologies as a component of ethnic identity (cf. Lechtman and Steinberg 1979).

On an inter-regional level, while many of the diffusion models that were proposed in the past have been refuted, there is still an underlying assumption that the Indus artisans were the recipients of knowledge from the highlands to the west, which they then passed on eastwards. Current investigations of information exchange and control of knowledge are gradually breaking away from the bounds of directionality that dominated the diffusion models of the past. In addition, the use and production of copper objects by groups in Rajasthan, some of which may still have been heavily dependent on hunting and/or shifting agriculture, reminds us that technological "stages" like the working of metal cannot be equated with other levels of social complexity, such as urban/non-urban. The comparison of metal processing techniques between culturally very distinct regions known to have communicated to some degree will add considerably to our knowledge of technological change and information exchange.

ACKNOWLEDGMENTS

As much of our experience and understanding of Indus Valley Tradition metallurgy derives from our recent firsthand experience at the site of Harappa, we would both like to dedicate this paper to the memory of Dr. George F. Dales who made it possible for us to work at Harappa and was always encouraging and supportive in our metallurgical research. We would also like to thank the Department of Archaeology and Museums, Pakistan for allowing us to work at Harappa and allowing Kenoyer to study the reserve collections in Harappa, Mohenjo-daro, and Karachi.

Other aspects of this paper are the result of considerable cooperation from colleagues and experimentation by both authors. We would especially like to thank colleagues who have allowed us to examine their unpublished materials and papers, including the late W. A. Fairservis, R. Hooja, J. Haquet, J.-F. Jarrige, M. Jansen, M. Tosi, M. Vidale, and the Museum of Fine Arts, Boston, where the Chanhu-daro collection is presently being curated and restudied. Special thanks are due M. Vidale for his comments and new information.

In terms of our own expertise it is important to note that Kenoyer has only recently begun to delve into the details of metallurgical research with the study of the metal objects from Harappa and through experimental research in smelting, melting, casting, and forging.

Miller acquired much of her practical knowledge of metalworking as the result of a year of part-time apprenticeship with Eleanor Moty and Fred Fenster of the Department of Art, University of Wisconsin-Madison, and at Max and Ruth Frölich's seminar on Ashante lost-wax casting at the Haystack Mountain School of Crafts; many thanks to the many metalsmiths associated with these programs for their enthusiasm and advice.

Please contact us with any suggestions or comments about our presentation of the data or our interpretations, especially if we have overlooked data (both authors are at: Dept. of Anthropology; 1180 Observatory Drive, No. 5240; Madison, WI 53706-1393; Email: kenoyer@macc.wisc.edu, heatherm@macc.wisc.edu). Finally, we would both like to thank Dr. Vince Pigott for inviting us to prepare this paper, R. Meadow and R. Wright for their comments on earlier drafts, and an anonymous reviewer for his/her painstaking reading and thoughtful suggestions, which significantly improved the clarity of this paper.

Note: We regret that we only received a copy of Chakrabarti and Lahiri's book after this article was in production. We highly recommend this seminal volume for more information on sites and ore sources in India, the use of copper after the Indus Tradition period, textual evidence for copper working, and ethnographic work on Indian coppersmiths.

Chakrabarti, D. K., and Lahiri, N.
    1996   *Copper and its Alloys in Ancient India.* New Delhi: Munshiram Manoharlal Publ. Pvt. Ltd.

APPENDIX A

Dashes (–), "tr," "0.00" recorded exactly as in original published tables. Blanks indicate that these elements were not reported.

MOHENJO-DARO: COPPER AND COPPER ALLOYS (SANAULLAH 1931:484)

Object no.	Description	% Cu	% Sn	% As	% Pb	% Fe	% Ni	% Sb	% Zn	% S	% Ag	% Co	% Cl	% Oxygen by difference	% Acid-insoluble residue	TOTAL %
	slab	82.71	13.21	1.17	0.11	0.42	0.56	0.33		0.00				1.49		100.00
	lump	83.92	12.13	0.00	0.17	0.00	0.17	tr		0.00				3.61		100.00
	chisel	85.37	11.09	0.07	tr	0.18	0.16	tr		0.11				3.02		100.00
	chisel	86.22	12.38	–	0.70	0.35	0.00	0.35		–				–		100.00
	buttons	88.05	8.22	tr	0.00	0.29	tr	2.60		0.84				–		100.00
	rod	91.90	4.51	1.96	0.17	0.15	–	1.15		0.16				–		100.00
	chisel (?)	92.41	0.00	3.42	3.28	0.59	0.15	0.10		0.05				–		100.00
	copper lump	92.49	0.37	1.30	tr	1.51	1.06	tr		2.26				1.01		100.00
	celt	94.76	0.09	4.42	0.26	0.15	0.14	–		–				–		99.82
	frag. of implement	95.80	0.00	0.74	1.58	0.12	0.25	0.72		0.61				0.18		100.00
	copper lump	96.42	0.00	0.00	0.09	0.00	0.35	–		0.36				2.78		100.00
	copper lump	96.67	0.00	0.15	0.02	0.03	1.27	0.88		0.98				–		100.00
	copper lump	97.07	0.00	0.98	tr	0.49	0.31	tr		1.15				–		100.00

MOHENJO-DARO: COPPER AND COPPER ALLOYS (HAMID IN MACKAY 1938:479–480 EXCEPT AS NOTED)

Object no.	Description	% Cu	% Sn	% As	% Pb	% Fe	% Ni	% Sb	% Zn	% S	% Ag	% Co	% Cl	% Oxygen by difference	% Acid-insoluble residue	TOTAL %
Dk 7856	chisel	75.25	7.84	0.00	0.39	0.51	0.61	1.25		0.00						85.85
Dk 5486	axe	80.56	1.76	2.10	0.20	0.34	0.58	tr		0.00						85.54
Dk 5360	pan with handle	81.94	0.37	0.80	0.05	0.33	0.21	0.18		0.14						84.02
Dk 6043	chisel	86.92	8.56	1.58	tr	0.02	0.68	0.54		0.07						98.37*
Dk 7861	portion of axe	88.49	9.88	0.00	0.22	0.10	0.30	0.14		0.06						99.19
Dk 7854	axe/celt	90.18	7.66	0.00	0.95	0.50	0.20	0.43		0.07						99.99
Dk 7535	axe/celt	91.01	6.14	0.66	0.59	0.33	0.48	0.25		0.12						99.58
Dk 7853	axe/celt	94.64	0.31	0.40	0.71	0.28	0.33	0.06		0.69						97.42
Dk 7859	spear/knife	95.23	0.00	0.24	0.82	0.56	0.41	0.00		0.48						97.74
Dk 7343	chisel/square rod	97.23	0.00	0.24	0.81	0.29	0.89	0.00		0.10						99.56
Pit-Dk 5316?**	copper ore	76.15		0.37	tr		0.23			1.12					4.8	77.87

*Addition error in original.
**Published in Pascoe in Mackay 1938:600.

APPENDIX A (CONTINUED)

HARAPPA: COPPER AND COPPER ALLOYS (SANAULLAH 1940:378)

Object no.	Description	% Cu	% Sn	% As	% Pb	% Fe	% Ni	% Sb	% Zn	% S	% Ag	% Co	% Cl	% Oxygen by difference	% Acid-insoluble residue	TOTAL %
X	chisel	87.42	10.45	1.10	0.52	0.34	0.17	–	–	*						100.00
Af150	awl	88.38	9.16	0.40	0.10	1.37	0.17	0.42	–	*						100.00
—	dagger	90.05	0.00	6.58	2.80	0.39	0.18	–	–	*						100.00
1208	dagger	91.00	6.76	0.04	0.88	0.74	0.14	0.44	–	*						100.00
Lot 277a/21	celt	91.10	7.85	0.42	tr	0.41	0.22	tr	–	*						100.00
4255	dagger	91.87	6.42	0.26	0.98	0.47	tr	–	–	*						100.00
11859	needle	92.55	0.29	2.96	3.72	0.20	0.21	–	0.07	*						100.00
B1754	chisel	92.61	6.43	0.36	tr	0.09	0.20	0.31	–	*						100.00
Lot 277k/3	chisel	94.92	3.60	0.60	0.20	0.39	0.29	–	–	*						100.00
5133	rectangular rod	97.20	0.84	0.70	0.00	0.98	0.09	0.19	–	*						100.00
J125	spearhead	97.66	0.33	0.06	0.70	1.11	0.14	–	–	*						100.00
Lot 277	folded sheet	97.69	0.15	1.19	0.85	0.07	0.05	tr	–	*						100.00
Lot 277g/2	saw	98.12	0.33	0.65	0.10	0.41	0.39	–	–	*						100.00
Lot 277a/19	celt	98.37	0.00	1.40	0.11	0.02	0.10	–	–	*						100.00
Lot 277	fragment	98.60	0.07	0.66	tr	0.41	0.26	–	–	*						100.00
Lot 277e/2	lancehead	98.69	0.10	0.68	tr	0.13	0.40	–	–	*						100.00

*SanaUllah notes that for the items from Harappa, "Sulphur is frequently present in minute quantities," but gives no particulars.

HARAPPA: COPPER AND COPPER ALLOYS (PIGOTT ET AL. 1989)

Object no.	Description	% Cu	% Sn	% As	% Pb	% Fe	% Ni	% Sb	% Zn	% S	% Ag	% Co	% Cl	% Oxygen by difference	% Acid-insoluble residue	TOTAL %
H88/4441	mirror	77.20	<.038	0.93	0.5	0.051	0.25	<.057	<1.1	0.38	<.029	<.022	18.9			98.21
H88/346-27	spherical bead	85.30	12.7	0.59	0.27	0.11	0.38	0.042	<.84	0.041	0.035	0.028	0.05			99.55
H88/751-21	spear point	93.40	2.6	1.91	0.17	0.15	0.3	<.053	<1.2	0.18	<.028	<.033	1.08			99.79
H88/529-120	blade/rod frags	93.70	<.015	4.71	0.051	0.13	0.15	<.022	<.052	0.074	<.012	<.046	<.95			98.82
H86.a	rod fragment	96.40	<.018	2.26	0.39	0.34	0.24	<.019	<.63	0.096	<.015	<.039	0.008			99.73
H86.c	rod fragment	96.50	<.014	2.08	0.32	0.34	0.23	<.018	<.72	0.12	0.02	<.02	0.052			99.66
H87/515	sheet metal/scoop	96.70	<.014	1.17	0.65	0.31	0.52	<.015	<.53	0.29	<.012	0.039	0.022			99.70
H88/325-08E	rod fragment	97.00	<.021	1.95	0.19	0.2	0.27	<.029	<.74	0.074	<.016	<.039	0.019			99.70
H88/325-08D	rod fragment	97.00	<.024	1.86	0.17	0.26	0.26	<.032	<.78	0.094	<.018	<.038	<.011			99.64
H88/761-1B	needle (point)	97.20	<.013	1.95	<.015	0.14	0.23	<.016	<.52	0.018	<.011	0.034	0.14			99.71
H88/761-1A	needle (eye end)	97.40	<.011	1.91	<.017	0.1	<.23	<.015	<.56	0.014	<.013	0.036	0.037			99.50
H88/374-1	large hook/handle	98.60	<.013	0.57	0.032	0.3	0.11	<.017	<.81	0.035	0.036	<.032	0.036			99.72
H88/413-A	finger ring	99.40	<.015	<.01	<.016	0.05	0.091	<.017	<.67	0.01	<.012	<.029	0.016			99.57

APPENDIX A (CONTINUED)

LOTHAL: COPPER AND COPPER ALLOYS (LAL 1985)

Object no.	Description	% Cu	% Sn	% As	% Pb	% Fe	% Ni	% Sb	% Zn	% S	% Ag	% Co	% Cl	% Oxygen by difference	% Acid-insoluble residue	TOTAL %
15217	spear/knife	39.07	2.27	*	–	tr	tr		–		–			36.08	12.58	90.00
—	revetted jar	43.00	–	*	–	1.00	–		–		–			53.77	2.23	100.00
14211	spear head	43.38	–	*	–	0.99	–		–		–			51.18	1.28/3.17 **	100.00
15176	knife handle	46.57	–	*	–	–	–		–		–			49.23	4.20	100.00
15139	ear ornament	48.24	tr	*	tr	0.70	tr		–		–			34.70	16.36	100.00
8480	spear head	48.26	–	*	–	tr	tr		–		–			38.09	13.65	100.00
15079	engraver	49.64	3.96	*	–	tr	tr		–		–			35.68	10.72	100.00
15155	pin	51.89	13.80	*	–	tr	–		–		–			28.91	5.40	100.00
15030	mirror	54.78	5.47	*	–	tr	–		–		–			31.58	8.17	100.00
15209	bangle	55.32	11.82	*	–	–	–		–		–			28.97	3.89	100.00
15251	lump	57.68	–	*	–	–	–		–		–			35.38	6.94	100.00
13886	rod with grooves	57.75	9.02	*	–	tr	tr		–		–			29.92	3.31	100.00
13134	pin	57.76	–	*	–	–	–		–		–			28.92	13.32	100.00
14855	ring from burial	58.36	–	*	–	tr	tr		–		–			28.60	13.04	100.00
15073	fragment	59.00	–	*	0.95	1.56	tr		–		–			25.38	13.11	100.00
10842	dagger/chisel	59.64	–	*	–	0.85	1.62		–		–			37.89	–	100.00
15137	rod	60.30	tr	*	–	tr	tr		–		–			26.26	13.44	100.00
5578	fish hook	60.65	–	*	1.30	tr	tr		–		–			32.16	5.99	100.10
14841	ornament	63.58	–	*	–	tr	tr		–		–			21.18	15.24	100.00
15085	fragment	66.60	–	*	–	0.80	tr		–		–			32.60	tr	100.00
12378	axe	70.00	–	*	–	–	–		–		–			7.00	23.00	100.00
3872	arrow head	70.30	–	*	–	tr	1.50		–		–			27.70	0.50	100.00
15210	engraver	70.45	–	*	–	tr	tr		–**		–			12.45	17.10	100.00
4189	object	70.69	–	*	–	0.90	tr		–		–			22.37	6.04	100.00
12264	bangle	72.26	–	*	–	–	tr		–		–			19.85	7.89	100.00
12432	fragment	72.50	–	*	–	0.95	2.48		–		–			16.65	7.42	100.00
15196	lump	72.83	–	*	–	0.55	tr		–		–			24.09	2.53	100.00
5971	bangle/ring	72.95	–	*	–	–	–		–		–			17.09	9.96	100.00
11893	chisel	74.28	9.62	*	–	tr	tr		–		–			13.02	3.06	99.98
12143	bangle	74.34	11.20	*	–	tr	tr		–		–			11.24	3.22	100.00
4148	fragment	74.84	–	*	–	0.61	tr		–		–			20.10	4.43	100.00
15169	fragment	75.60	–	*	–	tr	tr		–		–			23.90	0.50	100.00
15208	pin	77.40	–	*	–	0.50	tr		–		–			5.90	16.20	100.00
14302	hook	77.44	–	*	3.60	–	tr		–		–			22.56	–	100.00
4759	fragment	79.89	–	*	–	tr	tr		–		–			16.31	0.20	100.00
15194	lump	81.07	–	*	–	0.64	tr		–		–			15.74	2.55	100.00
15063	fish hook	82.90	–	*	–	tr	tr		–		–			15.90	1.20	100.00
11971	figurine	83.31	–	*	–	tr	tr		–		–			13.58	–	96.89
15036	lump	83.90	–	*	–	0.32	tr		–		–			12.46	3.32	100.00

APPENDIX A (CONTINUED)

LOTHAL: COPPER AND COPPER ALLOYS (LAL 1985) (CONTINUED)

Object no.	Description	% Cu	% Sn	% As	% Pb	% Fe	% Ni	% Sb	% Zn	% S	% Ag	% Co	% Cl	% Oxygen by difference	% Acid-insoluble residue	TOTAL %
5590	object	87.08	–	*	–	tr	tr		–					2.92	–	100.00
13140	hook	87.34	–	*	–	0.40	tr		–		–			11.48	0.78	100.00
5373	spear head	87.34	–	*	–	–	–		–		–			12.66	–	100.00
3091	shaft hole axe	88.27	–	*	–	tr	0.19		–		–			10.75	0.79	100.00
8110	chisel	88.53	–	*	–	tr	tr		–		–			10.99	0.48	100.00
5501	rod	88.60	–	*	–	tr	tr		–		–			11.10	0.30	100.00
14087	piece	88.76	–	*	–	tr	tr		–		–			10.66	0.58	100.00
12455	pin	89.95	–	*	–	tr	–		–		–			1.04	9.01	100.00
7700	chisel	90.44	–	*	–	tr	tr		–		–			9.56	–	100.00
12125	lump	90.68	–	*	–	4.02	–		–		–			5.03	0.28	100.00
6030	nail/pin	90.79	–	*	–	tr	0.63		–		–			8.58	–	100.00
SRC-324	fish hook	91.60	–	*	–	–	tr		–		–			8.40	–	100.00
SRC-B (20)	hand axe/celt	91.87	–	*	–	tr	0.80		–		–			7.11	0.20	100.00
4813	lump	93.65	–	*	1.51	tr	1.55		–		–			2.80	0.49	100.00
3091	hand axe/celt	94.33	–	*	–	–	tr		–		–			5.30	0.37	100.00
1344	bangle	94.90	–	*	–	2.14	0.45		–		–			2.51	–	100.00
4744	ring	95.31	–	*	–	–	–		–		–			4.69	–	100.00
10918	axe with sleeves†	96.27	–	*	2.51	tr	tr		–		–			1.22	–	100.00
15295	pin	96.76	0.57	*	–	tr	1.92		–		–			0.75	–	100.00
5957	celt	97.18	tr	*	–	tr	0.31		–		–			1.17	1.34	100.00
10918	axe/celt†	97.20	–	*	0.91	–	–		–		–			1.89	–	100.00
12147	arrow head	97.21	–	*	–	tr	tr		–		–			2.79	–	100.00
625	arrow head	97.70	–	*	–	–	tr		–		–			2.3	–	100.00
6040	fish hook	99.01	–	*	–	tr	0.28		–		–			0.71	–	100.00
14535	ingot	99.81	–	*	–	–	tr		–		–			0.19	–	100.00

*As was tested for, but no traces were found in any object.
**Typographical error in original.
†Two different analyses published.

APPENDIX A (CONTINUED)

RANGPUR: COPPER AND COPPER ALLOYS (RAO 1963:153; REPRINTED IN AGRAWAL 1971:167, TABLE 28).*

Object no.	Description	% Cu	% Sn	% As	% Pb	% Fe	% Ni	% Sb*	% Zn*	% S	% Ag*	% Co	% Cl	% Oxygen by difference	% Acid-insoluble residue	TOTAL %
RP 169	bangle	57.70	6.94	–	–	tr	tr		–					35.46		100.10**
RP 526	knife	59.00	5.28	–	–	tr	tr		–					35.72		100.00
RP 525	knife	59.60	2.69	–	–	1.08	–		–					36.63		100.00
RP 141	pin, needle	65.40	6.78	–	–	0.24	0.51		–					27.08		100.01**
RP 170	amulet	77.60	tr	–	–	0.57	0.1		–					21.73		100.00
RP 437	bangle	86.40	11.07	tr	tr	–	1.8		–					0.73		100.00
RP 324	celt	91.20	2.6	–	tr	–	2.1		–					4.1		100.00
RP 663	celt	91.35	4.09	tr	tr	–	tr		–					4.6		100.04**
RP 417	knife	94.80	0.7	–	–	tr	0.4		–					4.1		100.00
RP 635	ring, finger	96.10	tr	–	–	0.45	0.2		–					3.25		100.00
RP 442	pin, rolled head	96.60	tr	–	–	1.86	0.8		–					0.74		100.00
RP 260	bead	96.66	tr	–	–	1.4	0.38		–					1.56		100.00

*Zn tests, with no traces found, were only reported by Rao. Agrawal seems to indicate that Ag, Bi, and Sb were all tested for, with no traces found, but there is no indication of this in Rao's original report.
**Addition error in original (Rao 1963:153).
Note: A thirteenth object, RP 330, was reported by both Rao and Agrawal, but is not included here since it is from Period III, dating to after the Harappan Phase. This object had a high nickel content, 5.88%.

APPENDIX B

Dashes (–), "tr," "0.00" recorded exactly as in original published tables. Blanks indicate that these elements were not reported.

TIN BRONZE OBJECTS: MOHENJO-DARO, HARAPPA, LOTHAL, RANGPUR

Object no.	Description	Reference	% Cu	% Sn	% As	% Pb	% Fe	% Ni	% Sb	% Zn	% S	%Ag	% Co	% Cl	% Oxygen by difference	% Acid-insoluble residue	TOTAL %
Mirrors																	
Lothal 15030	mirror	Lal 1985	54.78	5.47	–	–	tr	–	–	–					31.58	8.17	100.00
Ornaments																	
Lothal 15155	pin	Lal 1985	51.89	13.89	0.00	–	tr	–	–	–					28.91	5.40	100.00
H88/34627	bead	Pigott et al. 1989	85.30	12.70	0.59	0.27	0.11	0.38	0.042	<.84	0.041	–	0.035	0.028	0.05		99.55
Lothal 15209	bangle	Lal 1985	55.32	11.82	0.00	–	–	–	–	–					28.97	3.89	100.00
Lothal 12143	bangle	Lal 1985	74.34	11.20	–	–	tr	tr	–	–					11.24	3.22	100.00
RP 437	bangle	Agrawal 1971; Rao 1963	86.40	11.07	–	tr	–	1.8	2.60	–					0.73		100.00
MD	buttons	SanaUllah 1931	88.05	8.22	tr	0.00	0.29	tr	–	–	0.84				–		100.00
RP 169	bangle	Agrawal 1971; Rao 1963	57.70	6.94	–	–	tr	tr	–	–					35.46		100.10
RP 141	pin, needle	Agrawal 1971; Rao 1963	65.40	6.78	–	–	0.24	0.51	–	–					27.08		100.01
Tools																	
MD Dk 7861	axe	Hamid in Mackay 1938	88.49	9.88	0.00	0.22	0.10	0.30	0.14	–	0.06						99.19
H Lot 277a/21	celt	SanaUllah 1940	91.10	7.85	0.42	tr	0.41	0.22	tr	–	*						100.00
MD Dk 7854	axe/celt	Hamid in Mackay 1938	90.18	7.66	0.00	0.95	0.50	0.20	0.43	–	0.07						99.99
MD Dk 7535	axe/celt	Hamid in Mackay 1938	91.01	6.14	0.66	0.59	0.33	0.48	0.25	–	0.12					4.6	99.58
RP 663	celt	Agrawal 1971; Rao 1963	91.35	4.09	tr	tr	–	tr	–	–					4.1		100.04
RP 324	celt	Agrawal 1971; Rao 1963	91.20	2.60	–	tr	–	2.1	–	–							100.00
MD Dk 5486	axe	Hamid in Mackay 1938	80.56	1.76	2.10	0.20	0.34	0.58	tr	–	0.00						85.54
H 1208	dagger	SanaUllah 1940	91.00	6.76	0.04	0.88	0.74	0.14	0.44	–	*						100.00
H 4255	dagger	SanaUllah 1940	91.87	6.42	0.26	0.98	0.47	tr	–	–	*						100.00
RP 526	knife	Agrawal 1971; Rao 1963	59.00	5.28	–	–	tr	tr	–	–					35.72		100.00
RP 525	knife	Agrawal 1971; Rao 1963	59.60	2.69	–	–	1.08	–	–	–					36.63		100.00
H88/751-21	spear point	Pigott et al. 1989	93.40	2.60	1.91	0.17	0.15	0.3	<.053	<1.2	0.18	<.028	<.033	1.08			99.79
Lothal 15217	spear/knife	Lal 1985	39.07	2.27	–	–	tr	tr	tr	–					36.08	12.58	90.00
MD	chisel	SanaUllah 1931	86.22	12.38	–	0.70	0.35	0.00	0.35	–	–				–		100.00
MD	chisel	SanaUllah 1931	85.37	11.09	0.07	tr	0.18	0.16	tr	–	0.11				3.02		100.00
HX	chisel	SanaUllah 1940	87.42	10.45	1.10	0.52	0.34	0.17	–	–	*						100.00
Lothal 11893	chisel	Lal 1985	74.28	9.62	1.58	tr	tr	tr	0.54	–	0.07				13.02	3.06	99.98
MD Dk 6043	chisel	Hamid in Mackay 1938	86.92	8.56	0.00	tr	0.02	0.68	1.25	–	0.00						98.37
MD Dk 7856	chisel	Hamid in Mackay 1938	75.25	7.84	0.00	0.39	0.51	0.61	0.31	–	*						85.85
H B1754	chisel	SanaUllah 1940	92.61	6.43	0.36	tr	0.09	0.20	–	–							100.00
Lothal 15079	engraver	Lal 1985	49.64	3.96	–	–	tr	tr	–	–					35.68	10.72	100.00
H Lot 277k/3	chisel	SanaUllah 1940	94.92	3.60	0.60	0.20	0.39	0.29	–	–	*						100.00
MD	rod	SanaUllah 1931	91.90	4.51	1.96	0.17	0.15	–	1.15	–	0.16				–		100.00
Lothal 13886	rod with grooves	Lal 1985	57.75	9.02	–	–	tr	tr	–	–					29.92	3.31	100.00
H Afl50	awl	SanaUllah 1940	88.38	9.16	0.40	0.10	1.37	0.17	0.42	–	*						100.00
Miscellaneous																	
MD	slab	SanaUllah 1931	82.71	13.21	1.17	0.11	0.42	0.56	0.33	–	0.00				1.49		100.00
MD	lump	SanaUllah 1931	83.92	12.13	0.00	0.17	0.00	0.17	tr	–	0.00				3.61		100.00

*SanaUllah notes that for the items from Harappa, "Sulphur is frequently present in minute quantities," but gives no particulars.

APPENDIX B (CONTINUED)

ARSENICAL BRONZE: MOHENJO-DARO, HARAPPA

Object no.	Description	Reference	% Cu	% Sn	% As	% Pb	% Fe	% Ni	% Sb	% Zn	% S	%Ag	% Co	% Cl	% Oxygen by difference	% Acid-insoluble residue	TOTAL %
Mirrors																	
H88/444-1	mirror	Pigott et al. 1989	77.20	<.038	0.93	0.5	0.051	0.25	<.057	<1.1	0.38	<.029	<.022	18.9	–		98.21
Tools																	
MD	celt	SanaUllah 1931	94.76	0.09	4.42	0.26	0.15	0.14	–	–	–				–		99.82
MD Dk 5486	axe	Hamid in Mackay 1938	80.56	1.76	2.10	0.20	0.34	0.58	tr	–	0.00						85.54
H Lot 277a/19	celt	SanaUllah 1940	98.37	0.00	1.40	0.11	0.02	0.10	–	–	*						100.00
H	dagger	SanaUllah 1940	90.05	0.00	6.58	2.80	0.39	0.18	–	–	*						100.00
H88/529-120	blade/rod frags	Pigott et al. 1989	93.70	<.015	4.71	0.051	0.13	0.15	<.022	<.052	0.074	<.012	<.046	<.95			98.82
H88/751-21	spear point	Pigott et al. 1989	93.40	2.6	1.91	0.17	0.15	0.3	<.053	<1.2	0.18	<.028	<.033	1.08			99.79
MD	chisel (?)	SanaUllah 1931	92.41	0.00	3.42	3.28	0.59	0.15	0.10		0.05				–		100.00
MD Dk 6043	chisel	Hamid in Mackay 1938	86.92	8.56	1.58	tr	0.02	0.68	0.54		0.07						98.37
H X	chisel	SanaUllah 1940	87.42	10.45	1.10	0.52	0.34	0.17	–	–	*						100.00
H86.a	rod	Pigott et al. 1989	96.40	<.018	2.26	0.39	0.34	0.24	<.019	<.63	0.096	<.015	<.039	0.008			99.73
MD	rod	SanaUllah 1931	91.90	4.51	1.96	0.17	0.15	–	1.15	–	0.16				–		100.00
H88/325-08E	rod	Pigott et al. 1989	97.00	<.021	1.95	0.19	0.2	0.27	<.029	<.74	0.074	<.016	<.039	0.019			99.70
H11859	needle	SanaUllah 1940	92.55	0.29	2.96	3.72	0.20	0.21	–	0.07	*						100.00
H88/761-1	needle	Pigott et al. 1989	97.20	<.013	1.95	<.015	0.14	0.23	<.016	<.52	0.018	<.011	0.034	0.14			99.71
Miscellaneous																	
MD	copper lump	SanaUllah 1931	92.49	0.37	1.30	tr	1.51	1.06	tr	–	2.26				1.01		100.00
H Lot 277	folded sheet	SanaUllah 1940	97.69	0.15	1.19	0.85	0.07	0.05	tr	–	*						100.00
H87/515	sheet metal/scoop	Pigott et al. 1989	96.70	<.014	1.17	0.65	0.31	0.52	<.015	<.53	0.29	<.012	0.039	0.022			99.70
MD	slab	SanaUllah 1931	82.71	13.21	1.17	0.11	0.42	0.56	0.33		0.00				1.49		100.00
MD	copper lump	SanaUllah 1931	97.07	0.00	0.98	tr	0.49	0.31	tr		1.15				–		100.00

*SanaUllah notes that for the items from Harappa, "Sulphur is frequently present in minute quantities," but gives no particulars.

REFERENCES CITED

Agrawal, D. P.
1971 *The Copper Bronze Age in India.* New Delhi: Munshiram Manoharlal.
1984 Metal Technology of the Harappans. Pp. 163–167 in *Frontiers of the Indus Civilization,* eds. B. B. Lal and S. P. Gupta. New Delhi: Books and Books.

Agrawala, R. C.
1984a Aravalli, The Major Source of Copper for the Indus Civilization and Indus Related Cultures. Pp. 157–162 in *Frontiers of the Indus Civilization,* eds. B. B. Lal and S. P. Gupta. New Delhi: Books and Books.
1984b Ganeshwar Culture—A Review. *Journal of the Oriental Institute,* Baroda 34(1–2):89–95.

Agrawala, R. C., and Kumar, V.
1982 Ganeshwar-Jodhpura Culture: New Traits in Indian Archaeology. Pp. 125–134 in *Harappan Civilization: A Contemporary Perspective,* ed. G. L. Possehl. Reprinted in 1993 in *Harappan Civilization: A Recent Perspective.* New Delhi: Oxford and IBH Publishing.

Allchin, B., and Allchin, R.
1982 *The Rise of Civilization in India and Pakistan.* Cambridge World Archaeology Series. Cambridge: Cambridge University Press.

Asthana, S.
1982 Harappan Trade in Metals and Minerals: A Regional Approach. Pp. 271–285 in *Harappan Civilization: A Contemporary Perspective,* ed. G. L. Possehl. Reprinted in 1993 in *Harappan Civilization: A Recent Perspective.* New Delhi: Oxford and IBH Publishing.

Bayley, J.
1989 Non-metallic Evidence for Metalworking. Pp. 291–303 in *Archaeometry. Proceedings of the 25th International Symposium,* ed. Y. Maniatis. Amsterdam: Elsevier Science Pub.

Berthoud, T., and Cleuziou, S.
1983 Farming Communities of the Oman Peninsula and the Copper of Makkan. *Journal of Oman Studies* 6(2):239–245.

Besenval, R.
1992 Le peuplement ancien du Kech-Makran. Travaux récents. *Paléorient* 18(1):103–107.

Bisht, R. S.
1982 Excavations at Banawali: 1974–77. Pp. 113–124 in *Harappan Civilization: A Contemporary Perspective,* ed. G. L. Possehl. Reprinted in 1993 in *Harappan Civilization: A Recent Perspective.* New Delhi: Oxford and IBH Publishing.
1984 Structural Remains and Town-planning of Banawali. Pp. 89–97 in *Frontiers of the Indus Civilization,* eds. B. B. Lal and S. P. Gupta. New Delhi: Books and Books.

Bondioli, L., and Vidale, M.
1986 Architecture and Craft Production across the Surface Palimpsest of Moenjodaro: Some Processual Perspectives. Pp. 115–138 in *Arqueologia Espacial 8: Coloquio sobre el microespacio-2. Del Paleolitico al Bronce Medio.* Teruel: Seminario de Arqueologia y Etnologia Turolense, Colegio Universitario de Tereul.

Casal, J. M.
1961 *Fouilles de Mundigak.* Paris: Librairie C. Klincksieck.

Cleland, J. H.
1977 *Chalcolithic and Bronze Age Chipped Stone Industries of the Indus Region.* Ph.D. dissertation, University of Virginia.

Cleuziou, S.
1984 Oman Peninsula and its Relations Eastwards During the Third Millennium. Pp. 371–394 in *Frontiers of the Indus Civilization,* eds. B. B. Lal and S. P. Gupta. New Delhi: Books and Books.
1989 The Chronology of Protohistoric Oman as seen from Hili. Pp. 47–78 in *Oman Studies,* Vol. LXIII, eds. P. M. Costa and M. Tosi. Rome: Serie Orientale.

Cleuziou, S., and Tosi, M.
1989 The Southeastern Frontier of the Ancient Near East. Pp. 15–48 in *South Asian Archaeology 1985,* eds. K. Frifelt and P. Sørensen. London: Curzon Press.

Costin, C. L.
1991 Craft Specialization: Issues in Defining, Documenting, and Explaining

the Organization of Production. Pp. 1–56 in *Archaeological Method and Theory*, Vol. 3, ed. M. B. Schiffer. Tucson: University of Arizona Press.

Craddock, P. T.
1989   The Scientific Investigation of Early Mining and Metallurgy. Pp. 178–212 in *Scientific Analysis in Archaeology and its Interpretation*, ed. J. Henderson. Oxford: Oxford University Press.
1991   Mining and Smelting in Antiquity. Pp. 57–73 in *Science and the Past*, ed. S. Bowman. Toronto: University of Toronto Press.

Craddock, P. T.; Freestone, I. C.; Gurjar, L. K.; Middleton, A.; and Willies, L.
1989   The Production of Lead, Silver and Zinc in Early India. Pp. 51–69 in *Old World Archaeometallurgy*, eds. A. Hauptmann, E. Pernicka, and G. Wagner. Bochum: Selbstverlag des Deutschen Bergbau-Museums.

Dales, G. F.
1979   The Balakot Project: Summary of Four Years of Excavations in Pakistan. *Man and Environment* 3:45–53.
1992   A Line in the Sand: Explorations in Afghan Sistan. Pp. 227–240 in *Archaeological Studies: Walter A. Fairservis, Jr. Feschrift*, ed. G. L. Possehl. New Delhi: Oxford and IBH Publishing.

Dales, G. F., and Flam, L.
1969   On Tracking the Woolly Kullis and the Like. *Expedition* 12(1):15–23.

Dales, G. F., and Kenoyer, J. M.
1989   Preliminary Report on the Fourth Season of Research at Harappa, Pakistan. University of California-Berkeley and University of Wisconsin-Madison. Unpublished manuscript.
1990   Excavation at Harappa—1988. *Pakistan Archaeology* 24:68–176.

Desch, C. H.
1931   Analyses of Copper and Bronze Specimens made for the Sumer Committee of the British Association. Pp. 486–488 in *Mohenjo-daro and the Indus Civilization*, Vol. 2, ed. J. Marshall. London: A. Probsthain.

Dupree, L.
1963   Deh Morasi Ghundai: A Chalcolithic Site in South-central Afghanistan. *Anthropological Papers of the American Museum of Natural History* 50:59–135.
1972   Prehistoric Research in Afghanistan 1959–1966. The American Philosophical Society, ns 62(4). Philadelphia.

Durrani, F. A.
1988   Excavations in the Gomal Valley: Rehman Dheri Excavation Report No. 1. *Ancient Pakistan* 6:1–232.

Dyson, R. H., and Remsen, W. C. S.
1989   Observations on Architecture and Stratigraphy at Tappeh Hesar. Pp. 69–109 in *Tappeh Hesar: Reports of the Restudy Project, 1976*, eds. R. H. Dyson and S. M. Howard. Florence: Casa Editrice Le Lettre.

Emmerich, A.
1965   *Sweat of the Sun and Tears of the Moon: Gold and Silver in Pre-Columbian Art.* Seattle: University of Washington Press.

Fairservis, W. A.
1952   Preliminary Report on the Prehistoric Archaeology of the Afghan-Baluchi Areas. *Novitates* No. 1587. New York: American Museum of Natural History.
1961   Archaeological Studies in the Seistan Basin of Southwestern Afghanistan and Eastern Iran in the Zhob and Loralai Districts, West Pakistan. *Anthropological Papers of the American Museum of Natural History* 48(part 1):1–128.

Fentress, M. A.
1977   *Resource Access, Exchange Systems and Regional Interaction in the Indus Valley: An Investigation of Archaeological Variability at Harappa and Moenjo Daro.* Ph.D. dissertation, University of Pennsylvania. Ann Arbor, MI: University Microfilms #77-10, 163.

Flam, L.
1993   Excavations at Ghazi Shah, Sindh, Pakistan. Pp. 457–468 in *Harappan Civilization: A Recent Perspective*, 2nd rev. ed., ed. G. L. Possehl. New Delhi: Oxford and IBH Publishing.

Fox, C.
1988   *Asante Brass Casting: Lost-wax Casting of Gold-weights, Ritual Vessels and Sculptures, with Handmade Equipment.* Cambridge: Cambridge University African Studies Center.

Francfort, H.-P.
1989 *Fouilles de Shortugaï recherches sur L'Asie centrale protohistorique.* Paris: Diffusion de Boccard.

Frifelt, K.
1991 *The Island of Umm An-Nar: Third Millennium Graves,* Vol. 1. Århus: Århus University Press.

Fröhlich, M.
1979 *Gelbgiesser im Kameruner Grasland.* Zürich: Rietberg Museum.

Hall, M. E., and Steadman, S. R.
1991 Tin and Anatolia: Another Look. *Journal of Mediterranean Archaeology* 4(1):217–234.

Haquet, J.
1994 Problématique de la métallurgie dans la plaine de Kachi (Pakistan) du néolithique à la fin de la civilisation de l'Indus au IIIe millénaire avant J.-C. Mémoire de D.E.A., Université de Paris I—U.E.R. d'Art et d'Archéologie, unpublished.

Hargreaves, H.
1929 *Excavations in Baluchistan, 1925.* Memoirs of the Archaeological Survey of India No. 35. Calcutta: Government of India.

Hauptmann, A.
1985 *5000 Jahre Kupfer in Oman. Bd. 1: Die Entwicklung der Kupfermetallurgie vom 3. Jahrtausend bis zum Neuzeit.* Montanhistorische Zeitschrift *Der Anschnitt,* Beiheft 4, Nr. 33. Bochum: Deutschen Bergbau-Musum.

Hauptmann, A., and Weisgerber, G.
1980a The Early Bronze Age Copper Metallurgy of Shahr-i-Sokhta (Iran). *Paléorient* 6:120–127.
1980b Third Millennium B.C. Copper Production in Oman. Pp. 131–138 in *Revue d'Archeometrie: Actes du XX Symposium International d'Archéométrie, Symposium for Archaeometry, Paris 26–29 Mars 1980,* Vol. III. Bulletin de Liaison du Groupe des Méthodes Physiques et Chimiques de l'Archéologie.

Hegde, K. T. M.
1965 *Chalcolithic Period Copper Metallurgy in India.* Ph.D. dissertation, Maharaja Sayajirao University, Baroda.
1969 Appendix I: Technical Studies in Copper Artifacts Excavated from Ahar. Pp. 225–228 in *Excavations at Ahar (Tambavati) 1961–62,* eds. H. D. Sankalia, S. B. Deo, and Z. D. Ansari. Poona: Deccan College.

Hegde, K. T. M., and Ericson, J. E.
1985 Ancient Indian Copper Smelting Furnaces. Pp. 59–70 in *Furnaces and Smelting Technology in Antiquity,* eds. P. T. Craddock and M. J. Hughes. British Museum Occasional Papers No. 48. London: British Museum Research Laboratory.

Herman, C. F.
1984 *De Harappa-Cultuur: De Koperen en Bronzen Wapens en Werktuigen. Eeen Morfologische Benadering met Aandacht voor de Stratigrafische Context.* Vol. 1: I–IX, 1–121; Vol. 2: 1–369; Vol. 3: 1–6; P1–P43; T1–T22; F1–F24. Katholieke Universiteit Leuven. Licentiate's Paper.

Hoffman, M. A., and Cleland, J.
1977 *Excavations at Allahdino II, The Lithic Industry.* New York: American Museum of Natural History.

Hooja, R.
1988 *The Ahar Culture and Beyond: Settlements and Frontiers of 'Mesolithic' and Early Agricultural Sites in South-eastern Rajasthan c. 3rd–2nd Millennia B.C.* BAR International Series 412. Oxford: British Archaeological Reports.

Hooja, R., and Kumar, V.
1995 Aspects of the Early Copper Age in Rajasthan. Paper presented at the European Association of South Asian Archaeologists Conference (Cambridge). To be published in *South Asian Archaeology 1995.*

Horne, L.
1990 A Study of Traditional Lost-wax Casting in India. *Journal of Metals* (October):46–47.

Jacobson, J.
1986 The Harappan Civilization: An Early State. Pp. 137–174 in *Studies in the Archaeology of India and Pakistan,* ed. J. Jacobson. New Delhi: Oxford and IBH Publishing.

Jarrige, J.-F.
1983 New Trends in the Archaeology of Greater Indus in the Light of Recent Excavations at Mehrgarh, Baluchistan. Paper presented at the International Seminar of Peshawar.
1985a A propos d'un foret a tige hélicoïdale en cuivre de Mundigak. Pp. 281–292 in *De l'Indus aux Balkans, Recueil Jean Deshayes,* eds. J.-L. Huot, M. Yon, and Y. Calvet. Paris: Éditions Recherche sur les Civilisations.
1985b Continuity and Change in the North Kachi Plain (Baluchistan, Pakistan) at the Beginning of the Second Millennium B.C. Pp. 35–68 in *South Asian Archaeology 1983,* eds. J. Shotsmans and M. Taddei. Naples: Istituto Universitario Orientale.
1990 Excavation at Nausharo 1987–88. *Pakistan Archaeology* 24:21–67.

Jarrige, J.-F., and Lechevallier, M.
1979 Excavations at Mehrgarh, Baluchistan: Their Significance in the Prehistoric Context of the Indo-Pakistan Borderlands. Pp. 463–536 in *South Asian Archaeology 1977,* ed. M. Taddei. Naples: Istituto Universitario Orientale.

Jarrige, J.-F., and Santoni, M.
1979 *Fouilles de Pirak.* Paris: Diffusion de Boccard.

Jarrige, J.-F., and Tosi, M.
1981 The Natural Resources of Mundigak. Pp. 115–142 in *South Asian Archaeology 1979,* ed. H. Härtel. Berlin: Dietrich Reimer.

Joshi, J. P., and Parpola, A.
1987 *Corpus of Indus Seals and Inscriptions.* 1: *Collections in India.* Helsinki: Suomalainen Tiedeakatemia.

Kenoyer, J. M.
1983 *Shell Working Industries of the Indus Civilization: An Archaeological and Ethnographic Perspective.* Ph.D. dissertation, University of California-Berkeley.
1989 Socio-economic Structures of the Indus Civilization as reflected in Specialized Crafts and the Question of Ritual Segregation. Pp. 183–192 in *Old Problems and New Perspectives in the Archaeology of South Asia,* ed. J. M. Kenoyer. Wisconsin Archaeological Reports No. 2. Madison: Dept. of Anthropology, University of Wisconsin-Madison.
1991 The Indus Valley Tradition of Pakistan and Western India. *Journal of World Prehistory* 5(4):331–385.
1992a Harappan Craft Specialization and the Question of Urban Segregation and Stratification. *Eastern Anthropologist* 45(1–2):39–54.
1992b Ornament Styles of the Indus Tradition: Evidence from Recent Excavations at Harappa, Pakistan. *Paléorient* 17(2):79–98.
1994 Experimental Studies of Indus Valley Technology at Harappa. Pp. 345–362 in *South Asian Archaeology 1993,* eds. A. Parpola and P. Koskikallio. Annales Academiae Scientiarum Fennicae; Series B, Vol. 271. Helsinki: Suomalainen Tiedeakatemia.

Kenoyer, J. M., and Vidale, M.
1992 A New Look at Stone Drills of the Indus Valley Tradition. Pp. 495–518 in *Materials Issues in Art and Archaeology,* III, eds. P. Vandiver, J. R. Druzick, G. S. Wheeler, and I. Freestone. Materials Research Society Monograph 267. Pittsburgh, PA.

Khan, F.; Knox, J. R.; and Thomas, K. D.
1989 New Perspectives on Early Settlement in Bannu District, Pakistan. Pp. 281–291 in *South Asian Archaeology 1985,* eds. K. Frifelt and P. Sørensen. London: Curzon Press.

Kumar, V.
1986 Excavations at Ganeshwar, District Sikar. *Indian Archaeology—A Review* 1983–84:71–72.

Kuppuram, G.
1989 *Ancient Indian Mining, Metallurgy, and Metal Industries.* Delhi: Sundeep Prakashan.

Lahiri, N.
1993 Some Ethnographic Aspects of the Ancient Copper-Bronze Tradition in India. *Journal of the Royal Asiatic Society* ser. 3, 3(2):219–231.

Lal, B. B.
1985 Report on the Chemical Analysis and Examination of Metallic and Other Objects from Lothal. Pp. 651–666 in *Lothal: A Harappan Port Town (1955–62),* Vol. 2, ed. S. R. Rao. Memoirs of the Archaeological Survey of

India No. 78. New Delhi: Archaeological Survey of India.

Lal, B. B., and Thapar, B. K.
1967 Excavation at Kalibangan: New Light on the Indus Civilization. *Cultural Forum* 9(4):78–88.

Lechtman, H., and Steinberg, A.
1979 The History of Technology: An Anthropological Point of View. Pp. 135–160 in *The History and Philosophy of Technology,* eds. G. Bugliarello and D. B. Doner. Urbana: University of Illinois Press.

Lowe, T. L.
1989a Refractories in High-Carbon Iron Processing: A Preliminary Study of the Deccani Wootz-Making Crucibles. Pp. 237–251 in *Ceramics and Civilization IV: Cross-Craft and Cross-Cultural Interactions in Ceramics,* eds. P. E. McGovern and M. D. Notis. Westerville, OH: The American Ceramic Society, Inc.
1989b Solidification and the Crucible Processing of Deccani Ancient Steel. Pp. 729–740 in *Principles of Solidification and Materials Processing,* Vol. 2, eds. R. Trivedi, J. A. Sekhar, and J. Mazumdar. New Delhi: Oxford and IBH Publishing.

Mackay, E. J. H.
1931 Technique and Description of Metal Vessels, Tools, Implements, and Other Objects. Pp. 488–508 in *Mohenjodaro and the Indus Civilization,* ed. J. Marshall. London: A. Probsthain.
1938 *Further Excavations at Mohenjodaro,* Vol. I text, Vol. II plates. Delhi: Government of India.
1943 *Chanhu-Daro Excavations 1935–36.* New Haven, CT: American Oriental Society.
1948 *Early Indus Civilizations,* 2nd ed. rev. and enlarged by D. Mackay. London: Luzac and Co., Ltd.

Marshall, J.
1931 *Mohenjo-daro and the Indus Civilization,* Vol. I, II text, Vol. III plates. London: A. Probsthain.

McCreight, T.
1982 *The Complete Metalsmith: An Illustrated Handbook.* Worcester, MA: Davis Publishing, Inc.

Meadow, R. H. (ed.)
1991 *Harappa Excavations 1986–1990: A Multidisciplinary Approach to Third Millennium Urbanism.* Monographs in World Archaeology 3. Madison, WI: Prehistory Press.

Meadow, R. H., and Kenoyer, J. M.
1992 Harappa Archaeological Project, 1992. Harvard University and University of Wisconsin-Madison. Unpublished manuscript.
1994 Excavations at Harappa 1994. Harvard University and University of Wisconsin-Madison. Unpublished manuscript.

Miller, H. M.-L.
1994a Indus Tradition Craft Production: Research Plan and Preliminary Survey Results Assessing Manufacturing Distribution at Harappa, Pakistan. Pp. 81–103 in *From Sumer to Meluhha: Contributions to the Archaeology of South and West Asia in Memory of George F. Dales, Jr.,* ed. J. M. Kenoyer. Wisconsin Archaeological Reports Vol. 3. Madison: Dept. of Anthropology, University of Wisconsin-Madison.
1994b Metal Processing at Harappa and Mohenjo-daro: Information from Nonmetal Remains. Pp. 497–510 in *South Asian Archaeology 1993,* eds. A. Parpola and P. Koskikallio. Annales Academiae Scientiarum Fennicae; Series B, Vol. 271. Helsinki: Suomalainen Tiedeakatemia.

Misra, V. N.
1969 Early Village Communities of the Banas Basin, Rajasthan. Pp. 296–310 in *Anthropology and Archaeology: Essays in Commemoration of Verrier Elwin,* eds. M. C. Pradhan, R. D. Singh, P. K. Misra, and D. B. Sastry. Oxford: Oxford University Press.

Moorey, P. R. S.
1985 *Materials and Manufacture in Ancient Mesopotamia: The Evidence of Archaeology and Art.* BAR International Series No. S237. Oxford: British Archaeological Reports.
1994 *Ancient Mesopotamian Materials and Industries.* Oxford: Clarendon Press.

Mughal, M. R.
1980 New Archaeological Evidence from Bahawalpur. *Man and Environment* 4:93–98.
1982 Recent Archaeological Research in the Cholistan Desert. Pp. 86–95 in

Harappan Civilization: A Contemporary Perspective, ed. G. L. Possehl. Reprinted in 1993 in *Harappan Civilization: A Recent Perspective.* New Delhi: Oxford and IBH Publishing.

1990 Further Evidence of the Early Harappan Culture in the Greater Indus Valley: 1971–1990. *South Asian Studies* 6:176–199.

Parpola, A.
1992 Copper Tablets from Mohenjo-daro and the Study of the Indus Script. In *Proceedings of the Second International Conference on Moenjodaro,* ed. I. M. Nadiem. Karachi: Department of Archaeology. In press.
1994a Deciphering the Indus Script: A Summary Report. Pp. 571–586 in *South Asian Archaeology 1993,* eds. A. Parpola and P. Koskikallio. Annales Academiae Scientiarum Fennicae; Series B, Vol. 271. Helsinki: Suomalainen Tiedeakatemia.
1994b *Deciphering the Indus Script.* Cambridge: Cambridge University Press.

Pascoe, E.
1931 Minerals and Metals. Pp. 674–685 in *Mohenjo-daro and the Indus Civilization,* ed. J. Marshall. London: A. Probsthain.

Pigott, V. C.
1989 Bronze in Pre-Islamic Iran. Pp. 457–471 in *Encyclopaedia Iranica,* Vol. IV. London: Routledge and Kegan Paul Ltd.

Pigott, V. C.; Fleming, S. J.; and Reistroffer, E.
1989 Preliminary Research on Copper-base Metallurgy at Harappa: The PIXE Analysis and Metallography. MASCA, University of Pennsylvania. Unpublished document.

Pigott, V. C.; Howard, S. M.; and Epstein, S. M.
1982 Pyrotechnology and Culture Change at Bronze Age Tepe Hissar (Iran). Pp. 215–236 in *Early Pyrotechnology,* eds. T. A. Wertime and S. F. Wertime. Washington, DC: Smithsonian Institution.

Possehl, G. L.
1990 Revolution in the Urban Revolution: The Emergence of Indus Urbanization. *Annual Review of Anthropology* 19:261–282.

Potts, D. T.
1990 *The Arabian Gulf in Antiquity: From Prehistory to the Fall of the Achaemenid Empire.* Oxford: Clarendon Press.

Pracchia, S.; Tosi, M.; and Vidale, M.
1985 On the Type, Distribution and Extent of Craft Industries at Mohenjo-daro. Pp. 207–247 in *South Asian Archaeology 1983,* eds. J. Shotsmans and M. Taddei. Naples: Istituto Universitario Orientale.

Rao, S. R.
1963 Excavations at Rangpur and Other Explorations in Gujarat. *Ancient India* 18–19:5–207.
1979 *Lothal: A Harappan Port Town (1955–62),* Vol. 1. Memoirs of the Archaeological Survey of India No. 78. New Delhi: Archaeological Survey of India.
1985 *Lothal: A Harappan Port Town (1955–62),* Vol. 2. Memoirs of the Archaeological Survey of India No. 78. New Delhi: Archaeological Survey of India.

Reeves, R.
1962 *Ciré Perdue Casting in India.* New Delhi: Crafts Museum.

Sali, S. A.
1986 *Daimabad 1976–79.* Memoirs of the Archaeological Survey of India No. 83. New Delhi: Archaeological Survey of India.

SanaUllah, M.
1931 Sources and Metallurgy of Copper and Its Alloys, *and* Notes and Analyses. Pp. 481–488 and 686–691 in *Mohenjo-daro and the Indus Civilization,* ed. J. Marshall. London: A. Probsthain.
1940 The Sources, Composition, and Techniques of Copper and its Alloys. Pp. 378–382 in *Excavations at Harappa,* ed. M. S. Vats. Delhi: Government of India Press.

Sankalia, H. D.; Deo, S. B.; and Ansari, Z. D.
1969 *Excavations at Ahar (Tambavati) 1961–62.* Poona: Deccan College.

Scheel, B.
1989 *Egyptian Metalworking and Tools.* Aylesbury, Bucks, UK: Shire Publications Ltd.

Schmandt-Besserat, D.
1980 Ocher in Prehistory: 300,000 Years of the Use of Iron Ore as Pigments. Pp.

127–150 in *The Coming of the Age of Iron,* eds. T. A. Wertime and J. D. Muhly. New Haven, CT: Yale University Press.

Shaffer, J. G.
1982 Harappan Culture: A Reconsideration. Pp. 41–50 in *Harappan Civilization: A Contemporary Perspective,* ed. G. L. Possehl. Reprinted in 1993 in *Harappan Civilization: A Recent Perspective.* New Delhi: Oxford and IBH Publishing.
1984 Bronze Age Iron from Afghanistan: Its Implications for South Asian Protohistory. Pp. 41–62 in *Studies in the Archaeology and Palaeoanthropology of South Asia,* eds. K. A. R. Kennedy and G. L. Possehl. New Delhi: Oxford and IBH Publishing.
1992 The Indus Valley, Baluchistan and Helmand Traditions: Neolithic Through Bronze Age. Pp. 441–464 in *Chronologies in Old World Archaeology,* 3rd ed., Vol. 1, ed. R. Ehrich. Chicago: University of Chicago Press.

Shah, S. G. M., and Parpola, A.
1991 *Corpus of Indus Seals and Inscriptions.* 2: *Collections in Pakistan.* Helsinki: Suomalainen Tiedeakatemia.

Sollberger, E.
1970 The Problem of Magan and Meluhha. *Bulletin of the London Institute of Archaeology* 8–9:247–250.

Stech, T., and Pigott, V.
1986 The Metals Trade in Southwest Asia in the Third Millennium B.C. *Iraq* 48:39–64.

Tosi, M.
1982 A Possible Harappan Seaport in Eastern Arabia: Ra's Al-Junayz in the Sultanate of Oman. Paper presented at the First International Conference on Pakistan Archaeology (Peshawar, March 1–4).
1984 The Notion of Craft Specialization and its Representation in the Archaeological Record of Early States in the Turanian Basin. Pp. 22–52 in *Marxist Perspectives in Archaeology,* ed. M. Spriggs. New Directions in Archaeology Series. Cambridge: Cambridge University Press.

Tosi, M.; Bondioli, L.; and Vidale, M.
1984 Craft Activity Areas and Surface Survey at Moenjodaro: Complementary Procedures for the Revaluation of a Restricted Site. Pp. 9–37 in *Interim Reports Volume 1,* eds. M. Jansen and G. Urban. Aachen and Rome: Rheinisch-Westfälische Technische Hochschule (RWTH) Aachen and Istituto Italiano per il Medio ed Estremo Oriente (IsMEO).

Tylecote, R. F.
1980 Furnaces, Crucibles, and Slags. Pp. 183–223 in *The Coming of the Age of Iron,* eds. T. A. Wertime and J. D. Muhly. New Haven, CT: Yale University Press.

Untracht, O.
1975 *Metal Techniques for Craftsmen.* New York: Doubleday.

Vats, M. S.
1940 *Excavations at Harappa,* Vol. I text, Vol. II plates. Delhi: Government of India Press.

Vidale, M.; Kenoyer, J. M.; and Bhan, K. K.
1992 A Discussion of the Concept of "Chaîne Opératoire" in the Study of Stratified Societies: Evidence from Ethnoarchaeology and Archaeology. Pp. 181–194 in *Ethnoarcheologie: Justification, Problémes, Limites,* ed. A. Gallay. Juan-Le-Pins, France: Centre de Recherches Archéologiques.

Weisgerber, G.
1981 Mehr als Kupfer in Oman. *Der Anschnitt* 33(5–6):174–263.
1983 Copper Production During the Third Millennium B.C. in Oman and the Question of Makkan. *Journal of Oman Studies* 6(2):269–276.
1984 Makkan and Meluhha: Third Millennium B.C. Copper Production in Oman and the Evidence of Contact with the Indus Valley. Pp. 196–201 in *South Asian Archaeology 1981,* ed. B. Allchin. Cambridge: Cambridge University Press.

Weisgerber, G., and Yule, P.
1989 The First Metal Hoard in Oman. Pp. 60–61 in *South Asian Archaeology 1985,* eds. K. Frifelt and P. Sørensen. London: Curzon Press.

Wraight, E. A.
1940 Report on the Metallography of Two

Ancient Bronze Specimens Found at Harappa. Pp. 382–383 in *Excavations at Harappa,* ed. M. S. Vats. Delhi: Government of India Press.

Wright, R. P.
1991   Patterns of Technology and the Organization of Production at Harappa. Pp. 71–88 in *Harappa Excavations 1986–1990: A Multidisciplinary Approach to Third Millennium Urbanism,* ed. R. H. Meadow. Madison, WI: Prehistory Press.

Yule, P.
1985a   *Figuren, Schmuckformen und Täfelchen der Harappa-Kultur,* Abteilung I. Munich: C. H. Beck'sche Verlagsbuchhandlung.
1985b   *Harappazeitliche Metallgefäße in Pakistan und Nordwestindien,* Abteilung II. Munich: C. H. Beck'sche Verlagsbuchhandlung.
1985c   *Metalwork of the Bronze Age in India.* Prähistorishe Bronzefunde, Abteilung XX, Band 8. Munich: C. H. Beck'sche Verlagsbuchhandlung.

The Early Iron Age in South Asia

Gregory L. Possehl and Praveena Gullapalli

ABSTRACT Traditional frameworks for the Early Iron Age in South Asia have emphasized diffusion and external stimuli as major factors in understanding the transition to iron production. However, scholars are now looking within the subcontinent for an understanding of the development of iron technology. Four principal archaeological assemblages document the Early Iron Age: the Gandharan Grave Culture, the Painted Grey Ware Assemblage, the Pirak Assemblage, and the Megalithic Complex. These regional manifestations are seen as possible outgrowths of a series of local Bronze Age traditions that seemed to have an awareness of iron. An adequate understanding of the technological processes involved in the production of early iron will yield much information regarding the transition to the Iron Age, but such an understanding has yet to be reached. A survey of the literature on metallurgical analyses reveals that although some promising work has been done, there is still much left to do if we are to reach a fuller understanding of the South Asian Iron Age. [Final ms. received 12/98.]

INTRODUCTION

There have been many theories and hypotheses about the beginnings of the Iron Age in ancient India, that is, the modern nation states of Afghanistan, Bangladesh, Bhutan, India, Nepal, Pakistan, and Sri Lanka. The most significant statements have been given by Gordon (1958), Kosambi (1963), Banerjee (1965), Ray (1969), Joshi (1970), Chakrabarti (1974, 1976, 1977, 1979, 1984, 1985, 1992), Tripathi (1976), Shaffer (1978), Allchin and Allchin (1982), and Agrawal (1982). In Asia the earliest human pyrotechnological discoveries resulted in the making of plaster and ceramics (Kingery et al. 1988). This innovation was followed by metallurgy, first of copper, then of iron. There are inter-regional parallels to this line of development which suggest either diffusion or a deep, internal logic to the process of discovery. We believe that there are good reasons for it to be postulated that there were a number of regions in Asia within which independent pyrotechnological innovation took place. While diffusion plays an important role within these regional contexts, it seems not to be the key to understanding the technological innovation of the Iron Age and its multiple occurrences. Although the development of iron metallurgy in South Asia had been seen as a relatively late development, with the subcontinent being the recipient of iron technology, scholars are now looking within the area for other explanations.

The advent of iron in the South Asian subcontinent has been caught up in diffusionist discourses, such as those, for example, that have attempted to explain the Painted Grey Ware and Megalithic cultures as intrusive into North and South India, respectively. Indian archaeological theory has often used migrations of people as explanations for changes in material culture (Chakrabarti 1988). However, subsequent research in these and other areas has begun to demonstrate greater time depth and cultural continuity within the subcontinent than previously envisioned. The "dark age" between the Harappan Civilization and the Painted Grey Ware Assemblage has been closed (Joshi 1978), with continuity and even an overlap demonstrated between the two; and excavations in South India have demonstrated cultural continuity from the Neolithic to Megalithic occupations (Kennedy 1975; Shaffer 1995). Consequently, there seems to be little archaeological evidence to support the view that these cultural phenomena are radical breaks from what came before them.

However, the role of iron technology in these developments is still unclear, and questions remain regarding its origins. In many cases demonstration of cultural continuity or of chronological priority has been deemed sufficient for understanding the technological development of iron (e.g., Chakrabarti 1977; Tandon 1967–68). Culturally the stage seems to be set for indigenous development of iron technology; however the technological environment also needs to be examined. To this end, after an introduction to the archaeological evidence, this paper addresses three interrelated issues: the evidence for "Bronze Age Iron" in the subcontinent and possible origins of ancient South Asian iron technology; the present understanding of technological practice during the Iron Age; and the theoretical framework within which research is being structured.

THE EARLY IRON AGE IN SOUTH ASIA

There are four principal archaeological assemblages that document the Early Iron Age in South Asia: the Gandharan Grave Culture, the Painted Grey Ware Assemblage, the Pirak Assemblage, and the Megalithic Complex (although Chakrabarti [1992] posits a six region division). The early iron from these assemblages has been dated to the first half of the first millennium B.C., with some dates circa 1000 B.C. (Nagaraja Rao 1971; Gaur 1983). A brief bibliographic essay on each of these cultural manifestations follows, to illustrate the regional patterns and character of the early Iron Age in the subcontinent. Other Iron Age assemblages are present in the subcontinent, but they are beyond the scope of this paper for various reasons. For example, the cairn graves of Baluchistan and the Northwest Frontier do not appear to fit the chronological horizon of an "Early Iron Age," as they seem to be dated from the first century B.C. to the early centuries A.D. (see Chakrabarti 1992:36–45 for a survey). Also, there is not yet enough known of the junction between the Late Bronze Age of Central India and the Early Iron Age there (see Agrawal 1982:257 for a brief synopsis) for those assemblages to be discussed here.

THE GANDHARAN GRAVE CULTURE

A series of cemeteries and associated habitation sites in northern Pakistan have produced evidence for Late Bronze Age and Early Iron Age life in the eastern portion of ancient Gandhara (Fig. 6.1). Most of the excavations have been conducted in Dir and Chitral Districts at the cemetery site of Timargarha and the settlement site of Balambat (Dani and Durrani 1964; Dani 1967; Stacul 1969a). Other excavations have taken place in Swat, at places like Aligrama, Butkara II, Loebanr I, Katelai, and Birkot-ghundai (Antonini 1963; Stacul 1966a, b, 1967a, b, 1969b; Antonini and Stacul 1972). Sites are distributed within the southern valleys of the Hindu Kush and possibly Ladakh (Francke 1914:71–74) and through to Peshawar and the Potwar Plateau (Khan 1979, 1983, 1988), possibly extending as far south as the Dera Jat and Thal Plain of the Indus-Jhelum Doab.

The details of ceramics and other small finds, along with continuity in the use of cemeteries and settlements, link these sites together into a complex, not fully understood archaeological assemblage. The earliest of the graves, dating to Period I of the Gandharan Grave Culture of Dani (1967:48), can be assigned to the Late Bronze Age and consist of pits lined and covered by stones. Flexed burials with handmade red and grey pottery and bronze artifacts are found at Timargarha. In Period II, there is evidence of cremation.

Graves in the latest series, dating to Period III of the Gandharan Grave Culture, contain iron as well as artifacts of other metals. At Timargarha seven artifacts of iron are reported, including a spoon, spearhead, and nail (Rahman 1967a:189 and 194–195). More iron artifacts come from the habitation site of Balambat in the same region (Rahman 1967b). The most interesting iron artifact is a cheek piece from a snaffle bit that comes from a Timargarha grave. It was examined by Karl Jettmar (1967:207), who noted:

> It is clear that it belongs typologically to those groups which played a great role in the Steppe-belt between the 10th and the 6th centuries BC. Its shape is rather similar to late pieces in Eastern Europe (6th century BC), also in iron.... With attention to all other objects in the Steppes, a dating in the 7th or 6th centuries BC could be tentatively proposed.

There is a shift in burial practices between the Late Bronze and Early Iron Ages, which is described in Dani (1978). This is accompanied by changes in ceramics and the like. But continuity between the Late Bronze Age and the Iron Age graves is implied by the continued use of the cemeteries. Both Dani and Stacul have proposed a three-phase sequence for the Gandharan Grave Culture. Dani terms the phases Period I, II, and III. Stacul has called them Archaic, Middle, and Late (1969b) and takes general disagreement with the name "Gandharan Grave Culture" (1981:91). Added to this is the position of the graves and habitation sites in the general Swat Valley sequence as defined by the Italian Archaeological Mission there (see Table 6.1).

The chronology of the Gandharan Grave Culture is not yet precise. Jettmar's opinion on the date of the cheek piece from the horse bit has already been quot-

TABLE 6.1
PHASES OF THE GANDHARAN GRAVE CULTURE

DANI 1967	STACUL 1969B	SWAT VALLEY SEQUENCE
Period III	Late	Period VII
Period II	Middle	Period VI
Period I	Archaic	Period V

Figure 6.1   Map of South Asia with sites mentioned in the text.

ed. Dani and Stacul have offered the following general chronologies (Table 6.2).

One can see the general agreement in these chronologies, which have been developed through the use of artifact comparison and comparative stratigraphy. They are not entirely congruent with the radiocarbon dates that are available (Appendix, Table 6.A).

A small amount of iron was found in grave 126 at Katelai Graveyard, which is assigned to Period V in the general Swat Valley sequence (Stacul 1979:341), but there is no iron in any of the 159 Period VI graves that were excavated up to 1979 (Stacul 1979:342). Iron

TABLE 6.2
CHRONOLOGY OF THE
GANDHARAN GRAVE CULTURE

DANI 1967	SWAT VALLEY SEQUENCE STACUL 1981
Period III, 800–500 B.C.	Period VII, 400–300 B.C.
Period II, 1100–800 B.C.	Period VI, 800–400 B.C.
Period I, 1500–1100 B.C.	Period V, 1400–800 B.C.

does occur in the upper levels of Aligrama, which are assigned to the late part of Period VI (Stacul 1981:90). There is an abundance of iron in Period VII graves (Stacul 1979:342). The transition to the Iron Age in ancient Gandhara can be reasonably placed at ca. 1000 B.C., or possibly a century later.

THE PAINTED GREY WARE ASSEMBLAGE

Painted Grey Ware was first defined as a ceramic type by Krishna Deva and R. E. M. Wheeler in the course of their analysis of the pottery from Ahichchhatra (Krishna Deva and Wheeler 1946). The significance of the type, and its widespread occurrence in northern India, came into focus as a result of B. B. Lal's (1954–55:138–143) excavations at Hastinapura.

Iron is widely associated with the Painted Grey Ware, at times in considerable quantity. R. C. Gaur (1983:219) notes of Atranjikhera in the Painted Grey Ware levels: "The presence of iron objects in such profusion and the discovery of furnaces, slags and certain specific tools used by blacksmiths, suggest that not only were iron goods manufactured at the site, but that the smelting of iron ore was also carried out there." Slag in some quantity also comes from Hulas in Saharanpur District of the Ganga-Yamuna Doab and from Jakhera (Sahi 1978).

Painted Grey Ware sites extend from the Bahawalpur region of Pakistan east across the Punjab into Uttar Pradesh in India. The number of sites by state is given in Table 6.3. There is an unpainted grey ware associated with the Painted Grey Ware. This occurs alone at some sites, but the two archaeological assemblages are thought to be one, and contemporaneous. Site counts for the settlements without the painted variety are given in Table 6.4. Painted Grey Ware, or Grey Ware, sites have recently been reported in the vicinity of Islamabad (Salim 1991a, b). While the largest Painted Grey Ware site is Satwadi in Pakistani Cholistan at 13.7 ha (Mughal 1984:501), most of the sites are in the 1- to 2-ha range (Joshi 1993:8).

The ceramic wares of the Painted Grey Ware assemblage are quite different from those of the Harappan and Post-urban Harappan that precede the assemblage in northern India and Cholistan. This difference, and the association with iron, tended to set the Painted Grey Ware apart, causing some archaeologists to think of it as intrusive, even foreign to the subcontinent. But Jagat Pati Joshi and his team from the Archaeological Survey of India have done much to dispel this impression through their explorations and excavations in the Indian Punjab. Their excavations at Bhagwanpura and Dadheri have found evidence for Post-urban Harappan occupations followed by an overlap between the Post-urban Harappan and the Painted Grey Ware (Joshi 1976, 1978, 1993). Thus the historical gap that once existed separating the Early Iron Age cultures from those of the preceding Bronze Age in northern India and Pakistan has now been closed and there is good evidence for cultural continuity between the two periods. Radiometric dates for Painted Grey Ware are presented in Appendix, Table 6.B.

With the exception of TF-191 from Atranjikhera (and the "sport" from Ganwaria) all of the dates for Painted Grey Ware are in the first millennium B.C. Radiocarbon seems to give a reasonable estimate for Painted Grey Ware to be dated in the 900–300 cal B.C. range, close to the estimate suggested by D. P. Agrawal (1982:255). One might push it closer to 1000 B.C. by using TF-191 from Atranjikhera, the Hulas TL date, and the radiocarbon date from Alamgirpur (TF-51) (Possehl 1994). This chronology is largely in agreement with the next Early Iron Age assemblage to be discussed: that from Pirak.

THE PIRAK ASSEMBLAGE

The site of Pirak on the Kachi Plain of the Greater Indus Valley in Pakistan was first reported by Robert Raikes in 1963 following observations he first made in 1956. Its mostly bichrome and grey pottery was intriguing from the start, since Pirak appeared to be a unique site in Pakistan. Since that time four more sites have been reported (Table 6.5) and there are apparently other small sherd scatters in Kachi (J.-F. Jarrige, pers. comm.).

TABLE 6.3
NUMBER OF PAINTED GREY WARE SITES BY STATE
(FROM MUGHAL 1982 AND JOSHI 1993:8)

STATE	NO. OF SITES
Bahawalpur	14
Punjab	108
Haryana	258
Rajasthan	101
Uttar Pradesh	218

TABLE 6.4
NUMBER OF SITES WITH ONLY THE UNPAINTED GREY WARE, AT TIMES ASSOCIATED WITH PAINTED GREY WARE, BY STATE (FROM JOSHI 1993:8)

STATE	NO. OF SITES
Jammu	1
Punjab	109
Haryana	46
Rajasthan	8
Uttar Pradesh	24

Excavations at Pirak took place between 1968 and 1974 by the French Archaeological Mission. This was Jean-Marie Casal's final excavation and the report was prepared in his honor by his colleagues (Jarrige and Santoni 1979; Enault 1979). Period I at the site has the characteristic Pirak pottery and well-built houses. Stray sherds of Wet Ware, Faiz Mohammad Grey Ware, and other third millennium wares indicate some continuity with the Baluchi Bronze Age, as does a compartmented copper stamp seal of Period II. The subsistence regime was based on rice, millets, and other cultivars. There is evidence in the form of figurines for the Bactrian camel and equid, possibly the domesticated horse. A final report on the faunal material has not yet been published, but there are bones of both animals reported, though not assigned directly to the first occupations (Meadow 1979). Artifacts of copper/bronze were found with a rich bone industry. Period II has continuity with Period I and is defined on the basis of new ceramic shapes and fabrics. Copper/bronze tools continue as does the original subsistence regime.

In Period III, evidence of iron production appears, in a large, basically quadrangular set of buildings constructed directly upon the remains of Period II. The iron artifacts, mostly points and "bars," are complemented by finds of crucibles and iron slag, as well as copper/bronze, lead, silver, and gold artifacts. The appearance of certain grey wares in Period III is important in the definition of this phase of occupation.

The chronology for Pirak is not entirely clear. The excavators propose that the site was first settled in the eighteenth century B.C. (Jarrige and Santoni 1979:19), ignoring a mid-third millennium radiocarbon date from Period I (Jarrige and Santoni 1979:352; see Appendix, Table 6.C). The radiocarbon dates for Periods II and III have a remarkable overlap. The excavators believe that Period II directly follows Period I and falls in the thirteenth and twelfth centuries B.C. (Jarrige and Santoni 1979:35). The radiocarbon dates for Period III are considered reliable, and this occupation is thought to begin as early as the twelfth century B.C. and to have been centered on the ninth century (Jarrige and Santoni 1979:44, 373). It is interesting that the Pirak sequence has no evidence for iron prior to Period III. The theory that it begins as early as 1200 or 1300 B.C. may not be as justified as positing a somewhat later date, perhaps approximately 900 B.C., for which there are three very consistent radiocarbon dates.

THE MEGALITHIC COMPLEX

The Megaliths of South Asia are an immense field of study, far too large to be adequately covered in this short survey. Burial monuments incorporating large stone constructions, associated with habitation sites, are known from prehistoric times in many regions of the subcontinent, principally the northwestern mountainous areas of Kashmir and Almora, the Vindhyas (Sharma 1985; Singh 1985), the Deccan, and the deep south of Peninsular India. Outliers in southwestern Sindh have been reported as well (Wheeler 1959:159–160; Qamar 1983). Aspects of this tradition are still in existence in the subcontinent, among the Nagas and Khasis of the northeast, and other tribes of

TABLE 6.5
SITES WITH PIRAK CERAMICS

SITE	REGION	WARE REPRESENTED AT SITE
Chashma Murad	Jhalawan	Complex B
		Pirak?
		Early Kulli
		Amri-Nal
		Kechi Beg
		Togau
Durkhan	Dhadar	Pirak III
Malazai	Quetta-Pishin	Islamic
		Early Historic
		Pirak II
		Quetta
Pathani Damb Two	Kachi	Pirak I
		Mature Harappan
Pirak	Kachi	Pirak III
		Pirak II
		Pirak I
Sulaimanzai	Quetta-Pishin	Pirak II
		Pirak I

the Chota Nagpur and Bastar regions. The connections have not yet been sorted out for this complex, historically deep set of relationships (Fürer-Haimendorf 1945). Cultural relations outside the subcontinent have been seen with regions as far away as Oman (Gupta 1970–71) and Thailand (Loofs-Wissowa 1985). For an up-to-date review, see Deo 1985.

Archaeological stratigraphy, radiocarbon dates, and the study of associated materials inform us, however, that the earliest Megaliths in the subcontinent date from somewhat early in the first or in the very late second millennium, and are associated with the early mass production of iron.

Good descriptions of the varieties of Megalithic burial monuments in South India are in Krishnaswami (1949) and in Sir Mortimer Wheeler (1959:154–158; see also Allchin and Allchin 1982:331–333). Wheeler offers eight types: (1) dolmenoid cists, (2) slabbed cists, (3) shallow-pit burials, (4) deep-pit burials, (5) umbrella-stones and hat-stones, (6) hood-stones, (7) multiple hood-stones, and (8) menhirs. Most of these monuments contain human remains and associated burial furniture, sometimes in great quantity. Horses and vehicles are present, along with pottery and metal artifacts including tools, horse trappings, Roman coins, and other artifacts. Iron occurs in abundance. A selection of excavated sites, both burial and habitation, comprise: Amirthgamangalam (Banerjee 1973); Brahmagiri (Wheeler 1947–48); Chandravalli (Krishna 1931; Wheeler 1947–48); Khapa (Deo 1970); Mahurjhari (Deo 1973); Nagarjunakonda (Subrahmanyam et al. 1975); Naikund (Deo and Jamkhedkar 1982); Paunar (Deo and Dhavalikar 1968); Porkalam (Thapar 1952); Raipur (Deglurkar and Lad 1992); Sanur (Banerjee and Soundara Rajan 1959); Takalghat (Deo 1970). L. Leshnik has produced a general study of Megaliths (1974) as has J. R. McIntosh (1981). Human remains from the burials have been studied by Kenneth A. R. Kennedy (1966, 1975, 1985) and others (Walimbe 1987–88, 1992; Walimbe, Gambir, and Venkatasubbaiah 1991).

S. B. Deo (1985:89) notes that over 1400 Megalithic sites are known in the subcontinent, 1116 of them in Peninsular India. Estimates of the number of sites per state are given in Table 6.6. Deo (1985:89) also notes that 239 of these sites have been excavated in some way, but this record goes back to 1823 when the stone monuments were first recognized as archaeological remains. Furthermore, excavation has been biased towards the funerary monuments themselves, with the result that a relatively small number of habitation sites have been investigated (Moorti 1994).

TABLE 6.6
NUMBERS OF MEGALITHIC SITES IN
PENINSULAR INDIA BY STATE (FROM DEO 1985:89)

Tamil Nadu	388
Karnataka	300
Kerala	188
Andhra Pradesh	147
Maharashtra	90
Pondicherry	3

The earliest iron implements associated with the Peninsular Indian Megalithic are simple implements such as arrowheads, daggers, and domestic vessels. They are associated with the transitional times between the Megalithic Complex and the preceding South Indian Neolithic. There is solid evidence for cultural continuity in the region during this shift in metal technology (Shaffer 1995). The earliest date for iron in the region (ca. 1100 B.C.) is from the Neolithic/Megalithic Transition Period of Hallur on the Tungabadhra River in Karnataka (Nagaraja Rao 1971, 1981). This tends to be supported by two thermoluminescence dates from Kumaranahalli. This Megalithic Complex at Hallur is also a very long lived one, extending well into the first millennium A.D. All of the relevant dates are given in the Appendix, Table 6.D.

Excavations at Gufkral in the Vale of Kashmir by the Archaeological Survey of India (*Indian Archaeology, A Review* 1981–82:19–25) have revealed a "Megalithic" occupation associated with iron. Very early radiocarbon dates for this occupation are given in the Appendix, Table 6.E. These dates would push back the widespread use of iron into the beginnings of the second millennium B.C. There is an underlying Neolithic at Gufkral, however, and we know almost nothing about the nature and amount of iron found in Period II. Could we be documenting an instance of "Bronze Age Iron" in the subcontinent here? Not enough is known to answer these questions, but the dates from Gufkral bring these issues into reasonably sharp focus.

It is evident that the manifestation of the Early Iron Age in South Asia is regional in character. The unifying element seems to be a contemporaneity in the appearance of iron in the various parts of the subcontinent. Far from being a homogeneous development, it can instead be seen to emerge as several localized phenomena.

SOUTH ASIAN PYROTECHNOLOGY—BRONZE AGE IRON?

There is a long history of pyrotechnology in the region involving sophisticated ceramic production, the smelting and manipulation of copper, bronze, lead, gold, and silver and a well-developed faience industry associated with the Harappan Civilization (ca. 2500–1900 B.C.). Thus, the technological and industrial context within which iron technology could have been discovered and developed was very much a part of ancient India. The earliest dates for iron in South Asia (from South India) are now almost contemporary with early iron in other parts of the Old World. Furthermore, iron ores are abundant in virtually every region of South Asia and raw material availability may have enhanced technological innovation, a theme very thoroughly reviewed by Chakrabarti (1977:168–171; 1991:23–51).

It has been argued that a well-developed copper/bronze technology may facilitate the transition to iron technology (Cooke and Aschenbrenner 1975; Charles 1980; Pigott 1981). Not only is there already a pyrotechnological tradition with specialists, but copper smelting can lead to accidental encounters with iron, either through use of iron ore as a flux or through smelting iron-rich copper ore. Such instances can illustrate familiarity with iron as a material. Several pieces of iron have been recovered from eight Bronze Age sites in South Asia. To our knowledge none has been analyzed to determine their technical properties and we do not know which of them is meteoritic and which (if any) were smelted. Some of these finds have also been reviewed by Jim Shaffer (1984). What follows is a compilation of evidence that demonstrates an early knowledge of iron in at least its mineral form. Taken together with a well-developed pyrotechnological tradition and cultural continuity, such early knowledge sets a stage amenable to the development of iron technology.

MUNDIGAK

Five iron items were found at Mundigak in Afghanistan associated with Period IV (2600–2100 B.C.):
1. a copper/bronze bell with an iron clapper (Fig. 6.2; Casal 1961:242, fig. 138, no. 17);
2. a fragment of a copper/bronze rod with two iron "buttons" affixed to it (Fig. 6.3; Casal 1961:249, fig 139, no. 14);
3. an iron button affixed to a copper/bronze mirror (Fig. 6.4; Casal 1961:250, fig. 140, no. 21);
4., 5. two indistinct lumps of "carbonates of iron" which were not illustrated by Casal (1961: 237–238) but weigh 1 and 3 kg, respectively.

SAID QALA TEPE

Said Qala Tepe is located about 100 km southeast of Mundigak (Shaffer 1978). Twenty-four indistinct ferrous "lumps" were recovered from reasonably secure, excavated "pre–Iron Age" contexts at the site, that is, Periods II and III. Shaffer (1984) refers to these nodules as "hammerstones"; they are not formed of smelted iron but are naturally occurring iron minerals. They would be contemporary with the end of Mundigak III and the beginnings of Mundigak IV (ca. 2700–2300

Figure 6.2 Copper/bronze bell with an iron clapper from Mundigak (Casal 1961:fig. 138, no. 17).

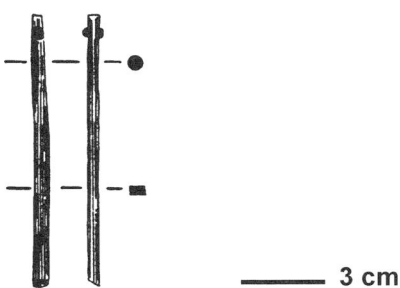

Figure 6.3 A fragment of a copper/bronze rod with two iron "buttons" from Mundigak (Casal 1961:fig. 139, no. 14).

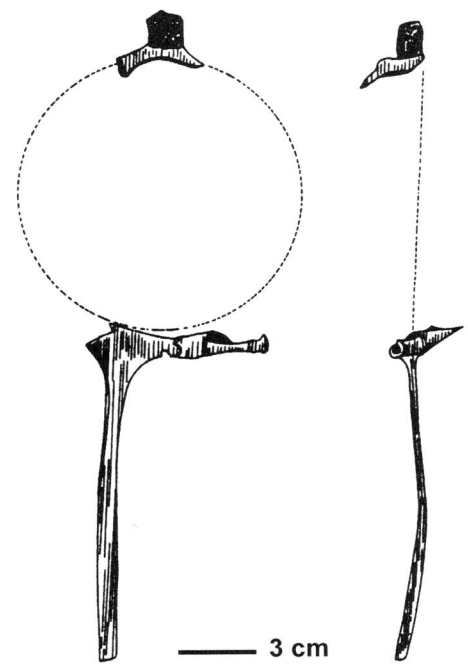

Figure 6.4 Iron button affixed to a copper/bronze mirror from Mundigak (Casal 1961:fig. 140, no. 21).

B.C.) Shaffer makes two interesting observations about the Said Qala iron. First, while copper/bronze artifacts are found in trash pits at the site none of the iron nodules came from this depositional environment. This may indicate that "such nodules possessed some cultural value and were not casually abandoned" (1984:46). Second, a number of the nodules at Said Qala were associated with a large "public" or "monumental" architectural feature, analogous to the findspots for some iron in Mundigak IV (Shaffer 1984:46).

DEH MORASI GHUNDAI

In the course of his excavations at Deh Morasi Ghundai in southern Afghanistan approximately 12 km west of Said Qala Tepe, Louis Dupree found a nodule of magnetite that had been utilized by the occupants of Period IIA (Dupree 1963:99). This was associated with the so-called "Shrine Complex" and would also be contemporary with Mundigak IV, although the precise chronology of the site is debated (Shaffer 1984:48).

AHAR

The site of Ahar, near the modern city of Udaipur (Sankalia et al. 1969), is located in southern Rajasthan. A small body of literature has been built up around the notion of Bronze Age iron at Ahar (Sahi 1979; Sankalia 1979; Lal 1987–88). Sahi encourages us to believe that the iron implements listed in Table 6.7 are in

TABLE 6.7
IRON IMPLEMENTS FROM AHAR (SANKALIA ET AL. 1969:202–203)

CONTEXT	IRON OBJECTS	FIGURE REFERENCE
	Period Ic	
Trench C, stratum 2	1 broken arrowhead	
	1 tanged arrowhead	Fig. 6.5, nos. 1, 2
Trench C, stratum 3	nail or wire	Fig. 6.5, no. 3
Trench D, stratum 1	1 tanged arrowhead?	
	1 chisel	
	nail or punch	Fig. 6.5, nos. 4, 5, 6
Trench D, stratum 2	tube or socket	Fig. 6.5, no. 7
Trench L, stratum 1	chisel	Fig. 6.5, no. 8
Trench X, stratum 5	2 arrowheads	Fig. 6.5, nos. 9, 10
	Period Ib	
Trench L, stratum 2	tanged arrowhead	Fig. 6.5, no. 11
Trench E, stratum 3	ring	Fig. 6.5, no. 12

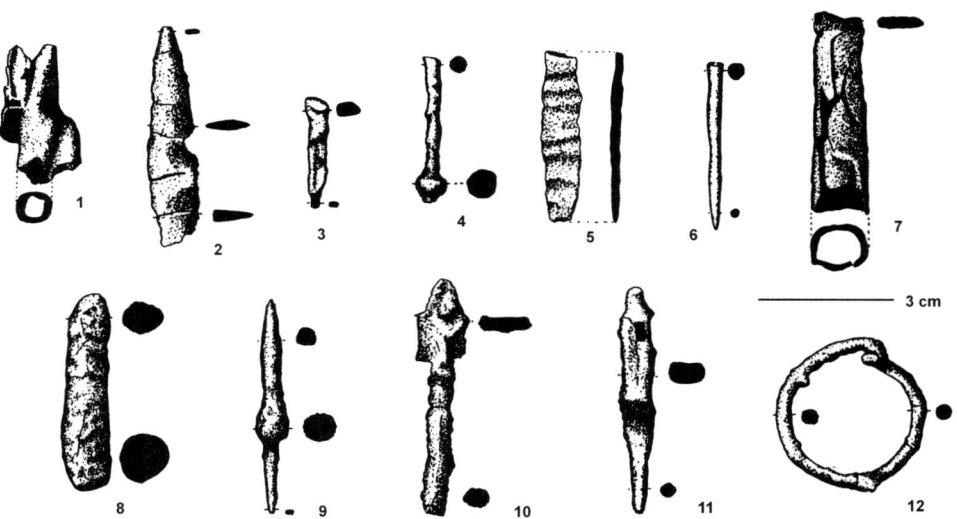

Figure 6.5   Iron implements from Ahar listed in Table 6.7 (after Sankalia et al. 1969:202–203).

their correct stratigraphic association, pointing out (1979:366) in reference to trench X, layer 5 that "from this deposit were obtained dish-on-stand and the lustrous red ware, besides the painted black-and-red ware. Charcoal sample TF-31 was also collected from this layer and it is this very deposit which had yielded two iron arrow-heads." TF-31 comes from Ahar Ic and is 1275±110 B.C., based on the 5730 half-life. The painted black-and-red ware is characteristic of the Bronze Age Period I at the site. However, Sankalia, who led the excavation of the site, pointed out in a review of Sahi's paper (1979:1–2) that Ahar had many pits and that there was "a great chance for contamination, the later objects getting mixed-up with the earlier." Makkhan Lal (1987–88), in a thoughtful paper, also points out that the strata in which the early iron was found were mixed, most likely due to redeposition from later periods. This leaves the artifacts in questionable stratigraphic context.

A chronology for Ahar I can be reasonably reconstructed as follows: Period Ia, 2600–2150 B.C.; Period Ib, 2150–1950 B.C.; and Period Ic, 1950–1500 B.C. (Possehl and Rissman 1992).

CHANHU-DARO

Chanhu-daro produced one iron artifact which looks like a broad, flat "Harappan"-type pointed blade, possibly an "arrow head" (Fig. 6.6; Mackay 1943:pl. LXXII, no. 24). Mackay attributes it to the Jhukar(?) levels on the plate caption and to Jhangar(?) in the text (1943:189, 304); however it was found just below the surface, raising questions about its context. It was the only piece of iron found in the Chanhu-daro excavations.

MOHENJO-DARO

At Mohenjo-daro (Marshall 1931:32, 485, 690) lollingite (an arsenic- and iron-bearing mineral) was found, suggesting that it may have been used in the smelting of arsenical copper or preparation of arsenic.

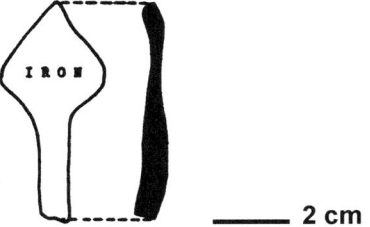

Figure 6.6   *"Harappan"-type pointed iron blade from Chanhu-daro (Mackay 1943:pl. LXXII).*

LOTHAL

A fragmentary piece of metal (object no. 15112) was recovered from Lothal, which is 66.10% iron and 9.30% copper (Lal 1985:table XXVIII, no. 11, p. 658), a high level of copper for an iron object. The occupational phase that this artifact came from is not known; however, it would have to be either Lothal A, Urban Harappan (2500–2000 B.C.), or Lothal B, the Post-urban Harappan (2000–1800 B.C.).

KATELAI GRAVEYARD

A single piece of iron has been found in the Swat Valley Period V cemetery site of Katelai (Stacul 1979:341). This artifact is highly corroded and has not been subjected to metallurgical testing. Period V in the Swat Valley sequence would date to 1500–800 B.C. and is associated with an abundance of copper-base artifacts.

SUMMARY OF BRONZE AGE IRON IN SOUTH ASIA

The eight sites with Bronze Age iron that have been very briefly outlined above are a most provocative data set. What emerges from the data is evidence that there existed, during the Bronze Age, an awareness of iron ores, of their hardness, and the possibility of encountering iron during copper smelting. The evidence from Said Qala Tepe especially documents an exploitation of the hardness of iron nodules in their use as hammerstones. The bimetallic artifacts from Mundigak have not been analyzed to determine whether or not the iron "buttons," or small nodules, were smelted. They can just as easily be iron ore nodules which were chosen for a variety of reasons. If bimetallic artifacts are an indication of the transitional phase between Bronze and Iron Ages (Pigott 1981), then the evidence from Mundigak does suggest such a phase in the area. However, to date there has not been any conclusive evidence for smelted iron during the Bronze Age. What is evident is an awareness of iron ore as a part of the natural environment, and willingness to exploit one or more of its characteristics.

It should also be noted that Chattopadhyay (1991) has referred to iron artifacts from the late Chalcolithic levels at Pandurajar Dhibi and Mangalkot (West Bengal), and has proposed the term "ferrochalcolithic" for this period. Furthermore, Datta (1998) also notes that iron artifacts have been recovered from the Chalcolithic levels at several sites in West Bengal, and argues that they constitute an Iron Age. This material has not been completely reported, however, and needs further investigation.

EARLY IRON AGE METALLURGY IN SOUTH ASIA

Knowledge of production techniques is essential to understanding the origins and development of iron technology in South Asia. Traditions of metalworking vary from region to region, shaped by cultural conditions, and the innovation of iron has to be locally situated. Metallurgical and metallographic analyses to determine composition and treatment can aid in determining how various regimes of metalworking interacted with each other within the subcontinent as well as with those in neighboring regions.

METALLURGICAL AND METALLOGRAPHIC ANALYSES

The analysis of Early Iron Age metals has been done by a number of scholars, including Agrawal (1983), Agrawal et al. (1983), Agrawal et al. (1990), De and Chattopadhyay (1991), Ghosh et al. (1987–88), Gogte (1982a, b), Gogte et al. (1981–82), Joshi (1973), and Munshi and Sarin (1970).

While a large number of iron artifacts have been recovered in India, the metallurgical analysis has not always kept pace. Much of the analysis that has been performed has not gone beyond compositional analysis and determination of steeling. This is partly due to the poor state of preservation, as a great number of artifacts are completely corroded. Furthermore, not all better-preserved specimens are accessible for analysis (Chattopadhyay 1984). Unfortunately, compositional analysis is rarely enough to illuminate metallurgical practice clearly—to understand how ancient smiths were producing and using their implements. The following summary of work highlights analyses of artifacts from the Early Iron Age that have included metallographic analyses, and does not address investigations of Early Historic and later artifacts.

Analyses from the excavation reports of Mahurjhari, Takalghat, and Khapa (Megalithic sites) deal with one artifact each of iron. From Mahurjhari (Joshi 1973), an iron axe from Megalith 4/Locality III (no date) was identified as containing 99.1% Fe and 0.9% C, with a trace of chromium. The axe was said to be a steel. From Khapa (Munshi and Sarin 1970) an iron spear from Megalith 7, dated from the eighth to the fourth centuries B.C., was analyzed for composition only. It was found to contain 99.6% Fe, and may have been "a variety of steel." Two major points arise from these analyses: first, one artifact does not provide a clear picture of metallurgical practice, and limiting analysis to compositional analysis does not take full advantage of the information contained within an artifact.

Agrawal et al. (1983) analyzed, both metallurgically and metallographically, three iron artifacts—two implements and an axe—from Tadakanahalli (ca. 1000 B.C.). Multiple sections were taken from each sample, allowing the reconstruction of the production process. The investigators concluded that each artifact was composed of two definite layers; that the edge was mainly martensite, meaning that the implements had undergone quenching; that layers of ferrite and pearlite are present; and that martensite decreases and pearlite/ferrite increases toward the interior. Furthermore, it was evident that layers of wrought iron that had been carburized were forged with those of high carbon content.

Metallographic examinations of iron artifacts from a series of sites from various time periods have produced evidence for an apparently long tradition of lamination techniques (Agrawal et al. 1990). A spear from Kumaranhalli (ca. 1200–1100 B.C.) and an axe from Tadakanhalli (ca. 1000 B.C.), both Megalithic sites in Karnataka, were subjected to metallographic analysis. The results revealed alternating layers of carburized (characterized as a hypoeutectoid steel) and uncarburized iron which had been hammered and welded together. There was no evidence for hardening or tempering. This lamination technique seems to continue, as revealed by analyses of later artifacts, including an axe from Jajamau dated to ca. 600–300 B.C., the Northern Black Polished Ware/Mauryan Phase. Examination of the cutting edge revealed two high carbon (medium steel) layers around a low carbon (mild steel) layer. In the final stages of preparation the axe had been quenched, tempered, and air cooled. It should be noted that of all the artifacts examined from these two sites, the latter axe was the only one which evinced any evidence of quenching.

Other evidence for the Northern Black Polished Ware Phase comes from Rajghat. Bhardwaj (1973) metallographically examined four iron tools dated to ca. 600 B.C., and found that while there was evidence for carburization, it was not consistent throughout the sample. O. P. Agrawal's (1983) examination of four iron artifacts from the Painted Grey Ware levels at Atranjikhera (ca. 900 B.C.) revealed evidence of surface carburization in three of the four implements, with the fourth being wrought iron.

These results indicate that specific properties of iron were at least beginning to be exploited by the smiths, who were able to manipulate high carbon and low carbon sheets to their advantage. It is results like these that allow not only reconstruction of metalworking practices, but also formulation of a coherent definition for the "Iron Age," one that is based on technological competence of prehistoric populations (i.e., the ability to manipulate the properties of iron to suit their needs).

Ghosh et al. (1987–88) performed metallographic examination of an iron dagger from Hatigra, West Bengal (ca. 1000 B.C.) as well as a chemical analysis and found the object to be a low carbon hypoeutectoid steel. De and Chattopadhyay (1991) carried out compositional and metallographic analyses of three iron artifacts (sword, shapeless bit, nail) from Dhuliapur and one (nail/rod) from Kankrajhar (West Bengal, no date provided) in order to determine similarities and differences in iron production between the neighboring sites. The results of the elemental analysis showed a difference between the samples from the two sites (lead was present in the artifact from Kankrajhar while absent in the other three). Metallographic analyses revealed pearlite and ferrite present in the Dhuliapur artifacts and martensite in the one from Kankrajhar.

Also from eastern India comes the metallographic analysis of a sickle (ca. 800 B.C.) from Neolithic deposits. Ghosh and Chattopadhyay (1982) sampled the artifact in three places and determined that it was a low carbon steel that had been forged at about 900°C and then allowed to air-cool slowly. However, as the investigators point out, the Neolithic character of the deposits and the fact that the sickle was a singular find do seem to indicate that it may have been imported rather than locally produced.

Chattopadhyay (1990) delineates for eastern India a three-stage sequence of early iron technology based on compositional and metallographic analyses of four iron artifacts. Stage I is illustrated by a dish from Pandurajar Dhibi (ca. 900 B.C., West Bengal) and is characterized by a lack of carburization and incomplete separation of slag and metal. Stage II sees the beginning of carburization, and is defined as the beginning of the Iron Age by Chattopadhyay. A dagger from Hatgira (ca. 900 B.C.) and a sickle-shaped agricultural implement from Barudih (ca. 850 B.C.) reveal pearlite and ferrite and evidence for slow cooling. Carbon content in the interior is 0.4–0.5%, while at the surface it is 0.6–0.8%. The final stage is characterized by quenching and tempering and is illustrated by a sickle from Pandurajar Dhibi dating to the third century B.C.

Technical studies such as these are able to exploit the wealth of information contained within iron artifacts, information that is extremely useful in helping archaeologists reconstruct the practices of ancient smiths. Such information, once situated within larger contexts of production and specialization, can illuminate social, political, and economic processes.

EVIDENCE FOR PRODUCTION

Another important aspect of iron technology that can be revealed archaeologically is production evidence. Despite the abundance of iron artifacts that have been recovered from archaeological contexts, direct evidence for the production of iron (e.g., workshops, furnaces) is rare.

The best-described furnace comes from Naikund (Maharashtra) (Fig. 6.7; Deo and Jamkhedkar 1982; Gogte 1982a, b), a Megalithic site with associated habitation area dated from the sixth to the fourth centuries B.C. The furnace was located through a three probe resistivity survey of the habitation area. The center of an area of greater resistivity was chosen as the site for a 4×4 m trench. Though actually in the habitation area,

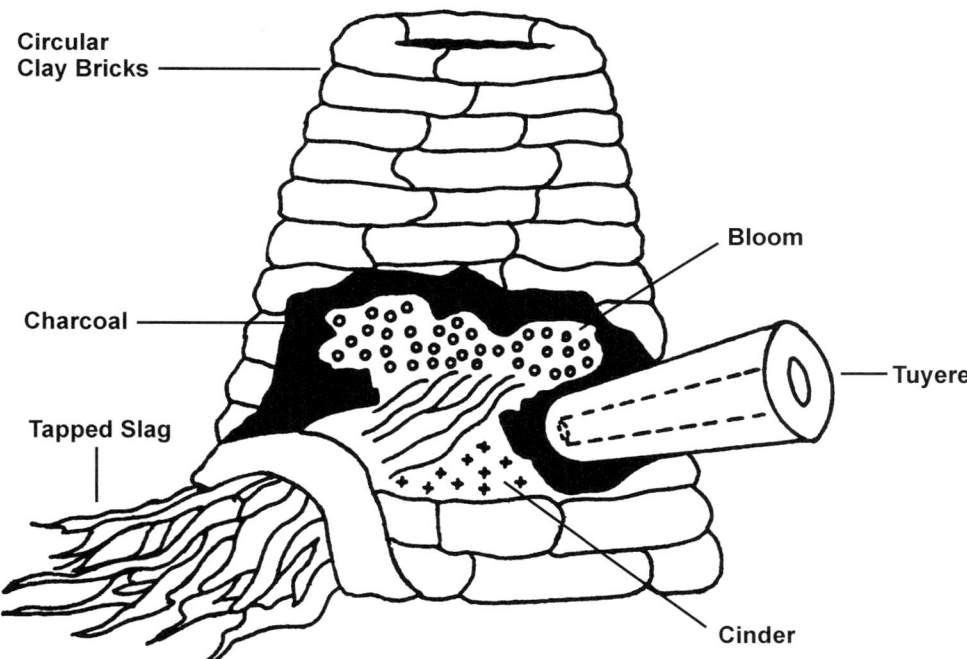

Figure 6.7 *Reconstruction of iron smelting furnace from excavated evidence, Naikund (Deo and Jamkhedkar 1982).*

this trench was located away from the main habitation excavations. The furnace was circular, about 30 cm in diameter and 25 cm in height, built up of interlocking clay bricks. The bottom of the furnace was paved with bricks. A taphole for the slag was detected, and two (vitrified) tuyères were recovered, as well as 40 kg of slag. Only one piece of iron was recovered, corroded and approximately 5 cm long, 0.5 cm thick. A few pieces of iron and manganese ores were also recovered, indicating the possible exploitation of the nearby manganeferrous belt. Despite the relatively detailed description of the furnace and the evidence for iron production at the site, however, there does not seem to be any greater architectural or settlement context. However, if the zone (ca. 66 m²) of greater resistivity is any indication, then perhaps we are seeing evidence for localized pyrotechnological or industrial activity within the settlement.

At Naikund, Gogte (1982a, b) also attempted to evaluate the efficiency of iron smelting carried out at the site. He determined that using about 10–12 kg of ore per operation, the Megalithic smelters were able to produce 3.0–4.2 kg of pure iron. Additionally, Gogte also determined that the ore being smelted was local, with rubble evidence located about 1 km southeast of the site.

Excavations at Atranjikhera (UP) (Gaur 1983) revealed a furnace and a "fire pit" (a possible furnace?). The furnace came from the Upper Phase of Period II (associated with the Painted Grey Ware culture) and was associated with a mud floor, a mud wall, a hearth, and the fire pit. A pair of tongs were recovered next to it (Fig. 6.8). Although Gaur does not elaborate further on the furnace, he does indicate that what he calls "fire pits" may also have functioned as furnaces, citing openings for bellows and the presence of iron tools within (1983:129 and pl. XXVII).

From Phase A of Period IV of the same site (associated with the Later Northern Black Polished Ware Culture) comes another fire pit/possible furnace. Associated with it were a mud floor, two domestic hearths, a drain, and a portion of a mud wall. The pit was well preserved, and from within it were recovered three finished iron artifacts as well as iron slag.

Even from the few examples mentioned above, it is apparent that a great deal of information about metal producing practices can be uncovered from metallurgical analyses and consideration of production facilities. It is this type of data that can allow more interesting discussion of the Iron Age, in South Asia and elsewhere.

IMPLICATIONS FOR THE SOUTH ASIAN IRON AGE

The full sociocultural and historical implications of the South Asian Iron Age seem to be subtle. It can be demonstrated that the iron tools (of the Iron Age) did not clear the forest of the great South Asian river valleys for farming (Lal 1986). Nor did they lead to the subjugation of indigenous populations by invading warriors. However, the exact nature of their impact is poorly understood, as is the nature of their innovation. The principal advantage in India and Pakistan of iron over copper/bronze would seem to be its broad and ready availability. If there was going to be a "metal for the masses" in South Asia it could not be based on copper. Surely this is something which must be considered as a historical implication, as a notion that has much historical appeal in South Asia. Two questions suitable for discussion and debate emerge from considerations of the South Asian Iron Age.

The first issue deals with the very nature and definition of an Iron Age. During the 1973 symposium "Radiocarbon and Indian Archaeology" there was a panel discussion on periodization. The diversity of opinion reflected the ambiguity and arbitrariness inherent in the demarcations of time so necessary to archaeology. The term "Iron Age" suffers from this same ambiguity. During that same discussion, M. C. Joshi proposed the following as a definition of an "age": the "span of time within which the dominant trait of a culture or cultural groups is determined by the basic material or materials, like copper, iron, etc., used for production by one or more societies" (quoted in Agrawal and Ghosh 1973:517). This definition seems to imply that the presence of a trait (such as iron) would qualify a site or a culture as pertaining to an "Iron Age." And indeed, the presence or absence of iron implements seems to have been the defining criterion by which the "Iron Age" label has been applied. The crucial point is the assessment of dominance, and it merits further discussion. The dominance of iron in the ar-

Figure 6.8   *Pair of tongs from Painted Grey Ware levels (Period III) at Atranjikhera (Gaur 1938:223, fig. 65).*

chaeological record has more often than not been associated with concomitant cultural change. Consequently, to refer to the "Iron Age" has begun to have connotations of social complexity and cultural evolution—carry-overs from the ideals of "technological stages" (Childe 1944).

Chakrabarti (1985:74) rightly notes that "not enough thought has been given to a precise definition of the Indian Iron Age." He questions whether or not what is thought of as Iron Age material really yields a new cultural pattern in contrast with the earlier copper/bronze or Neolithic/Chalcolithic background. He notes that at Pirak and Malwa, the Iron Age is more of a continuation of the past than a break with it, except for the use of iron. In the south, with the Megalithic association with iron and the Iron Age, little has been done beyond enumeration of burial types and assemblages. He further questions whether the first well-defined iron bearing levels in the Gangetic valley—the Painted Grey Ware levels—delineate an Iron Age: "To what extent can the beginning of iron in the Gangetic valley be related to something fundamentally innovative in the archaeological record?" (1985:76). The cultural change comes with the Northern Black Polished Ware period and not that of the Painted Grey Ware. Finally he points out that a region-by-region analysis shows that the existence of a pan-Indian Iron Age is highly doubtful.

If it is no longer feasible to define the Iron Age in terms of cultural and social change, then how can it be defined in a way that is meaningful to archaeologists? This point regarding the need for common ground is intimately linked to the first point discussed above, i.e., how are we going to define the Iron Age? Is it even possible to define the term in such a way that it will have cross-cultural validity? Snodgrass (1980) proposes that the definition of Iron Age be more closely tied to technical capabilities rather than other cultural factors. His criterion is that of "working iron," meaning iron that is "used to make the functional parts of the real cutting and piercing implements that form the basis of early technology.... [F]unctional parts may be defined as those parts which come into direct contact with the material to be cut or pierced, whether inanimate or (as in the case of weapons) animate" (1980:336). Indeed, this is the distinction that Thomsen employed in his formulation of the three-age system: "Thomsen, however, knew from his wide experience that bronze weapons and implements with cutting edges never occurred in the finds together with iron objects, and he made this observation the basis of his definition of the metal ages" (Graslund 1981:48). It may be this type of definition, based on ideas of technological competence, divorced from notions of cultural progress and evolution, that may allow a broader application of the term "Iron Age," one that may serve as a common starting point. If not, one can at least wonder whether such a typology—Stone, Bronze, Iron—remains useful and productive for continuing research.

The seemingly coincidental nature of the coming of the Iron Age near the end of the second millennium B.C. from the eastern Mediterranean across Asia to the South China Sea presents us with a need to measure the implications of this and consider common historical processes. We cannot make progress in understanding such inter-regional processes if we narrow our sights to our own regions and define an Iron Age in a way that fits only the local scene. If the Iron Age of India was an independent development, how is it to be seen in relationship to the other Iron Ages of surrounding regions, especially if they are all emerging within a similar time frame?

While the most obvious issue may be continuing imprecision in local chronologies, a more important one is the lack of a systematic approach to the Iron Age, in theory and in excavation. There has not been enough systematic research done with the goal of understanding the innovation of iron technology to make a definitive statement for either diffusion or indigenous invention. Although greater understanding of cultural change is emerging, the same cannot be said of the shift to widespread iron technology, and arguments and evidence for technological continuity or discontinuity need to be explicitly addressed if we are to reach a fuller understanding of the South Asian Iron Age.

APPENDIX

TABLE 6.A
RADIOCARBON DATES FOR THE GANDHARAN GRAVE CULTURE AND RELATED ARCHAEOLOGICAL ASSEMBLAGES

PERIOD AND SITE	LAB NO.	CALIBRATED DATE BC (1δ CAL) (CALIB-3 PROGRAM)
DATES FOR THE CHALCOLITHIC		
Aligrama Swat Valley Period IV Chalcolithic	P-2151A P-2151B	1681 (1625) 1529 1376 (1259, 1232, 1227) 1129
Kalako-deray	BM-0000	1617 (1527) 1518
Loebanr III	P-2584 P-2583 P-2585 P-2586	1443 (1410) 1320 1674 (1522) 1435 1598 (1516) 1435 1734 (1671, 1664, 1636) 1527
Timargara Period I Bronze Age Cemetery	H-XXX	1740 (1677) 1538
DATES FOR THE LATE BRONZE AGE		
Aligrama Swat Valley Period V Late Bronze Age	PRL-186 PRL-243 PRL-244 P-2150 PRL-246	1527 (1375, 1348, 1317) 994 1259 (1045) 916 916 (810) 768 1406 (1387, 1337, 1325) 1270 1516 (1381, 1342, 1321) 1063
Katelai	R-476 R-477 R-477A	1527 (1414) 1220 1121 (1009) 926 923 (897, 870) 826
Loebanr I	BM-196 BM-195	1257 (999) 827 1405 (1251, 1248, 1205) 943
DATES FOR THE EARLY IRON AGE		
Barama I Swat Valley Period VI Early Iron Age	R-196 R-195	812 (794) 569 400 (391) 374
Butkara II	R-194	752 (472, 467, 417) 404
Katelai	R-279 R-479	192 (157, 137, 125) 53 385 (365, 275, 264) 201
Loebanr I	R-278 R-276 R-474	481 (403) 393 763 (750, 746, 526) 409 751 (405) 391
Timargara Period III Iron Age Cemetery	H-XXY	1007 (924) 852

TABLE 6.B
RADIOMETRIC DATES FOR PAINTED GREY WARE

SITE AND PERIOD	LAB NO.	CALIBRATED DATE BC/AD (1δ CAL) (CALIB-3 PROGRAM)
RADIOCARBON DATES		
Ahichchhatra, Late Painted Grey Ware	TF-317	367 (185) 43 BC
Alamgirpur, Painted Grey Ware	TF-51	AD 889 (997) 1035
Allahapur, Painted Grey Ware	PRL-81	399 (372) 195 BC
Allahapur, Painted Grey Ware, Northern Black Polished Ware Overlap	PRL-83	371 (189) 43 BC
Atranjikhera, Painted Grey Ware	TF-194 TF-287 TF-291 TF-191	759 (409) 392 BC AD 381 (435) 591 764 (410) 390 BC 1255 (1034) 913 BC
Atranjikhera, Painted Grey Ware, Northern Black Polished Ware Overlap	UW-XXXX BM-194 BM-193	783 (408) 269 BC 790 (413) 372 BC 199 (36) BC AD 125
Bateshwar, Painted Grey Ware	PRL-198	793 (758, 679, 650, 547) 407 BC
Ganwaria, Painted Grey Ware	PRL-325	3508 (3360) 3107 BC
Hastinapura, Painted Grey Ware	TF-83 TF-90 TF-91 TF-112 TF-85	393 (353, 304, 208) 113 BC 403 (372) 189 BC 791 (519) 393 BC 398 (368) 191 BC 763 (404) 368 BC
Hastinapura, Painted Grey Ware (intrusive burial?)	UCLA-684	AD 119 (239) 382
Jodhpura, Painted Grey Ware	PRL-274 PRL-213 PRL-273 PRL-272	399 (365, 275, 264) 173 BC 391 (347, 316, 205) 101 BC 519 (389) 193 BC 976 (814) 765 BC
Jodhpura, Painted Grey Ware, Black and Red Ware Overlap	PRL-212	401 (372) 193 BC
Khalaua, Painted Grey Ware	PRL-68 PRL-67 TF-1228	772 (401) 202 BC 798 (519) 382 BC 764 (413) 392 BC
Mathura, Painted Grey Ware, Northern Black Polished Ware Overlap	PRL-342 PRL-340	395 (196) 2 BC 772 (405) 257 BC
Mathura, Sonkh Mound, Painted Grey Ware	I-6277	809 (790) 543 BC
Noh, Painted Grey Ware	UCLA-703B TF-994 TF-993 UCLA-703A	1112 (822) 533 BC 810 (785) 525 BC 895 (797) 522 BC 843 (755, 686, 540) 249 BC
Thapli, Painted Grey Ware	PRL-731 PRL-732	193 (31, 18, 9) BC AD 125 199 (50) BC AD 72
Tilaurakot, Painted Grey Ware	TF-737	393 (359, 285, 254) 173 BC
THERMOLUMINESCENCE DATES		
Hulas, Painted Grey Ware	PRL-HLS-TL-1	965 BC
Sanghol, Painted Grey Ware	PRL-TL-14	AD 227

TABLE 6.C
RADIOCARBON DATES FOR PIRAK

LAB NO.	CALIBRATED DATE BC/AD (1δ CAL) (CALIB-3 PROGRAM)
PERIOD III	
LY-1643	1396 (1196, 1181, 1165, 1141, 1139) 943 BC
TF-1108	898 (807) 777 BC
LY-1644	AD 59 (239) 424
PERIOD II	
LY-1642	1527 (1414) 1220 BC
TF-1202	1262 (1125) 1004 BC
PRL-391	987 (843) 803 BC
PRL-390	987 (843) 803 BC
PRL-388	994 (843) 801 BC
TF-1201	922 (807) 559 BC
PRL-389	823 (795) 546 BC
PERIOD I	
LY-1640	1884 (1731, 1728, 1686) 1520 BC
LY-1641	2924 (2586) 2197 BC
TF-1109	982 (829) 794 BC
TF-861	902 (810) 784 BC

TABLE 6.D
RADIOMETRIC DATES FOR MEGALITHIC IRON IN PENINSULAR INDIA

SITE AND PERIOD	LAB NO.	CALIBRATED DATE BC/AD (1δ CAL) (CALIB-3 PROGRAM)
RADIOCARBON DATES		
Alagankulam, Period II, Black and Red Ware	PRL-1297	361 (173) 36 BC
	PRL-1296	199 (90, 67) BC AD 19
	PRL-1298	401 (361, 282, 257) 113 BC
Appukullu, Megalithic?	BS-38	402 (359, 285, 254) 95 BC
Bhagimohari, Megalithic	XX-000	835 (801) 766 BC
	XX-000	757 (405) 382 BC
Hallur, Period IB, Late Neolithic	TF-586	1415 (1310) 1134 BC
	TF-576	1679 (1522) 1426 BC
Hallur, Period II, Neolithic/Iron Age Overlap	TF-575	1255 (1039) 918 BC
	TF-573	1116 (976, 965, 935) 837 BC
	TF-570	1378 (1196, 1181, 1165, 1141, 1139) 1007 BC
Hathinia Hill, Megalithic	TF-109	AD 1702 (1955*) 1955*
Kakoria, Megalithic	TF-179	AD 1646 (1673, 1779, 1797, 1945, 1953) 1954
	TF-183	AD 1643 (1672, 1781, 1795, 1946, 1953) 1954
Kilayur, Black and Red Ware	TF-207	385 (337, 324, 202) 101 BC
Korkai, Megalithic	TF-987	906 (818) 795 BC

TABLE 6.D (CONTINUED)

SITE AND PERIOD	LAB NO.	CALIBRATED DATE BC/AD (1δ CAL) (CALIB-3 PROGRAM)
Kotia, Megalithic?	TF-319	359 (169) 33 BC
	TF-322	AD 1301 (1403) 1437
	TF-320	AD 1401 (1436) 1480
	TF-321	AD 993 (1034) 1212
	TF-318	AD 1636 (1666) 1954
Naikund, Megalithic	BS-94	797 (759, 676, 658, 641, 550) 405 BC
	BS-92	782 (522) 398 BC
Paiyampalli, Megalithic	TF-828	333 (102) BC AD 9
	TF-824	AD 1183 (1271) 1294
	TF-823	800 (764, 614, 606) 410 BC
	TF-825	AD 1262 (1294) 1396
Paiyampalli, Neolithic/Megalithic Transition	TF-827	2033 (1892) 1747 BC
Paiyampalli, Late Southern Neolithic	TF-833	1735 (1499, 1481, 1456) 1223 BC
	TF-829	AD 978 (1027) 1176
	TF-832	AD 1193 (1278) 1300
	TF-829 (BS)	AD 787 (897, 910, 958) 1015
Palavoy, Neolithic/Megalithic Transition	TF-700	1855 (1680) 1526 BC
Polakonda, Megalithic	BS-97	170 (38) BC AD 70
Satanikota, Megalithic	BS-202	AD 544 (635) 672
	BS-201	AD 340 (427) 554
	BS-204	8081 (8018) 7927 BC
	BS-203	6458 (6372) 6182 BC
T. Narsipur, Megalithic	TF-414	AD 1638 (1666) 1954
Takalghat, Pre-Iron Megalithic	TF-783	796 (759, 676, 658, 641, 550) 406 BC
	TF-784	768 (487, 442, 424) 395 BC
	TF-784 (BS)	758 (401) 264 BC
Togarappalli, Megalithic	PRL-135	369 (181) 36 BC
	PRL-134	377 (196) 58 BC
Veerapuram, Megalithic	PRL-730	1525 (1414) 1227 BC
	PRL-728	1259 (1009) 842 BC
	PRL-729	1197 (987, 956, 944) 822 BC
	PRL-727	357 (90, 67) BC AD 72
	PRL-725	AD 83 (249) 420

THERMOLUMINESCENCE DATES

Kumaranhalli	PRL-TL-47	1290±90 BC
	PRL-TL-50	1270±170 BC

1995 denotes influence of bomb ^{14}C

TABLE 6.E
RADIOCARBON DATES FOR GUFKRAL, PERIOD II, MEGALITHIC IRON AGE

LAB NO.	CALIBRATED DATE BC (1δ CAL) (CALIB-3 PROGRAM)
BS-434	2195 (2035) 1900
BS-431	1885 (1747) 1677
BS-433	2131 (1945) 1779
BS-371	1888 (1747) 1674

REFERENCES CITED

Agrawal, D. P.
1982 *The Archaeology of India.* Scandinavian Institute of Asian Studies Monograph Series no. 40.

Agrawal, D. P., and Ghosh, A.
1973 *Radiocarbon and Indian Archaeology.* Bombay: Tata Institute of Fundamental Research.

Agrawal, O. P.
1983 Scientific and Technological Examination. In *Excavations at Atranjikhera*, ed. R. C. Gaur, pp. 487–498. Delhi: Motilal Banarsidas and Center for Advanced Study, Department of History, Aligarh Muslim University.

Agrawal, O. P.; Harinarain; and Bhatia, S. K.
1983 Technical Studies of Iron Implements from the Megalithic Site Tadakanahalli. *Puratattva* 12:97–100.

Agrawal, O. P.; Narain, H.; Prakash, J.; and Bhatia, S. K.
1990 Lamination Technique in Iron from Artifacts in Ancient India. *Journal of the Historical Metallurgy Society* 24(1):12–26.

Allchin, B., and Allchin, R.
1982 *The Rise of Civilization in India and Pakistan.* Cambridge: Cambridge University Press.

Antonini, S.
1963 Preliminary Notes on the Excavation of the Necropolises found in Western Pakistan. *East and West* 14:13–26.

Antonini, S., and Stacul, G.
1972 *The Protohistoric Graveyard of Swat (Pakistan).* Rome: IsMEO.

Banerjee, N. R.
1965 *The Iron Age in India.* Delhi: Munshiram Manoharlal.
1973 Amirthamangalam 1955: A Megalithic Urn-burial Site in District Chingleput, Tamilnadu. *Ancient India* 22:3–36.

Banerjee, N. R., and Soundara Rajan, K. V.
1959 Sanur 1950 & 1952: A Megalithic Site in District Chingleput. *Ancient India* 15:4–42.

Bhardwaj, H. C.
1973 Aspects of Early Iron Technology in India. In *Radiocarbon and Indian Archaeology*, eds. D. P. Agrawal and A. Ghosh, pp. 391–399. Bombay: Tata Institute of Fundamental Research.

Casal, J.-M.
1961 *Fouilles de Mundigak,* 2 vols. Memoires de la Delegation Archéologique Français en Afghanistan 17. Paris.

Chakrabarti, D. K.
1974 Beginning of Iron in India: Problem Reconsidered. Pp. 345–356 in *Perspectives in Palaeoanthropology: Professor D. Sen Festschrift*, ed. A. K. Ghosh. Calcutta: Firma K. L. Mukhopadhyay.
1976 The Beginnings of Iron in India. *Antiquity* 50:114–124.
1977 Distribution of Iron Ores and the Archaeological Evidence of Early Iron in India. *Journal of the Social and Economic History of the Orient* 20(2):166–184.
1979 Early Iron Age in the Indian Northwest. Pp. 347–364 in *Essays in Indian Protohistory*, eds. D. P. Agrawal and D. K. Chakrabarti. Delhi: B. R. Publishing Corporation.
1984 Study of the Iron Age in India. *Puratattva* 13–14:81–85.
1985 The Issues in the Indian Iron Age. Pp. 74–88 in *Recent Advances in Indian Archaeology*, eds. S. B. Deo and K. Paddayya. Poona: Deccan College Postgraduate and Research Institute.
1988 *A History of Indian Archaeology: From the Beginning to 1947.* Delhi: Munshiram Manoharlal.
1991 Aspects of Continuity in Indian Metalworking Technology: A Study of Geological and Ethnographic Sources. *Cultural Heritage of the Indian Village.* Occasional Papers of the British Museum no. 47. London.
1992 *The Early Use of Iron in India.* Delhi: Munshiram Manoharlal.

Charles, J. A.
1980 The Coming of Copper and Copper-base Alloys and Iron: A Metallurgical Sequence. In *The Coming of the Age of Iron*, eds. T. A. Wertime and J. D. Muhly, pp. 151–181. New Haven: Yale University Press.

Chattopadhyay, P. K.
1984 Archaeometallurgical Studies in Indian Subcontinent: A Survey on Metal-

1990   lography of Iron Objects. *Indian Journal of History of Science* 19(4):361–365.
On Early Iron Technology in Eastern India. In *Adaptation and Other Essays: Proceedings of the Archaeology Conference, 1988*, eds. N. C. Ghosh and S. Chakrabarti, pp. 252–258. Santiniketan: Visva-Bharati Research Publications.

1991   Iron Smelting Technology in West Bengal. In *Studies in Archaeology: Papers Presented in Memory of P. C. Dasgupta*, ed. A. Datta, pp. 167–172. Delhi: Books and Books.

Childe, V. G.
1944   Archaeological Ages as Technological Stages. *Journal of the Royal Anthropological Institute of Great Britain and Ireland* 74:7–24.

Cooke, S. B., and Aschenbrenner, S. E.
1975   The Occurrence of Metallic Iron in Ancient Copper. *Journal of Field Archaeology* 2(3):251–266.

Dani, A. H.
1978   Gandhara Grave Culture and the Aryan Problem. *Journal of Central Asia* 1(1):42–56.

Dani, A. H. (ed.)
1967   Timargarha and the Gandharan Grave Complex. *Ancient Pakistan* 3:1–407.

Dani, A. H., and Durrani, F. A.
1964   A New Grave Complex in West Pakistan. *Asian Perspectives* 8:164–165.

Datta, A.
1998   Iron Technology at the Chalcolithic Phase of West Bengal. Pp. 36–43 in *Archaeometallurgy in India*, ed. V. Tripathia. Delhi: Sharada Publishing House.

De, S., and Chattopadhyay, P. K.
1991   Archaeometallurgical Studies at Kankarjhor and Dhuliapur. *Puratattva* 20:115–117.

Deglurkar, G. B., and Lad, G. P.
1992   *Megalithic Raipur* (1985–1990). Poona: Deccan College Post-graduate and Research Institute.

Deo, S. B.
1970   *Excavations at Takalghat and Khapa*. Nagpur: Nagpur University.

1973   *Mahurjhari Excavation 1970–72*. Nagpur: Nagpur University.

1985   The Megaliths: Their Culture, Ecology, Economy and Technology. Pp. 89–99 in *Recent Advances in Indian Archaeology: Proceedings of the Seminar held in Poona in 1983*, eds. S. B. Deo and K. Paddayya. Poona: Deccan College Post-graduate and Research Institute.

Deo, S. B., and Dhavalikar, M. K.
1968   *Paunar Excavation (1967)*. Nagpur: Nagpur University.

Deo, S. B., and Jamkhedkar, A. P.
1982   *Excavations at Naikund 1978–80*. Bombay: Government of Maharashtra, Department of Archaeology and Museums.

Dupree, L.
1963   *Deh Morasi Ghundai: A Chalcolithic Site in South-Central Afghanistan*. Anthropological Papers of the American Museum of Natural History, no. 50(2). New York.

Enault, J.-F.
1979   *Fouilles de Pirak*, 2 vols. Publications de la Commission des Fouilles Archéologiques. Fouilles du Pakistan, No. 2, Vol. 2. Paris.

Francke, A. H.
1914 [1977]   *A History of Ladakh*. New Delhi: Sterling Publishers.

Fürer-Haimendorf, C. von
1945   The Problem of the Megalithic Culture in Middle India. *Man in India* 25:73–86.

Gaur, R. C.
1983   Excavations at Atranjikhera: Early Civilization of the Upper Ganga Basin. Delhi: Motilal Banarsidass and Center of Advanced Study, Department of History, Aligarh Muslim University.

Ghosh, A. K., and Chattopadhyay, P. K.
1982   An Early Steel Implement from Barudih, Bihar Province, India: Metallurgical Studies. *MASCA Journal* 2(2):63–64.

Ghosh, N. C.; A. K. Nag; and P. K. Chattopadhyay
1987–88   The Archaeological Background and Iron Sample from Hatigra. *Puratattva* 18:21–27.

Gogte, V. D.
1982a Megalithic Iron Smelting at Naikund (Part 1): Discovery by Three-probe Resistivity Survey. Pp. 52–55 in *Excavations at Naikund 1978–80,* eds. S. B. Deo and A. P. Jamkhedkar. Bombay: Government of Maharashtra, Department of Archaeology and Museums.
1982b Megalithic Iron Smelting at Naikund (Part 2): Efficiency of Iron Smelting by Chemical Analysis. Pp. 56–59 in *Excavations at Naikund 1978–80,* eds. S. B. Deo and A. P. Jamkhedka. Bombay: Government of Maharashtra, Department of Archaeology and Museums.

Gogte, V. D.; Kanetkar, S. M.; and Kshirsagar, A. A.
1981–82 Efficiency of Megalithic Iron Smelting by Chemical Analysis. *Bulletin of the Deccan College Post-graduate and Research Institute* 41:68–71.

Gordon, D. H.
1958 *The Prehistoric Background of Indian Culture.* Bombay: Bhulabhai Memorial Trust.

Graslund, B.
1981 The Background to C. J. Thomsen's Three Age System. In *Towards a History of Archaeology,* ed. G. Daniel, pp. 45–50. London: Thames and Hudson.

Gupta, S. P.
1970–71 Gulf of Oman: The Original Home of Indian Megaliths. *Puratattva* 4:4–18.

Indian Archaeology, A Review
1981–81 Excavations at Gufkral. *Indian Archaeology, A Review,* pp. 19–25. Delhi: Archaeological Survey of India.

Jarrige, J.-F., and Santoni, M.
1979 *Fouilles de Pirak,* 2 vols. Publications de la Commission des Fouilles Archéologiques. Fouilles du Pakistan, No. 2, Vol. 1. Paris.

Jettmar, K.
1967 An Iron Cheek-piece of a Snaffle Found at Timargarha. Pp. 203–209 in Timargarha and Gandhara Grave Culture. *Ancient Pakistan* 3, ed. A. H. Dani.

Joshi, A. P.
1973 Analysis of Copper and Iron Objects. P. 77 in *Mahurjhari Excavations 1970–72,* ed. S. B. Deo. Nagpur: Nagpur University.

Joshi, J. P.
1976 Excavations at Bhagwanpura. Pp. 238–239 in *Mahabharata: Myth and Reality, Differing Views,* eds. S. P. Gupta and K. S. Ramachnadran. Delhi: Agam Prakashan.
1978 Interlocking of Late Harappa Culture and Painted Grey Ware Culture in the Light of Recent Excavations. *Man and Environment* 2:98–101.
1993 *Excavation at Bhagwanpura 1975–76 and Other Explorations and Excavations 1975–81 in Haryana, Jammu & Kashmir and Punjab.* Delhi: Memoirs of the Archaeological Survey of India 89.

Joshi, S. D.
1970 *History of Metal Founding on the Indian Sub-Continent Since Ancient Times (4000 B.C.–1970 A.D.).* Ranch: Sushila S. Joshi.

Kennedy, K. A. R.
1966 A New Interpretation of Population Dynamics in Iron Age India. *American Journal of Physical Anthropology* 25:215.
1975 *The Physical Anthropology of the Megalith Builders of South India and Sri Lanka.* Canberra: Oriental Monograph Series of the National University of Australia.
1985 The Element of Racial Biology in Indian Megalithism: A Multivariate Analysis Approach. Pp. 455–464 in *Recent Advances in Indo-Pacific Prehistory,* eds. V. N. Misra and P. Bellwood. Delhi: Oxford & IBH.

Khan, G. M.
1979 Excavations at Zarif Karuna. *Pakistan Archaeology* 9:1–95.
1983 Hathial Excavation. *Journal of Central Asia* 6(2):35–44.
1988 New Elements of Chronology in Taxila Valley. *Pakistan Archaeology* 23:275–288.

Kingery, D. W.; Vandiver, P. B.; and Prickett, M.
1988 The Beginnings of Pyrotechnology, Part II: Production and Use of Lime and Gypsum Plaster in the Pre-Pottery Neolithic Near East. *Journal of Field Archaeology* 15:219–244.

Kosambi, D. D.
1963 The Beginnings of the Iron Age in India. *Journal of the Economic and Social History of the Orient* 6:309–318.

Krishna, M. H.
1931 *Excavation at Chandravalli*. Bangalore: Government of Mysore.

Krishna Deva, and Wheeler, R. E. M.
1946 Appendix B, Note on the Painted Grey Wares at Ahichchhatra. *Ancient India* 1:58–59.

Krishnaswami, V. D.
1949 Megalithic Types of South India. *Ancient India* 5:35–45.

Lal, B. B.
1954–55 Excavations at Hastinapura and Other Explorations in the Upper Ganga and Sutlej Basins 1950–52. *Ancient India* 10–11:5–151.
1985 Report on the Chemical Analysis and Examination of Metallic and Other Objects from Lothal. Pp. 651–666 in *Lothal: A Harappan Port Town (1955–62)*, Vol. 2, ed. S. R. Rao. Memoirs of the Archaeological Survey of India No. 78. New Delhi: Archaeological Survey of India.

Lal, M.
1986 Iron Tools, Forest Clearance and Urbanisation in the Gangetic Plains. *Man and Environment* 10:83–90.
1987–88 Iron at Ahar: A Comment. *Puratattva* 18:109–112.

Leshnik, L. S.
1974 *South Indian 'Megalithic' Burials: The Pandukal Complex*. Weisbaden: Franz Steiner Verlag GmbH.

Loofs-Wissowa, H. H. E.
1985 Traces of the South Indian Megalithic in Thailand: A 'Dolmen' in a 'Chedi' in U-Thong. Pp. 497–500 in *Recent Advances in Indo-Pacific Prehistory*, eds. V. N. Misra and P. Bellwood. Delhi: Oxford & IBH.

Mackay, E. J. H.
1943 *Chanhu-daro Excavations 1935–36*. American Oriental Society, American Oriental Series, Vol. 20. New Haven, CT.

Marshall, J. (ed.)
1931 Mohenjo-daro and the Indus Civilization, 3 vols. London: Arthur Probsthain.

McIntosh, J. R.
1981 The Megalith Builders of South India: A Historical Survey. Pp. 459–468 in *South Asian Archaeology 1979*, ed. H. Hartel. Berlin: Dietrich Reimer Verlag.

Meadow, R. H.
1979 A Preliminary Report on the Faunal Remains from Pirak. P. 334 in *Fouilles de Pirak*, eds J.-F. Jarrige and M. Santoni. Publications de la Commission des Fouilles Archéologiques. Fouilles du Pakistan, No. 2, Vol 1. Paris.

Moorti, U. S.
1994 *Megalithic Culture of South India: Socioeconomic Perspectives*. Varanasi: Ganga Kaveri Publishing House.

Mughal, M. R.
1982 Recent Archaeological Research in the Cholistan Desert. Pp. 85–95 in *Harappan Civilization: A Contemporary Perspective*, ed. G. L. Possehl. Delhi: Oxford & IBH and the American Institute of Indian Studies.
1984 The Post-Harappan Phase in Bahawalpur Distt., Pakistan. Pp. 499–503 in *Frontiers of the Indus Civilization*, eds. B. B. Lal and S. P. Gupta. Delhi: Books and Books.

Munshi, K. N., and Sarin, R.
1970 Analysis of Copper and Iron Objects from Takalghat and Khapa. Pp. 78–79 in *Excavations at Takalghat and Khapa (1968–69)*, ed. S. B. Deo. Nagpur: Nagpur University.

Nagaraja Rao, M. S.
1971 *Protohistoric Cultures of the Tungabhadra Valley: A Report on Hallur Excavations*. Dharwar: M. S. Nagaraja Rao.
1981 Earliest Iron-using People in India and the Megaliths. Pp. 25–32 in *Mandu: Recent Researches in Indian Archaeology and Art History, the Shri M. N. Deshpande Festschrift*, ed. M. S. Nagaraja Rao. Delhi: Agam Kala Prakashan.

Pigott, V. C.
1981 *The Adoption of Iron in Western Iran in the Early First Millennium B.C.: An Archaeometallurgical Study*. Ph.D. dissertation, Department of Anthropology, University of Pennsylvania, Philadelphia. Ann Arbor, MI: University Microfilms International.

Possehl, G. L. (compiler)

1994   *Radiometric Dates for South Asian Archaeology.* An Occasional Publication of the Asia Section. Philadelphia: The University of Pennsylvania Museum.

Possehl, G. L., and Rissman, P. C.
1992   The Chronology of Prehistoric India: From Earliest Times to the Iron Age. Pp. 465–490 and 447–474 in *Chronologies in Old World Archaeology,* 3rd ed., 2 vols., ed. R. W. Ehrich. Chicago: University of Chicago Press.

Qamar, M. S.
1983   Excavation of Megalithic Burial at Dumlotti, District Karachi. *Journal of Central Asia* 6:97–110.

Rahman, A.
1967a   Part II, Section 3: Graves of the 1965 Season. Pp. 70–95 in Timargarha and Gandhara Grave Culture. *Ancient Pakistan* 3, ed. A. H. Dani.
1967b   Part IV, Section 1: Report on the Small Finds from the Grave. Pp. 185–200 in Timargarha and Gandhara Grave Culture. *Ancient Pakistan* 3, ed. A. H. Dani.

Raikes, R. L.
1963   A New Prehistoric Bichrome Ware from the Plains of Baluchistan. *East and West* 14:56–67.

Ray, R. U.
1969   *Metals in Ancient India (From the Earliest Times to the Beginning of the Christian Era).* University of Gorakhpur.

Sahi, M. D. N.
1978   New Light on the Life of the Painted Grey Ware People as Revealed from Excavations at Jakhera (Dist. Etah). *Man and Environment* 2:101–103.
1979   Iron at Ahar. Pp. 365–368 in *Essays in Indian Protohistory,* eds. D. P. Agrawal and D. Chakrabarti. Delhi: B. R. Publishing Corporation.

Salim, M.
1991a   Painted Grey Ware Sites around Islamabad. *Pakistan Archaeology* 26(1):144–155.
1991b   Dhok Gangaal: A Painted Grey Ware Site in Islamabad. *Ancient Pakistan* 27:27–33.

Sankalia, H. D.
1979   Review of *Essays in Indian Protohistory,* edited by D. P. Agrawal and D. K. Chakrabarti. *Commerce* (Aug.):1–2.

Sankalia, H. D.; Deo, S. B.; and Ansari, Z. D.
1969   *Excavations at Ahar (Timbavati) 1961–62.* Poona: Deccan College Postgraduate and Research Institute.

Shaffer, J. G.
1978   *Prehistoric Baluchistan: With Excavation Report on Said Qala Tepe.* Delhi: B. R. Publishing Corporation.
1984   Bronze Age Iron from Afghanistan: Its Implications for South Asian Protohistory. Pp. 41–62 in *Studies in the Archaeology and Palaeoanthropology of South Asia,* eds. K. A. R. Kennedy and G. L. Possehl. Delhi: Oxford & IBH and the American Institute of Indian Studies.
1995   The Watgal Excavations: An Interim Report. *Man and Environment* 20:57–4.

Sharma, G. R.
1985   Megalithic Cultures of the Northern Vindhyas. Pp. 477–480 in *Recent Advances in Indo-Pacific Prehistory,* eds. V. N. Misra and P. Bellwood. Delhi: Oxford & IBH.

Singh, P.
1985   Megalithic Remains in the Vindhyas. Pp. 473–476 in *Recent Advances in Indo-Pacific Prehistory,* eds. V. N. Misra and P. Bellwood. Delhi: Oxford & IBH.

Snodgrass, A. M.
1980   Iron and Early Metallurgy in the Mediterranean. Pp. 335–374 in *The Coming of the Age of Iron,* eds. T. A. Wertime and J. D. Muhly, pp. 375–416. New Haven: Yale University Press.

Stacul, G.
1966a   Preliminary Report on the Pre-Buddhist Necropolis in Swat (W. Pakistan). *East and West* 16:37–79.
1966b   Notes on the Discovery of a Necropolis near Kherai in the Gorband Valley (Swat—West Pakistan). *East and West* 16(3–4):261–274.
1967a   Discovery of Four Pre-Buddhist Cemeteries near Pacha in Buner (Swat, W. Pakistan). *East and West* 17(3–4):220–232.
1967b   Excavations in a Rock Shelter near Ghaligai (Swat, Pakistan). *East and West* 17(3–4):185–219.
1969a   Discovery of Proto-historic Cemeteries in the Chitral Valley (West Pakistan).

	East and West 19:92–99.
1969b	Excavation near Ghaligai (1968) and the Chronological Sequence of Protohistorical Cultures in the Swat Valley (West Pakistan). *East and West* 19(1–2):44–91.
1979	The Sequence and Proto-historical Periods at Aligrama (Swat, Pakistan). Pp. 88–90 in *South Asian Archaeology 1975*, ed. E. van Lohuizen-de Leeuw. Leiden: E. J. Brill.
1981	On Periods and Cultures in the Swat Valley and Beyond. *Puratattva* 10:89–91.

Subrahmanyam, R.; Banerjee, K. D.; Khare, M. D.; Rao, B. V.; Sarkar, H.; Singh, R.; Joshi, R. V.; Lal, S. B.; Rao, V. V.; Srinivasan, K. R.; Totadri, K.
1975   *Nagarjunakonda (1954–60)*. Delhi: Memoirs of the Archaeological Survey of India, No. 75.

Tandon, O. P.
1967–68   Alamgirpur and the Iron Age in India. *Puratattva* 1:54–60.

Thapar, B. K.
1952   Porkalam 1948: Excavation of a Megalithic Urn-burial. *Ancient India* 8:3–16.

Tripathi, V.
1976   *The Painted Grey Ware: An Iron Age Culture of Northern India*. Delhi: Concept Publishing Company.

Walimbe, S. R.
1987–88   Human Skeletal Remains from Megalithic Vidarbha. *Puratattva* 18:61–71.
1992   Human Skeletal Evidence from Megalithic Raipur. Pp. 125–132 in *Megalithic Raipur (1985–1990)*, eds. G. B. Deglurkar and G. P. Lad. Poona: Deccan College Post-graduate and Research Institute.

Walimbe, S. R.; Gambhir, P. B.; and Venkatasubbaiah, P. C.
1991   A Note on Megalithic Human Skeletal Remains from Kanyathirtham, Cuddapah District, Andhra Pradesh. *Man and Environment* 16(1):99–101.

Wheeler, R. E. M.
1947–48   Brahmagiri and Chandravalli 1947: Megalithic and Other Cultures in the Chitaldrug District, Mysore State. *Ancient India* 4:180–310.

7

The Transition to Iron in Ancient China

Bennet Bronson

ABSTRACT The making of iron emerged late in China—later than in many other parts of Eurasia. Yet it flowered with exceptional quickness. Within a period of less than three centuries, China advanced from its first known experiments with smelted iron to a position of leadership. Perhaps by 200 B.C. and certainly by 100 B.C., the Chinese iron industry had become the largest and most technically innovative anywhere in the ancient world. The purpose of this paper is to explore that remarkable transformation.

The first section examines the evidence for iron in China before 400 B.C. The second and third sketch the phase of rapid development that occurred between 400 and 200 B.C., and then the emergence of a large-scale iron industry between 200 and 1 B.C. The fourth focuses on the distribution and dating of various Chinese technical innovations: blast furnace smelting, white iron casting, fining, malleabilizing, and solid-state decarburizing. The fifth section discusses the origin of ironmaking in China, the reasons for its unusual but surprisingly modern character, and why it was so long in spreading to other regions. [Final ms. received 10/96.]

THE EARLIEST IRON IN CHINA

Several bimetallic implements of meteoritic iron and bronze have been found at sites of the Shang and Western Zhou periods in eastern China, ca. 1500–771 B.C. (Hua et al. 1986; Gettens et al. 1971).[1,2] In the far northwest of China, in Xinjiang Province, smelted iron was fairly common by 800–600 B.C.[3] However, iron smelting and working did not make further progress in the rest of China or elsewhere in East Asia until much later. It is now generally agreed that the first smelted iron east of Xinjiang is not earlier than about 500 B.C., at the end of the Spring and Autumn period (770–476 B.C.). Iron objects that can plausibly be attributed to the late sixth and early fifth centuries remain very scarce. The most frequently cited examples are two pieces found in epigraphically dated graves at the site of Liuhe or Luhe in Jiangsu Province (see Fig. 7.1 for locations of ironworking sites in China; KG 1965[3]:113). One of these pieces, a ball-like object, is of high-carbon cast iron; the other, a rod, is of low-carbon wrought iron. Although neither piece is from an entirely satisfactory context (both were found in supposedly undisturbed parts of partially looted graves), the consensus is that they do indeed date to very late Spring and Autumn times.

A number of other early iron finds from central and northern China have been reported, but the dating of these is considered to be less convincing, either because the relevant radiocarbon or epigraphic dates have a wide margin of error or because the association between the dated materials and the iron is not secure.

Table 7.1 presents a list of iron objects to which dates earlier than 400 B.C. have been assigned by epigraphic or radiometric methods. The Mayang date was on timber in an ancient copper mine where an iron wedge was found; the association between the timber and the wedge was not tight. The Zengjiagou dates were on wooden coffin planks from two large tombs; the iron axe was found in a third, smaller, tomb considered by the excavators to be contemporary with the first two. The Xichuan iron sample was also from a smaller tomb in the same group as larger tombs that could be dated, in this case by epigraphic methods. The Jinan and Changsha dates, from major urban complexes of the historical Chu kingdom, are more convincing in terms of the association between iron and directly datable objects but pertain to cultural phases that have yielded a number of significantly later dates. The dated wood samples and the iron hoe at Tianxingguan came from a single tomb and thus were securely associated; the problem here is that another, inscriptional, date is so well established as to cast doubt on the validity of the radiocarbon determinations.

As far as I am aware, few other ancient iron objects have been found anywhere east of Xinjiang for which there is direct archaeometric or epigraphic evidence pointing to a date either in the late Spring and Autumn period or in the earliest stage, the first 75 years or so, of the succeeding Warring States period (475–221 B.C.). Moreover, as Huang (1976:64) and Yang (1982:17–24) have noted, there is no unambigu-

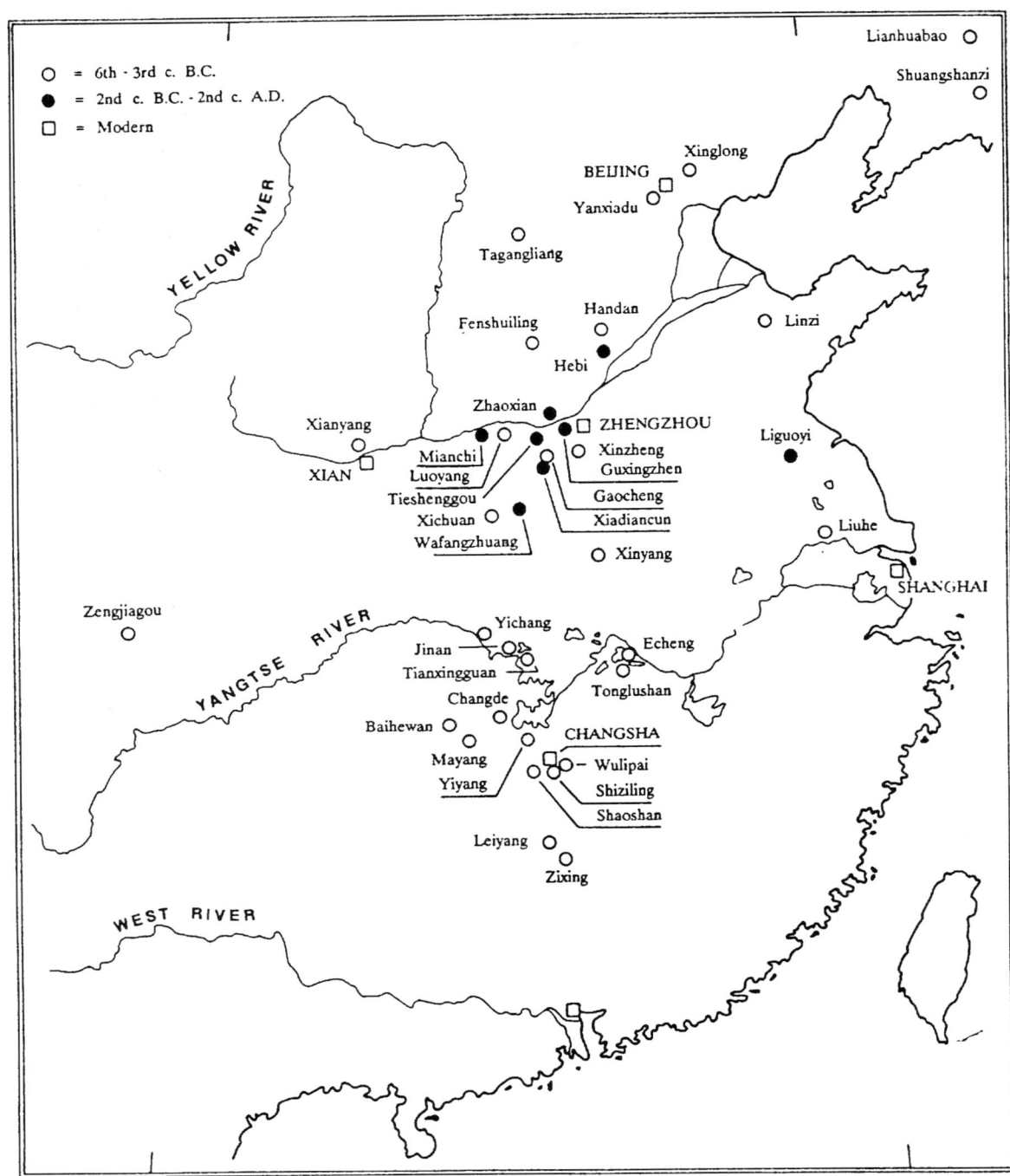

Figure 7.1 Ironworking sites in China, 500 B.C.–A.D. 200.

TABLE 7.1
DATED FINDS OF EARLY IRON IN CHINA

SITE	PROVINCE	OBJECT(S)	DATING METHOD	DATE B.C.	LAB. NO(S).	REF.
Mayang	Hunan	wedges	C-14	899–790	BK 79057	1
Zengjiagou	Sichuan	axe	C-14	508–392 795–433	ZK 1062 ZK 1063	2
Xichuan	Henan	dagger	inscription	ca. 550		3
Luoyang	Henan	pole fitting	C-14	755–392	Y 1513	4
Liuhe	Jiangsu	cast ball wrought rod	inscription	ca. 500		5
Jinan City	Hubei	14 objects, cast and	C-14	736–389	ZK 0243(2)	6
		wrought axe	C-14	408–233	ZK 0400	7
Changsha	Hunan	ring	C-14	507–263	ZK 0001	8
Tianxingguan	Hubei	cast (?) hoe	C-14	829–662 791–414 401–208	ZK 0564 BK 78037 WB 79-6	9
			inscription	340–361		
Wangshan (in Jinan City)	Hubei	belt hook	C-14	387–112 397–204	ZK 0026 PV 27	10

Notes and references:

(1) Locality 2202. KG 1985(2):120–124; AICASS 1991:202.
(2) Dates are on Tombs M11 and M12; iron is from Tomb M13. KG 1984(12):1084; AICASS 1991:226.
(3) Tomb M10. WW 1980(10):18.
(4) Carbon in cast iron sample. Merwe 1969:96; AICASS 1991:151.
(5) Tomb M1. KG 1965(3):113.
(6) Level 3, South Gate. KGXB 1982(3):334; KG 1980(4):376; AICASS 1991:187.
(7) Lining of Well J93. KGXB 1982(4):493; AICASS 1991:187.
(8) Tomb M406. Changsha Report 1957:66; KG 1977(4):231; AICASS 1991:201.
(9) Tomb M1. KGXB 1982(1):73; KGYWW 1980(4):134; KG 1980(4):376; AICASS 1991:187–188.

ous literary evidence for sixth and fifth century iron in China. And yet historians of technology often assume that Chinese ironmaking was well established by 400 or even 500 B.C. In some cases, the assumptions are based more on faith than evidence, for in China as elsewhere scholars may prefer early dates on the grounds of national and local pride. In other cases, the assumptions derive from misinterpretations of archaeological sequences at sites that cover time spans extending from the late Spring and Autumn period down to the fourth or third century B.C. Many such sites have produced a few pieces of iron. Excavators and commentators have naturally been tempted to assign that iron to the earliest dates at which the sites can have been occupied.

The temptation to push finds of iron as far back in time as possible would be minimized if the sites in question could be reliably divided into cultural phases. Unfortunately, however, doing so is not easy at the present stage of research. The great majority of the supposedly earliest iron-yielding sites are associated with the above-mentioned Chu culture and are located in Hunan and Hubei Provinces of central China. The fact that Chu excavations have been focused on rich but unconnected graves in cemetery sites has made it exceedingly difficult to derive a satisfactory relative chronology, and this in turn has rendered the great bulk of Chu iron essentially undatable. A conservative opinion on the subject is that of Wagner (1987). Based on a careful review of published data on cemeteries near Changsha and Jinan, the two principal Chu cities, he has concluded—convincingly in my opinion—that on current evidence we cannot be certain that any Chu iron object is earlier than 400 B.C. (ibid.:46).

The possibility that all iron in Chu may be as late as the fourth century is disconcerting, for most scholars believe that the use of iron in China first began within the territories of Chu, on the central Yangtse, and the neighboring kingdom of Wu, near the mouth of that river. Thus far, the claims of Wu to be a center of early ironmaking are based largely on the two above-mentioned iron objects from Liuhe. The claims of Chu, on the other hand, are based on numerous finds—so numerous, in fact, that Chu sites have yielded at least 90% of the non-Xinjiang Chinese iron objects to which early Warring States and Spring and Autumn dates have been ascribed. Is it conceivable that none of this iron is earlier than the fourth century? Does this mean that all smelted iron in China east of Xinjiang, with the possible exception of the two Liuhe pieces, is similarly late?

Perhaps it is best to leave the question open. Currently available evidence does not disprove a late date for the beginnings of Chinese iron production, but it does not prove it either. Several recent finds of iron at Chu sites—at Yangjiashan and Yaoling near Changsha, for example—are claimed to be from convincing Spring and Autumn period contexts (Huang 1984: 155). Iron-yielding sites believed to be very early are also known from northern as well as central China, including the Luoyang Cement Factory site (Li 1975:5) in Henan Province; Tagangliang (WW 1986[11]:21–26) in Shanxi Province; and possibly Jinjiazhuang or Lingtai in Gansu Province in the far northwest (KG 1981[4]:298–301; Li 1985:318–319). Some of these, like Jinjiazhuang, are controversial.[4] But others are quite promising: for instance, Tagangliang, with a cast iron axehead from a stylistically dated late Spring and Autumn period tomb, could prove to be as old as or even older than the iron-yielding sites of Chu and Wu in spite of the weight of evidence that Chinese iron was first smelted within the boundaries of those kingdoms. Moreover, as the following section shows, ironmaking advanced with astonishing rapidity during the fourth and third centuries B.C. It could reasonably be claimed that the sheer speed and scale of this development implies a lead time of at least a century or two during which ironmaking methods were worked out, perhaps on a scale too small to be noticed or perhaps in a part of China that is still poorly explored by archaeologists.

THE EXPANSION OF IRON USE IN THE FOURTH AND THIRD CENTURIES B.C.

The archaeological evidence for iron in the fourth century B.C., the middle of the Warring States period, is also unsatisfactory. This is chiefly because so much iron of that date has come from cemeteries, especially those associated with the Chu kingdom in Hunan and Hubei. Chu was a major center of ironworking and yet, as Table 7.2 shows, Chu graves are much more likely to contain bronze than iron. A similar situation exists in contemporary burial sites in other parts of China. The inhabitants of the numerous kingdoms into which China was divided during the early centuries of the Iron Age had a clear preference for grave furnishings made of bronze.

This preference was marked throughout the middle and late Warring States period (ca. 400–221 B.C.). Table 7.3 summarizes the contents of 343 graves from sites in the Changsha area and 558 graves from Yutaishan near the Chu capital, Jinan. It is evident that at these sites bronze implements fulfilled many of the functions that might otherwise be served by iron; the predominance of bronze weapons is especially striking. To the extent that these particular graves can be dated, there does appear to be a trend for iron to increase in popularity over time: less of it exists at Yutaishan, which went out of use in the early third century B.C., than at Changsha, where Chu traditions survived for a hundred years longer. And yet iron is much less common than bronze at both sites, and indeed at almost all other burial sites in China that antedate the Christian Era. One could well imagine from these data that in China the popularization of iron happened very late, and that it long remained a scarce and valuable metal.

Other data, however, show that the choice of materials for use by the dead did not reflect the actual importance of iron in everyday life. One indication of this is the fact that so much iron has been found in the backfill, rather than the burial chambers, of tombs. At Tianxingguan, for example, the furnishings of a single burial included 33 objects made of bronze, 660 of shell and bone, 28 of leather, 10 of gold leaf, 200 of tin and 110 of lacquer; the single iron object found was a hoe from the backfill in the stepped trench that led down into the tomb (KGXB 1982[1]:73). A tomb at Fenshuiling produced 400 objects of various kinds; the 3 iron tools recovered all came from the zone above the coffin and in the fill (WW 1972[4]:38–46). Similarly, at the Changsha localities Wulipai, Changyang, and Shiziling, the iron in 2 of the 13 iron-yielding tombs listed in Table 7.2 came from backfill rather than the

TABLE 7.2
IRON IN BURIAL SITES OF THE WARRING STATES PERIOD (475–221 B.C.)

SITE	PROVINCE	NUMBER OF TOMBS	TOMBS WITH BRONZE	TOMBS WITH IRON	REF.
Baihewan	Hunan	64	28	7	1
Changde	Hunan	83	4	3	2
Changsha	Hunan	77	(many)	13	3
Echeng	Hubei	15	11	1	4
Leiyang	Hunan	19	?	1	5
Shaoshan	Hunan	76	42	7	6
Tagangliang	Shanxi	4	3	1	7
Taihuiguan	Hubei	10	6	1	8
Wangshan	Hubei	3	3	1	9
Xichuan	Henan	25	?	1	10
Yiyang '80	Hunan	47	23	3	11
Yiyang '85	Hunan	93	43	19	12
Yichang	Hubei	6	4	1	13
Yutaishan	Hubei	558	>216	9	14
Zengjiakou	Sichuan	6	1	1	15
Zixing	Hunan	80	66	23	16

References:

(1) KGXB 1986(3):339–360
(2) KG 1963(9):461–473
(3) Changyang—WW 1978(10):44–47;
  Shiziling—KG 1977(1):62–64;
  Shiziling & Wulipai—Changsha Report 1957
(4) KGXB 1983(2):223–251; KG 1978(4):256–260
(5) WW 1985(6):1–15
(6) WW 1977(4):36–54
(7) WW 1986(11):21–26

(8) KG 1973(6):337–343
(9) WW 1966(5):36; KG 1977(3):203
(10) WW 1980(10):18
(11) KGXB 1981(4):519–549
(12) KGXB 1985(1):89–117
(13) KGXB 1976(2):115–148
(14) KG 1980(5):391–402
(15) KG 1984(12):1084
(16) KGXB 1983(1):93–124

tomb floor (Changsha Report 1957:66). At Taihuiguan near Jinan, iron from backfill accounted for 1 of 1 iron-yielding tombs (KG 1973[6]:338); at Changde, for 3 of 3 (KG 1963[9]:461–462); at Yiyang '85, 7 of 19 (KGXB 1985[1]:535); and at Zixing, 12 of 23, plus 2 tombs with iron both in the backfill and on the tomb floor (KGXB 1983[1]:115). In almost all cases, the iron objects found in backfill were parts of digging implements: spades, hoes, and the like. Hence, it seems that many tombs of the fourth and third centuries, though furnished mainly or exclusively with bronze, were dug out with iron tools: a datum which is not compatible with the thesis that iron was valuable or scarce.

Another indication that the furnishings of Warring States burials are misleading as indicators of the contemporary importance of iron is an inventory of weapons found in Grave M44, a mass burial of warriors, perhaps fallen in battle or executed as rebels, at Yanxiadu in Yi County, Hebei Province (KG 1975 [4]:228–243). Here, the weapons are presumably practical rather than ceremonial and, as Wagner (1988:176) points out, are preponderantly of iron. Of 77 weapons and identifiable weapon parts, 63 were of iron, and iron also accounted for 16 out of 20 metal accessories and tools.[5] Nine of the iron objects were submitted for analysis: one was of white cast iron, one of partially decarburized white iron, one of malleabilized white iron, and six—all weapons—of low-carbon iron partially carburized and pack-welded to form a heterogeneous steel. In view of the sophistication and variety of the ironworking processes involved, it is sig-

TABLE 7.3
NUMBERS OF BRONZE AND IRON ARTIFACTS IN CHU TOMBS, CHANGSHA (HUNAN) (343 TOMBS)
AND YUTAISHAN, NEAR JINAN (HUBEI) (558 TOMBS). DATA FROM WAGNER (1987:TABLE 3)

IMPLEMENT TYPE	BRONZE	BRONZE/IRON	IRON
Digging tools	0		6
Axeheads	0		4
Arrow points	217	12*	4
Swords and daggers	323		10
Halberd heads	7		1
Spear- and dagger-axeheads	157		0
Writing knives	7		17
Scraper blades	0		5
Tripod cauldrons	24	11+**	0
Rings	8		1

* iron tangs with bronze points
** iron legs with bronze bodies

nificant that the grave can be dated, by the presence of no fewer than 1,360 coins, to the early third century B.C.

Further archaeological evidence for the importance of iron in everyday life during the later Warring States period comes from reports of excavations at contemporary settlement sites. Substantial quantities of that metal were found in third and possibly fourth century contexts at the ancient cities of Linzi in Shandong Province, Handan in Hebei Province, and Xianyang in Shaanxi Province (Chang 1977:333–345). Jinan City in Hubei Province, the Chu capital destroyed by the state of Qin in 278 B.C., has proved to be especially rich in iron. For example, excavations of the ancient residential-commercial area near the South Gate of Jinan City have yielded a fair amount of iron, including 13 iron implements from the third level alone: 3 chisels, 2 spear- or arrowheads, 2 knives, 1 hoe, 1 spade, 1 adze, 1 axe, 1 fishhook, and 1 sickle (KGXB 1982[3]:341, 347–348). This is an impressive number when compared with the total of 2 all-iron objects and 10 bronze-iron objects found in the 558 graves excavated at Yutaishan (see Table 7.3; Wagner 1987:143), which is contemporary with Jinan and only one kilometer outside the walls of that city.

It is important to note that many fourth and third century non-burial sites in which iron is found are located in northern China rather than in the central region—the ancient Chu and Wu kingdoms—which, as noted above, is considered to be the cradle of Chinese iron metallurgy. In the kingdom of Yan, near modern Beijing, several sites have proved to be quite rich in iron: Yanxiadu, where iron tools have been found in workshops as well as in the above-mentioned mass burial (Li 1985:115, 324); Lianhuabao in Fushun County (ibid.:326–327) and Shuangshunzi in Kuandian County (WWZLCK 1980[3]:128–129), both in Liaoning Province; and Dafujianggou in Xinglong County, Hebei Province, thought to be the site of an official iron foundry of the Yan government (Li 1985:327; KG 1956[1]:29–37). The kingdom of Han (a small state south and west of modern Zhengzhou in Henan Province, not the homonymous dynasty of later times) has yielded two ironworking sites that probably date to the third century: Gaocheng in Dengfeng County (Yang 1982:109–110) and Xinzheng or Zhenghan Gucheng in Xinzheng County (Li 1987:203–204). The kingdom of Qin around modern Xian in Shaanxi Province, later to conquer all China under Emperor Shi Huangdi, is known from literary evidence to have had a state-controlled iron industry during the third century, and many iron artifacts have been found within the territory of Qin despite the complete absence of that metal from the equipment of the famous "terracotta army" that surrounds Shi Huangdi's tomb (Li 1985:327–329).

Thus, it seems clear that iron was in fairly common use, for utilitarian if not ceremonial purposes, during the third century and probably the late fourth century B.C. (see Figs. 7.2 and 7.3). It also seems clear that rather sophisticated iron processing techniques, including pack-welding, malleabilizing, iron-to-steel welding, and controlled quenching were already in existence by the early third century. True, skill in handling iron must have evolved unevenly in different areas, and it is quite possible that a few otherwise developed regions of China continued to use only bronze until the very end of the Warring States period. But there can be no doubt that many regions were already well into the Iron Age by the time that Shi Huangdi completed the conquest of the other states and brought most of the country under Qin Dynasty rule in 221 B.C. It had taken less than 300 years to reach this point from the time of the first known appearance of smelted iron in China.

THE FLORESCENCE OF IRON IN THE SECOND AND FIRST CENTURIES B.C.

The rise of iron continued to accelerate during the short-lived Qin Dynasty (221–206 B.C.) and the first decades of the Han Dynasty (206 B.C.–A.D. 220). Metal tomb furnishings were still mainly of bronze, and would remain so for at least another two hundred years. But iron was dominant in other areas of cultural and economic life (see Figs. 7.4–7.7). By 200 B.C., iron-making is known to have been organized on an industrial scale and to have been a source of very substantial profits to the organizers. The great historian Sima Qian (145–90 B.C.) enumerated nine families as illustrations of "the ways in which some of the worthy men of the present age, working within an area of a thousand miles, have managed to acquire wealth." Five of the nine were merchants or money-lenders. The other four were all smelters and casters of iron (Watson 1969:352–355). To quote one of Sima Qian's examples,

> The ancestors of the Kong family of Yuan were men of Liang [near Kaifeng in modern Henan] who made their living by smelting iron. When Qin overthrew the state of Liang [or Wei, in 225 B.C.], the Kong family was moved to Nanyang [also in Henan], where they began smelting iron with bellows and laying out ponds and fields. Soon they were riding around in carriages with a mounted retinue and visiting the feudal lords. (Watson 1969:353)

All four of Sima Qian's ironmaking families were active and well on their way to success by the beginning of the Han Dynasty in 206 B.C. The Chinese iron industry continued to be dominated by large-scale private enterprises until 117 B.C., when a policy of government monopoly of iron production was inaugurated (Gale 1931:xxv). Large ironmaking firms continued to exist, sometimes directly managed by the government and sometimes private but under government license, until at least the end of the Han period in A.D. 220. They were to reappear often in later centuries.[6]

The size and sophistication of large ironmaking enterprises during the early Han period has been confirmed by archaeological research. The four biggest sites thus far excavated, all of them in northern Henan Province, are Tieshenggou in Gong County (Henan Team 1962; Zhao et al. 1985), Hebi in Hebi County (KG 1963[10]:550–552), Wafangzhuang in Nanyang County (WW 1960[1]:58–60) and Guxing or Guxingzhen in Zhengzhou Municipality (WW 1978[2]:28–43). Tieshenggou, Guxingzhen, and probably Hebi were integrated plants, with a number of blast furnaces and facilities for secondary processing: 8 blast furnaces, 1 fining hearth ("puddling furnace"), 1 annealing furnace for solid-state decarburizing, 11 mold-baking furnaces, 5 general-purpose heating furnaces, and at least one forge at Tieshenggou;[7] 2 blast furnaces and numerous "kilns" for malleabilizing and

Figure 7.2 Spade head of cast iron. Henan Province, third century B.C. Field Museum # 127392. (Photograph courtesy of The Field Museum, Chicago, IL, neg. no. 76307.)

Figure 7.3 Digging tool edges of cast iron. Henan Province, third century B.C. Field Museum #s 127035 and 127034. (Photograph courtesy of The Field Museum, Chicago, IL, neg. no. 76310.)

solid-state decarburizing at Guxingzhen; and 13 blast furnaces at Hebi. Wafangzhuang is now thought to have specialized in foundry operations, using cast pig iron smelted elsewhere. At the smelting sites, blast furnaces with diameters of one to two meters were commonly built. Although the operation of these furnaces is not clearly understood (for one thing, no one knows how enough air pressure was generated to drive them), they evidently had a large capacity. The cupola furnaces used for remelting cast iron were also big; one of the 17 cupolas at Wafangzhuang still held 700 kilograms of pig iron at the time of excavation.

A few other early smelting and casting sites have also been at least cursorily investigated. Two of them, both foundries in Henan Province, are assigned to the third century B.C.: Xinzheng in Xinzheng County (WWZLCK 1980[3]:62–63), and Gaocheng or Guyangcheng in Dengfeng County (WW 1977[12]:52–65). Three others date, in part at least, to the second or first centuries: Xiadiancun in Linru County, Henan Province (WW 1960[1]:60); Liguoyi in Xuzhou County, Jiangsu Province (WW 1960[4]:46–47); and Yaaishan in Kuqa County, Xinjiang Province (WW 1960[6]:28). While some of these are smaller than Tieshenggou, Guxingzhen, and Wafangzhuang, they appear to have utilized a similar technology. Despite their early date, both Gaocheng and Xinzheng had large cupola furnaces and employed very complex molding systems. The blast or cupola furnaces at Xiadiancun and Liguoyi were about two meters in diameter.

THE CHARACTER OF CHINESE IRONWORKING TECHNOLOGY

Ancient Chinese iron processing employed a number of techniques that are unusual in the context of ancient ironmaking elsewhere in the world. Several comprehensive summaries are available, mostly in Chinese (Henan Museum 1978; Yang 1982:53–93; Li 1975:1–22; Hua et al. 1986), but a few in English (Needham 1958; Barnard and Sato 1975:60–68; He 1983; Wagner 1988). Here I shall try only to summarize those summaries and to bring a few details up to date.

The majority of iron objects found at early sites in China are of cast iron. This is true even at the very earliest sites, those datable to the fourth and perhaps fifth centuries. Although it has sometimes been suggested that some of this early cast iron was made by a two-step process, through the carburization of bloomery iron, no secure evidence for such a process exists. As far as is known, at least some ore was being smelted directly to a high-carbon liquid metal, probably in blast furnace–like structures, from the very beginning.

Low-carbon iron is also present in early sites: one of the two iron objects found at Liuhe (see Table 7.1), accepted as the earliest appearance of smelted iron in China, represents a metal with an apparently low carbon content. However, it is not clear whether this and similar metals are bloom iron (made in a single stage, in bloomery furnaces) rather than wrought iron (made in two stages, in blast furnaces and then fineries).

There would be no problem if bloom and wrought iron could be easily distinguished, but the two are essentially identical in terms of structure and chemistry. Some Chinese authorities (e.g., Yang 1982:220) appear to be confident that the products of fineries and bloomeries can be differentiated on the basis of slag content and other features. However, Western archaeometallurgists—who, due to the prevalence of direct-process ironmaking in their countries until quite recently, see a great deal of iron from bloomeries as well as from fineries—tend not to think that distinguishing the two kinds of iron is possible (Tylecote 1986:218, 221).

Figure 7.4 Steel knife with ring pommel. Shaanxi Province, third–second century B.C. Field Museum # 121000. (Photograph courtesy of The Field Museum, Chicago, IL, neg. no. 76309.)

BLOOMERY PROCESSES

The older literature on Chinese ironmaking contends that bloomeries were used extensively in China during the early Iron Age, and many archaeometallurgists still find it hard to believe that the Chinese could have begun smelting to cast iron without many years of prior experience in bloom iron production.[8] Yet archaeologists have not been able to find a single example of an actual bloomery, from the Warring States or any other period. The team that excavated the large second–first century B.C. smelting site of Tieshenggou in Henan Province in the late 1950s claimed initially that it had discovered no fewer than three bloomeries, along with a number of large blast furnaces and other iron processing installations (Henan Team 1962:8). These bloomeries have now become firmly embedded in the literature (see, for example, Yang 1982:71–72; and Barnard and Sato 1975:61). However, Zhao Qingyun's recent reinvestigation of Tieshenggou has demonstrated convincingly that none of the three could possibly be bloomeries and that they are in fact probably kilns for making molds.[9] Moreover, nothing like a bloomery has been reported either as a chance find from a smaller Warring States period site or in the course of systematic large-scale excavations of smelting sites of the Han period: for instance, Guxingzhen or Wafangzhuang.

It is also significant that there are no reports of ancient bloomery slags. Unlike the irons produced in bloomeries and fineries, bloomery slags are easily distinguished from blast furnace slags due to their high iron content. They can also be distinguished from finery slags, with which they have nearly identical chemical compositions, by the fact that they tend to be found in larger quantities. Massive deposits of high-iron bloomery slag occur at many early ironmaking sites in Europe, Africa, and South and Southeast Asia. One finds it hard to believe that such deposits could have remained unnoticed in China if bloomeries had ever been used there in significant numbers.

This leaves us with no actual examples of bloomeries or of identified bloomery slag from any archaeological site in China east of Xinjiang. Early fineries are known, as at Tieshenggou; bloomeries are not. It is true that smelting systems recorded during the early twentieth century in Shanxi Province (Tegengren 1923–24:319–333) and eastern Yunnan Province (Brown 1923:88–90) yielded a solid iron product with a carbon content that was sometimes quite low; these might validly be considered bloomery processes. However, in neither case was the product a normal bloom, and in both cases the associated equipment—long crucibles fired in open stalls in Shanxi and substantial blast furnace–like structures in Yunnan—was not at all like the equipment employed in the bloomery processes known to have been used by pre-modern Europeans, Africans, and Southeast, South, and Central Asians. There is no reason to think that either of the Shanxi or eastern Yunnan iron smelting systems represents a survival of an early bloomery tradition.

Figure 7.5 Iron/steel spears and swords. Shaanxi Province, second century B.C.–second century A.D. Field Museum #s (left-right): 120995, -996, -993, -994. (Photograph courtesy of The Field Museum, Chicago, IL, neg. no. A38928.)

FINERY PROCESSES

Making wrought iron by liquefying cast iron and then removing the alloyed carbon through oxidation is much better documented in China than is smelting in bloomeries. Often called "puddling" in translations and "steel frying" in Chinese, finery processes have been widely used by traditional Chinese ironmakers in recent centuries (see, e.g., Wagner 1985:60–69). Such processes were also used in antiquity. Twenty-eight pieces of fined wrought iron and steel were identified at the second century B.C. site of Tieshenggou, along with a fining hearth (Zhao et al. 1985:175).

Did fineries exist at an even earlier date, back in the fourth and third centuries B.C.? Most Chinese authorities (i.e., Li 1975:13–15; Hua et al. 1986:5; Yang 1982:220) think not. The consensus is that earlier artifacts made of low-carbon iron are likely to be from bloomeries even when the artifacts in question are very similar to slightly later objects known to be from

Figure 7.6   *Cast iron wine jar. Shaanxi Province, first–second century A.D. Field Museum # 120986. (Photograph courtesy of The Field Museum, Chicago, IL, neg. no. A98577.)*

fineries.[10] This consensus seems not to be affected by the above-mentioned fact that bloomeries and bloomery slag are undocumented in China whereas fineries there are common and quite old.

Is it possible that fineries existed as early as the middle Warring States period, and that the slaggy low-carbon iron found at sites like Tonglushan (WW 1975[2]:19–21) and Yanxiadu (KG 1975[4]:241–243) was made by fining cast iron rather than in a bloomery? I personally believe that this is indeed possible, and that the discovery of fining might well have followed quite closely on the first attempts to utilize cast iron. However, no such hypothesis can be directly proved or disproved until at least one iron smelting site of the relevant period has been excavated.

The point is that no one working with early Chinese metals has yet tried to make those distinctions, and thus the question of the relative age of fined and bloom iron in China must rest solely on *a priori* beliefs: first, that ironmaking was or was not introduced from another, necessarily bloomery-using, region; and second, that a two-stage process for making low-carbon iron is or is not too complex to have been in use at an early date. Both beliefs are considered in the concluding section of this paper.

FUEL

A number of recent commentators have stated that coal was used as a fuel for blast furnaces as early as the first century B.C., citing the initial reports of the excavations at Tieshenggou and Guxingzhen. However, the excavators appear to have been mistaken. Zhao and his associates (1985:171–172) observed that only char-

Figure 7.7 *Cast iron stove. Shaanxi Province, first–sixth century A.D. Field Museum # 120985. (Photograph courtesy of The Field Museum, Chicago, IL, neg. no. A98581.)*

coal was present in the smelting furnaces and slags at Tieshenggou, and that the coal found at that site and at Guxingzhen was used solely in kilns for firing tiles and molds. In support of this observation, Zhao notes that all cast iron of the Han and earlier periods is quite low in sulfur (see also Table 7.4). A high sulfur content has long been regarded as a normal trait of later Chinese cast iron, but such iron does not seem to become common before the fifth or sixth century A.D.[11]

WHITE IRON

The great bulk of analyzed iron specimens from Han and pre-Han sites are either of unmodified white cast iron or of white iron that has been subjected to various annealing treatments. As Table 7.4 shows, this metal is characteristically low in sulfur and silicon but rather high in phosphorus. The low silicon content shows that the blast furnaces involved were operated at relatively low temperatures. The wide use of iron molds for casting iron tools would have increased cooling rates and so encouraged the formation of white rather than gray iron.

GRAY IRON

Although gray cast iron becomes quite common by A.D. 500 (Pinel et al. 1938), it appears only sporadically in earlier periods. Zhao et al. (1985:175) claim that 7 out of 73 analyzed samples from Tieshenggou were gray; Qiu and Yu (1983:254) classify 14 out of 188 samples from Guxingzhen as gray but suggest that in some cases the gray color is due to graphite precipitated during the annealing of white iron. It seems possible that the same explanation applies to the occasional finds of "gray iron" reported from other early sites.

TABLE 7.4
CHEMICAL ANALYSES OF CHINESE IRON, WARRING STATES AND HAN PERIODS

SITE	OBJECT	TYPE OF IRON	C	P	S	Si	Mn	Ni	PERIOD	REF.
Tonglushan	digger	forged	0.10	0.11	0.004	0.10	0.06	0.02	WS	1
	wedge	forged "bloom"	0.06	0.12	0.009	0.06	0.05	0.01	WS	1
	hammer	cast, some graphite	4.3	0.152	0.019	0.19	0.05	0.02	WS	1
	axe	whiteheart malleable	0.7–2.5	0.108	0.016	0.13	0.05	0.1	WS	1
	hoe	cast, decarburized	0.07–2.98	0.10	0.006	0.08	0.01	0.02	WS	1
Mayang	2 hammers	cast	3.2–3.5	0.03	0.04	0.7	0.6		WS	2
	2 wedges	soft steel	0.1–0.15			0.3–0.5	0.5		WS	2
Changsha	sword	quenched	0.5						WS	3
	tripod	cast	4.3						WS	3
	tripod	cast	4.3						WS	3
Luoyang	casting	white cast	4.19	0.08	0.014	0.055	0.011	0.029	WS	4
Xian	stove	white cast	4.32	0.38	0.027	0.11	0.07	0.01	Han	4
Tieshenggou	sheet	cast	4.12	0.15	0.043	0.27	0.125		Han	5
	lump	high carbon	1.288	0.024	0.022	0.231	0.017		Han	6
	spade	—	2.57	0.489	0.024	0.13	0.16		Han	7
	spade	cast	3.55	0.40	0.022	0.09	0.12		Han	7
	lump	mottled	4.0	0.41	0.07	0.42	0.21		Han	7
	sheet	gray cast	3.80	0.48	0.04	0.22	0.09		Han	7
	digger	blackheart malleable	1.98	0.29	0.048	0.14	0.14		Han	7
	digger	cast, decarburized	3.30	0.24	0.03	0.09	0.10		Han	7
Mancheng	lump	cast	4.05	0.217	0.063	0.018	0.03		Han	8
Guxingzhen	—	cast	4.0	0.29	0.091	0.21	0.21		Han	9
	sheet	mottled	4.2	0.0	0.012	0.07	0.05		Han	10
	sheet	gray cast	3.8	0.292	0.020	0.12	0.05		Han	10

References:
(1) WW 1975(2):19–21
(2) KG 1985(2):124
(3) WW 1978(10):46
(4) Henger 1970:46
(5) KGXB 1978(1):10
(6) KGXB 1978(1):22
(7) Zhao et al. 1985:180
(8) Yang 1982:87
(9) KGXB 1978(1):10
(10) KJS 1985(13):111

MALLEABILIZED IRON

Both whiteheart and blackheart malleable were made, apparently by well-controlled annealing over long periods of time, at Tieshenggou and Guxingzhen (Qui and Yu 1983:234; Henan Team 1962; Zhao et al. 1985:175, 178). A few pieces of blackheart malleable at both sites exhibited a tendency for the graphite to assume a spheroidal form, although Chinese metallurgists do not consider that true nodular graphite iron appeared until very late in or after the Han Dynasty (i.e., at Mianchi—see He 1983:394–395). Several objects of the Warring States period are often cited as examples of early malleabilizing; one, an adze from a supposedly fifth century B.C. pit at the Luoyang Cement Factory site, has been well studied by Li Zhong (1975:5); because it has a white iron core with a pearlitic surface, he considers that it represents an early stage in the development of malleabilizing. One of the weapons from the warriors' grave at Yanxiadu, reliably dated to the late fourth or early third century B.C., is also reported to be of malleablized iron (KG 1975[4]:241–243).

SOLID-STATE DECARBURIZED IRON AND STEEL

A closely related technology based on controlled annealing served to remove most or all carbon from white cast iron, thus converting it to wrought iron or steel. Early efforts to soften and toughen cast iron objects seem often to have resulted in a steel or wrought iron skin over a white iron core: as in the case of the above-mentioned adze from Luoyang, the rate of carbon diffusion in solid metal made it impossible to remove the carbon completely from objects with thick cross-sections. By the late Warring States period, however, ironworkers seem to have overcome this problem by casting white iron into thin sheets and rods, subjecting them to a decarburizing anneal, and then forging the decarburized metal into a finished product. Molds for casting sheets and rods of this kind are said to have been found at the third century site of Xinzheng (Li 1987:203–204). Eighty-seven cast trapezoidal sheets, free from impurities and with carbon contents as low as 0.1%, were found at the Han period site of Guxingzhen (WW 1978[2]:36, 38; Qiu and Yu 1983:256–257). By the first century A.D., Chinese ironworkers had returned to the idea of casting objects to final shape (presumably only those with thin cross-sections) before decarburizing. Han and Yu (1983:239) report that several medium-carbon steel scissors from Dongshima in Zhengzhou, Henan Province, had been made in this way.

MOLDING SYSTEMS

Complex piece molds for casting single objects and bivalve molds for the side-by-side casting of multiple objects already had a long history in China by the time cast iron came into use. Molds made to be stacked so that several could be filled through a single gate were employed by bronze casters in the early or middle Warring States (He 1983:411; Hua et al. 1986:248). By the late Warring States period, ironworkers too had begun to use bivalve molds for multiple castings (at Gaocheng—Yang 1982:109–110) as well as molds made of cast iron (at Xinzheng and at Xinglong—Li 1985:327). Stackable molds for casting large numbers of iron objects were in wide use during the Han period. The best known examples are those discovered at a specialized mold-making kiln in Zhaoxian, Wen County, Henan Province. Hua (1983; see also WW 1976, 9:66–75 and Henan Museum Staff 1978) has reconstructed the systems by which the molds were made and used; he states that Zhaoxian alone yielded 500 sets of molds for making 16 products in 36 different sizes.

A number of important Chinese ironworking methods and equipment will not be discussed here, as they are not known to have appeared until the first millennium A.D. These include the "co-fusion" steelmaking process, whereby wrought iron and cast iron are heated together to form a medium-carbon metal (Needham 1958); crucible smelting, of great importance in Shanxi Province in later times but not convincingly attested during the early period; the use of water power for blowing furnaces; and the vertical fan bellows and double-acting piston bellows. With regard to water power and efficient bellows, it may seem reasonable to think that both would have been necessary to drive the large-diameter furnaces favored by early Han ironmasters. However, thus far we have no evidence that either was in existence before the first century A.D.[12]

THE ORIGINS AND EXPANSION OF CHINESE IRON

We have seen that iron in metropolitan China (that is, China east of Xinjiang Province) appeared late but that it developed with extraordinary rapidity. It was probably not used except in minuscule quantities until the fifth century B.C.; yet the making of iron was fully developed by 200 B.C., with most of the key techniques in place and production on a massive scale. At first glance, this might seem a *prima facie* case for technological diffusion. China, after all, was undergoing very rapid political, demographic, and economic expansion during those three centuries, and introduced technologies are often observed to flower rapidly when sown in fertile soil of this kind. One might be tempted to compare Chinese ironmaking in the late Warring States through early Han periods with American ironmaking in the 1830s through 1940s or

Japanese ironmaking in the 1870s through 1970s, as instances of a borrowed technological system reaching new heights due to social and economic demand in a new environment.

There is an apparently insuperable objection to a hypothesis of diffusion, however: the ironmaking of early China was strongly idiosyncratic, bearing little resemblance to ironmaking as practiced elsewhere in the world. With the probable exceptions of iron-to-steel welding and the pack- or "pattern"-welded blade,[13] and with the possible exception of the bloomery, it is difficult to find a single device or concept used in early Chinese ironmaking that is shared with any other country. Indeed, by the second century B.C., Chinese ferrous metallurgy had developed to such an extent that the direction of diffusion is likely to have been reversed: if one were to find the same device in use by Chinese and foreign ironmakers, the chances are that it would prove to have diffused from China, not to China.

No one doubts that iron came later to China than to many other regions. This paper will not attempt to review the evidence for very early iron elsewhere in Eurasia. It suffices to say that the effective use of that metal is much older in the Middle East and Mediterranean than in metropolitan China, and probably somewhat older in parts of Central and South Asia. Even Southeast Asia, which is not usually thought to be especially precocious in that regard, may well have begun making and using iron one or two centuries before China east of Xinjiang. For instance, thus far Thailand has yielded about fifteen radiocarbon dates for iron-associated contexts that are earlier than 500 B.C. (Bronson 1986:206); metropolitan China has yielded less than ten (see Table 7.1).

In view of the chronological priority of other regions with respect to ironmaking, we must reckon with the likelihood that certain general concepts—for instance, the idea that iron ores were not only smeltable but far more common than ores of copper and that iron was therefore in principle much cheaper—could have entered China from elsewhere. Stimulus diffusion of this kind is undoubtedly common enough in history, although in the absence of written evidence it is not likely to be provable in any given instance. I see no reason to think that a Central Asian visitor familiar with iron— perhaps a cultured traveler from Achaemenian Iran, perhaps a member of a metallurgically advanced tribal people like the Shors of more recent times[14]—could not have come to the Yellow River or Yangtse areas and convinced the Bronze Age people there that iron was a useful and easily made metal. Yet there is no evidence that this hypothetical traveler passed on anything more concrete than one or two abstract ideas and perhaps a few techniques for forging weapons. To repeat, it is hard to find a single specific process or item of metallurgical equipment that was used by early Chinese ironmakers and by ironmakers elsewhere in Eurasia.

One difficulty in assuming that Chinese iron technology represents an largely indigenous development is to devise a model or scenario whereby that technology could evolve from a preceding copper-based metallurgical system. Thus far we have no hard evidence that bears on this. To my knowledge, neither experimental iron castings nor accidentally produced samples of cast (or bloom) iron have been found at Bronze Age smelting and casting sites. It is true that, as Hua Jueming and associates (1986:284–285) observe, the Chinese of the Shang and early Zhou periods showed an interest in meteoritic iron and that there is some amount of technological carryover between the metal processing industries of the late Bronze and early Iron Ages. However, these data do not have a direct bearing on the problem of the initial stages of iron production in China. We must therefore have recourse to pure hypothesis while we wait for more evidence to accumulate.

A plausible model for the development of a cast iron–based industry from a copper-based industry can be assembled from the ideas of several specialists. Wagner (1986:7–10) has noted that large shaft furnaces were already in use for smelting copper during the Spring and Autumn period; although these could have been charged with dead-roasted sulfide ores, they appear to have been used mainly in areas where, as at Tonglushan in Henan Province, the ore minerals were copper oxides and carbonates. As Yang Kuan (1982:53–57) indicates, some of those shaft furnaces appear to have been run at very high temperatures. Such furnaces could easily have smelted iron. Since they would have been routinely charged with substantial quantities of iron ore as a flux during copper smelting, Wagner (1986:10) suggests that they probably did sometimes produce iron in the form of metal:

> It might be that metallic iron was well-known to the copper-smelters, a sign of incorrect operation of the furnace. Then the discovery of iron-smelting would simply have been the discovery that this undesirable by-product was useful.

Accidentally produced solid bloom iron might not have been interesting to Chinese bronze workers, who had long been accustomed to depend almost entirely on casting rather than forging for the fabrication of finished products. Accidentally produced liquid cast iron, on the other hand, would almost certainly have attracted their attention. The large melting furnaces and complex molding systems that had been in use by bronze casters since the Shang period would have been (and, in fact, proved to be) readily adaptable to iron casting. Iron ores were already familiar, as were the techniques of mining and preparing them as fluxes for copper-smelting furnaces. And a substitute for bronze was needed. The endemic warfare and rapid population growth of the early Warring States period must have created a demand for metal tools and weapons that was far too great to be satisfied by the available resources of copper and tin.

It might be objected that white iron, the form which accidentally produced cast iron would almost certainly take, would have been unattractive to potential ancient consumers. Both Trousdale (1977) and Keightley (1976) have suggested that the Chinese must have begun with bloom iron, arguing that cast iron weapons would have been so brittle as to be unusable. However, a recent experiment by Rostoker (1987) forces us to reevaluate that opinion. Specifically addressing himself to the problem of ancient Chinese cast iron, Rostoker fabricated several implements from unmodified white iron with a 3.5% carbon content and tested these as tools and weapons. A chisel, a penetrating point, a chopper blade, and an axe, all of white iron, were used for cutting wood. In all four shapes, the white iron proved to be surprisingly tough and, of course, far harder and more able to retain a cutting edge than any alloy of copper. Rostoker concludes that correctly used white iron implements are excellent tools and potentially effective weapons, and that they would perform even better if subjected to decarburizing treatments of the kind known to have been used by the Chinese in the later Warring States period.

A second idea of Rostoker's, proposed during conversations we had, points to a possible route by which finery processes could develop directly out of a cast iron–based metallurgical system. The key to that development, Rostoker suggests, might be the fact that shaft furnaces used as cupola furnaces for remelting cast iron pigs or scrap will produce solid low-carbon metal if supplied with an excess of air. Even in modern cupolas it occasionally happens that oxygen levels are allowed to rise to a point where the carbon in the melting cast iron becomes partially or wholly oxidized and passes off as a gas; the result is a solid lump of wrought iron that is similar to the product of a bloomery.[15] Inadvertent decarburization of this sort must have occurred at least occasionally in cupola furnaces of the Warring States and Han periods. In most cases, the foundrymen would not have been pleased at having to shut down their furnaces in order to extract the unwanted lump. But a few pioneers might have experimented with the lump anyway, and discovered it to be a forgeable and potentially useful metal. From there it would have been only a short step to designing a furnace that could convert cast to wrought iron in a predictable fashion. That step could have been taken within a few decades of the initial appearance of cupola furnaces for remelting, and in China those must be almost as early as the blast furnace itself.

It is therefore not absurd from a theoretical standpoint to maintain that white cast iron might have been the first type of iron known and used in China east of Xinjiang. The archaeological evidence points in that direction anyway, for the great majority of iron artifacts from the Warring States and Han periods are made either of white iron or of a metal derived from white iron. It is also not absurd to think that the first low-carbon iron used in China might have come from fineries rather than bloomeries. As noted above, the available archaeological data do not show that either method for making wrought iron was earlier than the other. Yet it may be significant that early objects of wrought iron were scarce and that they remained so down through and past the Han Dynasty. Bloom iron, after all, is an excellent metal and one that can be produced and fabricated on a small scale much more easily than cast or finery iron. If bloomery-made wrought iron had been introduced before cast iron or produced at all widely during the first years of the Iron Age, surely it would have assumed a much greater role in Chinese metallurgical technology. There would be more of it in Warring States period sites, and early Chinese ironworkers would not have been driven to experiment with such a variety of seemingly uneconomical methods for converting white cast iron into metals with properties similar to those of the bloom irons made routinely by ancient peoples in other countries.

The early experiments with malleabilizing, solid-state decarburizing, and fining did pay off in later periods, however, for Chinese ironmaking techniques proved to be capable of exceptionally rapid expansion. In Rome and India at about 100 B.C., ironmaking was carried on much as it had been five hundred years earlier. Larger numbers of slightly bigger bloomery furnaces may have been used in some places, and marketing systems were of course far more complex. Yet the technology involved in all ancient ironmaking outside China was conservative and intrinsically confining. Per-unit production was still limited to a few tens of kilograms of metal per day. Per-unit costs, measured in energy inputs or man days, cannot have fallen substantially since the dawn of the Iron Age.

China, in contrast, found itself in possession of an ironmaking technology that was flexible, readily geared to rationalized production methods, and potentially responsive to attempts to achieve economies of scale. While the biggest Roman or Indian bloomery was not much more efficient than a smaller one and was limited in any case to a maximum production of about 50 kilograms per day, blast furnaces of traditional Chinese types could be increased in size by at least an order of magnitude, with consequent gains in efficiency. And while European smithies in the days before water-powered hammers could increase production only by hiring more smiths, a large Chinese casting shop like the one at Wafangzhuang had numerous options for raising output and decreasing per-unit costs: enlarging the melting furnaces, improving the configuration of larger numbers of stacked and multiple molds, or simplifying processes to allow greater use of unskilled labor. The existence of very large ironworks during the early Han period shows that at least some of this potential was realized. The sheer quantity of iron implements in use by Han times shows that iron had become cheap and abundant—arguably, cheaper and more abundant than in any other ancient civilization.

An interesting question is why the Han system took so long to spread beyond the borders of China. In its developed big-industry form, as at Tieshenggou and Guxingzhen, that system might indeed have been too capital-intensive to be readily adopted in thinly populated "barbarian" areas. However, Chinese methods of making and handling cast iron can be employed on a much smaller scale,[16] and in that form should have been usable and useful to most iron-using peoples. Areas that had little or no iron prior to intensive contact with China, like Korea and Japan, did adopt a modified Chinese-style technology quite early. But cast iron came late to the parts of Southeast, Central, and North Asia that already knew the use of bloomery iron. These areas appear to have resisted the idea that iron could be melted and cast, rather than forged, into implements.

In Southeast Asia, we have no evidence that local peoples made cast iron before the seventeenth century A.D. (Bronson 1986), and that technology seems never to have penetrated to premodern India or Africa. Soviet archaeologists have found evidence of cast iron manufacture in several parts of Central Asia during the twelfth–fourteenth centuries A.D.—at Karakorum in Mongolia (Terekhova 1974), at Sayginskoye on the central Amur (Len'kov 1974), and at Great Bolgars and Sarai Berke near the mouth of the Volga (Rubtsov 1975:42–43). Pig iron smelting and casting may also have reached Afghanistan and Iran during this period,[17] and of course it spread to or was independently discovered by Europeans at just about that time. But before then, for more than a thousand years, the whole of the astonishing technology developed by Chinese ironmakers seems to have been confined to China and its eastern neighbors.

ACKNOWLEDGMENTS

I wish to thank Dr. Chuimei Ho for her advice on Chinese archaeological issues and for sharing her extensive research on Chinese-language sources. I also wish to thank the anonymous reviewer of this article for his or her detailed comments and for calling my attention to recent work in Xinjiang Province. I also am grateful to Dr. William Rostoker, now deceased, for his helpful suggestions, and to Dr. Donald B. Wagner for sending me several key articles by himself and other authors.

NOTES

1. The present paper was originally written in 1988 and has been revised only to the extent that references and radiocarbon dates have been corrected. I have not tried to cover data and ideas from more recent writings, including Wagner's authoritative new book, *Iron and Steel in Ancient China* (E. J. Brill, Leiden, 1993). Although Wagner and I differ in some areas and although he has made use of a wider range of sources, our conclusions are not dissimilar.

2. Chinese character versions of names mentioned in the text are supplied in an appendix following the article.

3. Sites include Yanbulake in Hami (KGXB 1989 [3]: 329–362), Subashi in Shanshan (KG 1984[1]:41–50), Chawuhu Pass Cemeteries 1 and 2 in Hejing (KGXB 1988 [1]:75–100; KG 1990[6]:511–518), Qunbake in Luntai (KG 1987 [11]:987–1001; KG 1991[8]:684–713), and tumuli near Taxkorgan in the Pamirs (KGXB 1981[2]:199–216). Ceramic and other finds from these sites, along the northern and western edge of the Takla Makan desert, do not resemble contemporary artifacts from the rest of China.

4. The bronze-handled iron sword from Jinjiazhuang in Lingtai County, Gansu, came from a tomb dated to the early Spring and Autumn period, the eighth or seventh century B.C. Li (1985) regards this sword as definitely of man-made iron in spite of the fact that it is too rusted to be examined metallographically. Some specialists think it is meteoritic. Huang (1984:155–156) implies that it might be an import from a non-Chinese culture.

5. The totals given by Wagner differ slightly from those in the original report in *Kaogu*. There were 12 rather than 20 iron halberd heads, 3 rather than 4 iron belt hooks, and 20 rather than 19 bimetallic arrowheads.

6. Such enterprises became a hallmark of traditional ferrous metallurgy in China; Hartwell (1962, 1966) has documented the existence of even larger ironworks in the vicinity of Kaifeng during the eleventh century.

7. The original site report for Tieshenggou (Henan Team 1962) states that the site contains 17 smelting furnaces, a low-temperature roasting oven, a melting furnace and a fining hearth. These totals must be revised in the light of the re-investigation carried out in 1980 by Zhao Qingyun and his associates (1985).

8. Li Xueqin (1985:323), for example, is quite emphatic in speaking of the "false impression" that "iron metallurgy was characterized by cast iron from the very beginning in China, unlike the rest of the world.... We believe this to be an unreasonable view. Unless cast iron technology came in from the outside, the appearance of cast iron must indicate that a very long process of development of iron metallurgy preceded it." Barnard is one of those who have consistently maintained the opposite point of view, that cast iron in China came before wrought iron (Barnard and Sato 1975:61).

9. According to the excavators, Furnaces #12, 13, and 14 functioned as bloomeries, "for sponge iron making." Zhao et al. (1985:169), however, have established that all three were actually mold-baking kilns with semicircular chambers.

10. Weapons with pack-welded laminated structures are a case in point: for swords supposedly of finery-made iron

from the Han period, see Hua et al. (1986:5) and Yang (1982:220); for Warring States period swords supposedly of bloomery-made iron, see the descriptions in *Kaogu* (1975[4]:241–243).

11. The pioneering analyses of Pinel and his associates (1938:178) showed an average sulfur content of about 0.5% for eight dated cast iron objects of the sixth through eleventh centuries A.D. Even higher levels of sulfur may be present in cast iron bells of the Ming (1368–1644 A.D.) and Qing (1644–1911 A.D.) Dynasties (Rostoker et. al. 1984).

12. Wagner (1988:184–185), noting that a number of early Han ironworks in the Nanyang area of Henan (i.e., near Wafangzhuang) are located on rivers rather than close to ore sources, makes the interesting suggestion that water power may have been in use as early as the third century B.C. He may be right. On the other hand, it took a much greater weight of fuel than of ore to produce a given quantity of iron. Hence one would expect many ironworks to have been located nearer fuel sources than mines, and fuel supplies often had to have been carried by boat.

13. Weapons made of wrought iron layers partially carburized through cementation and then pack-welded into a single blade existed during and probably before the third century B.C. (e.g., at Yanxiadu—KG 1975[4]:241–243). As "hundred-refined" swords (Sasaki 1982; Rostoker et al. 1985), they were quite common in the Han and later periods.

14. On the extensive iron smelting carried on during the seventeenth century by the nomadic Shor ethnic group of the Kuznets Basin in Siberia, see Levin and Potapov (1964:445–446).

15. Foundrymen of the preindustrial period were acutely conscious of this potential problem. The French cannon maker Pierre Grignon, for example, warned repeatedly of the danger that the melted cast iron in a cupola might "pass to the state of natural [i.e., wrought] iron, which will cause annoying impediments and the ruin of the furnace" (Grignon 1806:419).

16. For instance, the microindustries for casting and mending iron pans described in Hommel (1937:27–32); or the small smelting operations in McCaskey (1903) and Wagner (1985).

17. Descriptions of traditional cast iron smelting by Tadjiks in Badakshan, Afghanistan, and by Persians in the Elburz area, Iran, are given by Kussmaul and Schafer (1973) and by Böhne (1928); in both cases the techniques used are distinctive enough to show that they were not recently borrowed from either the Far East or the West.

APPENDIX
CHINESE CHARACTER VERSIONS OF NAMES MENTIONED IN THE TEXT

Baihewan	白鹤湾	Hunan	湖南
Changde	常德	Jiangsu	江苏
Changsha	长沙	Jinan	纪南
Changyang	长杨	Kaifeng	开封
Changzhi	长治	Kuandian	宽甸
Chu (State)	楚	Kuga	库车
Dafujianggou	大副将沟	Leiyang	耒阳
Dengfeng	登封	Liang	梁
Dongshima	东史马	Lianhuabao	莲花堡
Echeng	鄂城	Liaoning	辽宁
Fenshuiling	分水岭	Liguoyi	利国驿
Fushun	抚顺	Linru	临汝
Gaocheng	告城	Linzi	临淄
Gong County	巩县	Liuhe	六合
Guweicun	固围村	Luoyang Cement Factory	洛阳水泥厂
Guxingzhen	古荥镇	Mancheng	满城
Han (State)	韩	Mayang	麻阳
Handan	邯郸	Mianchi	渑池
Hebei	河北	Nanyang	南阳
Hebi	鹤壁	Qin State	秦
Henan	河南	Shaanxi	陕西
Hubei	湖北	Shandong	山东
Hui County	辉县	Shang Dynasty	商朝

Shanxi	山西	Xinglong	兴隆
Shaoshan	韶山	Xinjiang	新疆
Shi Huangdi	始皇帝	Xinyang	信阳
Shiziling	识字岭	Xinzheng	新郑
Shuangshanzi Commune	双山子公社	Xuzhou	徐州
Sichuan	四川	Yaaishan	亚艾山
Sima Qian	司马迁	Yan (State)	燕
Spring & Autumn	春秋	Yanxiadu	燕下都
Tagangliang	塔岗梁	Yichang	宜昌
Taihuiguan	太晖观	Yiyang	益阳
Tianxinguan	天星观	Yunnan	云南
Tonglushan	铜绿山	Yutaishan	雨台山
Wafangzhuang	瓦房庄	Zengjiagou	曾家沟
Wangshan	望山	Zhaoxian Commune	招贤公社
Warring States	战国	Zhengzhou	郑州
Wei (State)	魏	Zhou Dynasty	周朝
Wen County	温县	Zixing	资兴
Wu (State)	吴		
Wulipei	五里牌		
Xiadiancun	下店村		
Xian	西安		
Xianyang	咸阳		
Xichuan	淅川		

REFERENCES CITED

Note: The authorship of articles in Chinese journals is often credited to institutions or to numerous collaborators. In order to save space, only the most frequently used Chinese-language articles are listed below; these are cited in the text using the standard author-date-page format. Chinese-language articles not listed below are cited in the text in a journal-year-issue-page format: "KG 1971(2):23," for instance, means *Kaogu*, volume (and year) 1971, issue 2, page 23. The following abbreviations for the titles of journals are employed:

KG: *Kaogu* [Archaeology]
KGXB: *Kaogu Xuebao* [Archaeological Reports]
KGYWW: *Kaogu ye Wenwu* [Archaeology and Culture]
KJS: *Kejishi Wenji* [Journal of the History of Technology and Science]
WW: *Wen Wu* [Cultural Material]
WWZLCK: *Wenwu Ziliao Congkan* [Collection of Cultural Data]

AICASS (Institute of Archaeology, Chinese Academy of Social Sciences)
1991 *Radiocarbon Dates in Chinese Archaeology 1965–1991*. Beijing: Cultural Relics Publishing House. [in Chinese, with English abstract]

Barnard, N., and Sato T.
1975 *Metallurgical Remains of Ancient China*. Tokyo: Nichiosha.

Böhne, E.
1928 Die Eisenindustrie Masenderans. *Stahl und Eisen* 48(45):1577–1580.

Bronson, B.
1986 Notes on the History of Iron in Thailand. *Journal of the Siam Society* 73:205–225.

Brown, J. C.
1923 The Mines and Mineral Resources of Yunnan with Short Accounts of its Agricultural Products and Trade. *Memoirs of the Geological Survey of India* 47.

Chang K. C.
1977 *The Archaeology of Ancient China*, 3rd ed. New Haven, CT: Yale University Press.

Changsha Report
1957 *Changsha Excavation Report*. Beijing: Institute of Archaeology, Academia Sinica. [in Chinese]

Gale, E. M.
1931 *Discourses on Salt and Iron*. Leiden: E. J. Brill.

Gettens, R. J.; Clarke, R. S.; and Chase, W. T.
1971 *Two Early Chinese Bronze Weapons with Meteoric Iron Blades*. Occasional Papers 4(1), Freer Gallery of Art. Washington, DC.

Grignon, P. C.
1806 *L'art de fabriquer le fer, de fondre et de forger les pieces d'artillerie*. Paris: Auguste Delalain.

Han R., and Yu X.
1983 The Eastern Han Scissors from Dongshima, Zhengzhou, and Cast Iron Decarburized Steel. *Zhongyuan Wenwu* 1983 (special issue):239–241. [in Chinese]

Hartwell, R. M.
1962 A Revolution in the Chinese Iron and Coal Industries during the Sung, 960–1126 A.D. *Journal of Asian Studies* 21:153–162.
1966 Markets, Technology and the Structure of Enterprise in the Development of the Eleventh Century Chinese Iron and Steel Industry. *Journal of Economic History* 26:29–58.

He T.
1983 Iron Smelting. Pp. 392–418 in *Ancient China's Technology and Science*, ed. The Institute of the History of Natural Sciences. Beijing: Foreign Languages Press.

Henan Museum (Honan Provincial Museum, The Blast-furnace Plant of the Shih-ching-shan Steel Plant, Shutu Iron Steel Company, and the Chinese Archaeometallurgy Study Group)
1978 The Iron and Steel Making Techniques of the Han Dynasty in Honan. *Kaogu Xuebao* 1978(1):1–26. [in Chinese]

Henan Museum Staff
1978 *Han Period Stack Mold Casting of Cast Iron: The Excavation and Study of the Wenxian Cast Iron Mold Kiln*. Beijing: New China Publishing House. [in Chinese]

Henan Team (Archaeological Team of the Bureau of Culture, Honan Province)
1962 *T'ieh Sheng Kou, Kung Hsien*. Beijing: Wen Wu Press. [in Chinese]

Henger, G. W.
1970 The Metallography and Chemical Analyses of Iron-base Samples Dating from Antiquity to Modern Times. *Bulletin of the Historical Metallurgy Group* 4(2):45–52.

Hommel, R. P.
1937 *China at Work*. New York: John Day.

Hua J.
1983 The Mass Production of Iron Castings in Ancient China. *Scientific American* 248(1):106–114.

Hua J. et al. (eds.)
1986 *Collection of Articles on Ancient Chinese Metallurgy*. Beijing: Wen Wu Press. [in Chinese]

Huang Z.
1976 The Early Use of Iron in China. *Wen Wu* 1976(8):62–70. [in Chinese]
1984 A Discussion of Chu Iron. *Hunan Kaogu Jikan* 1984(2):142–157. [in Chinese]

Keightley, D. N.
1976 Where Have All the Swords Gone? Reflections on the Unification of China. *Early China* 2:31–34.

Kussmaul, F., and Schafer, F.
1973 *Tadschiken (Afghanistan, Badakshan) Formen und Eisengeissen*. Gottingen: Institut für den Wissenschaftlichen Film.

Len'kov, V. D.
1974 *Metallurgy and Metal Working of the Jurchens in the 12th Century A.D.* Novosibirsk. [in Russian]

Levin, M. G., and Potapov, L. P. (eds.)
1964 *The Peoples of Siberia*. Chicago: University of Chicago Press.

Li X. Q.
1985 *Eastern Zhou and Qin Civilizations*, trans. by K. C. Chang. New Haven: Yale University Press.

Li J. H.
1987 A Brief Review of Ancient Metallurgy in Henan. *Huaxia Kaogu* 1987(1):202–220. Zhengzhou: Henan Province Cultural Research Institute. [in Chinese]

Li Z. H.
1975 The Development of Iron and Steel Technology in China. *Kaogu Xuebao* 1975(2):1–22. [in Chinese]

McCaskey, H. D.
1903 Report on a Geological Reconnaissance of the Iron Region of Angat, Bulacan. *Philippines Mining Bureau Bulletin* No. 3. Manila: Bureau of Public Printing.

Merwe, N. J. van der
1969 *The Carbon-14 Dating of Iron*. Chicago: University of Chicago Press.

Needham, J.
1958 *The Development of Iron and Steel Technology in China*. London: The Newcomen Society.

Qiu L., and Yu X.
1983 Preliminary Studies of Iron Artifacts Excavated at Guxingzhen. *Zhongyuan Wenwu* 1983 (special issue):242–264. Zhengzhou: Henan Provincial Museum. [in Chinese]

Pinel, M. L., Read, T. T.; and Wright, T. A.
1938 Composition and Microstructure of Ancient Chinese Iron Castings. *Transactions of the American Institute of Mining, Metallurgy and Engineering* 131:174–184.

Rostoker, W.
1987 White Cast Iron as a Weapon and Tool Material. *Archeomaterials* 1(2):145–148.

Rostoker, W.; Bronson, B.; and Dvorak, J.
1984 The Cast Iron Bells of China. *Technology and Culture* 25(4):750–767.

Rostoker, W.; Notis, M. B.; Dvorak, J. R.; and Bronson, B.
1985 Some Insights on the "Hundred Refined" Steel of Ancient China. *MASCA Journal* 3(4):99–103.

Rubtsov, N. N.
1975 *History of Foundry Practice in USSR*. New Delhi: Indian National Scientific Documentation Centre.

Sasaki, M.
1982 A Seven-branched Sword Made by the Hundred Refined Method. *Tetsu-to-Hagane* 68(1):178–184. [in Japanese]

Tegengren, F. R.
1923–24 The Iron Ores and Iron Industry of China, Pt. 2. *Memoirs of the Geological Survey of China*, Series A, No. 2. Peking.

Terekhova, N. N.
1974　Technology of Cast Iron Production in Ancient Mongolia. *Sovetskaia Arkheologiia* 1974(1):69–78. [in Russian]

Trousdale, W.
1977　Where All the Swords Have Gone. Reflections on Some Questions Raised by Professor Keightley. *Early China* 3:65–66.

Tylecote, R. F.
1986　*The Prehistory of Metallurgy in the British Isles.* London: Institute of Metals.

Wagner, D. B.
1985　*Dabieshan, Traditional Chinese Iron-production Techniques Practised in Southern Henan in the Twentieth Century.* Monograph 52, Scandinavian Institute of Asian Studies. Copenhagen.
1986　Ancient Chinese Copper Smelting, Sixth Century BC: Recent Excavations and Simulation Experiments. *Historical Metallurgy* 20(1):1–16.
1987　The Dating of the Chu Graves of Changsha: The Earliest Iron Artifacts in China? *Acta Orientalia* 48:111–156.
1988　Swords and Ploughshares, Ironmasters and Officials: Iron in China in the Third Century B.C. Pp. 176–191 in *Analecta Hafniensa*, ed. L. Littrup. London: Curzon Press.

Watson, B. (trans. and ed.)
1969　*Records of the Historian, Chapters from the 'Shi Chi' of Ssu-ma Ch'ien.* New York: Columbia University Press.

Yang K.
1982　*Development of Ironworking in Ancient China.* Shanghai: Peoples Press. [in Chinese]

Zhao Q. et al.
1985　A Reinvestigation of the Remains of the Iron and Steel Works of the Han Dynasty at Tieshenggou, Henan. *Kaogu Xuebao* 1985(2):157–182. [in Chinese]

Appendix I

Avilova, L. I., and N. N. Terekhova,
Sovetsko-amerikanskiy simpozium "Drevneyshaya metallurgiya starogo sveta"
(The Soviet-American Symposium "Early Metallurgy in the Old World").
Sovetskaya arkheologiya 1989, no. 3, pp. 290–296.

[Any remarks in square brackets are those of the translator.]

The symposium was held in the cities of Tbilisi and Signakhi between September 28 and October 7, 1988. It was the fourth working session of Soviet-American archaeologists conducted within the framework of the cooperative agreement between the Institute of Archaeology of the USSR Academy of Sciences and the American Council of Learned Societies. Former symposia were dedicated to such problems as the origin of proto-urban and early urban civilizations and the role of cultural exchange in the development of human societies. The theme of this symposium was the development of metallurgy and the working of nonferrous and ferrous metals within the period between the appearance of metals and the end of the Bronze Age. Also included were some aspects of the development of early technologies, the problems of transfer of technological skills, and the formation of technological traditions. Because of the given theme, participants of the symposium made extensive use of the application of natural science methods to archaeological investigations.

The symposium opened with the presentation of Academician A. M. Apakidze of the Georgian Academy of Sciences in Tbilisi, entitled "On Some of the Principal Trends in the Archaeological Investigations in the Georgian Soviet Socialist Republic." In it were elucidated the principal accomplishments of Georgian archaeology in recent years as well as its central problems and prospects for its [future] development.

The report of G. E. Areshyan (Erevan), "The Acquisition of Copper and Iron in the Near East as an Object of a Comparative-Historical Investigation," was theoretical in nature. A comparison of models of social and cultural development was presented, detailing the acquisition of metals as new materials in early society. The author pointed out a number of sequential stages in the process: (1) complete absence of metal utilization; (2) rare and casual inclusion of metal in the works of man, at which time it becomes an object of human experimentation for the first time; (3) inclusion of metal in the production cycle, [its] distribution, and its use as a secondary component, at which time it becomes a natural resource of the community; (4) a stage during which the new material achieves a decisive importance as a productive force and undergoes considerable improvement in its working technology and the meaning of its social significance; (5) dependable application with a gradual development of technology; (6) decline of the social significance of the metal, stagnation, and then the loss of technological skills. G. E. Areshyan emphasized the regular, law-governed aspects of the change from the bronze industry to the iron industry. The change was associated with the separate stages of societal development, with the Iron Age societies being more "democratic" than those of the Bronze Age.

The report of C. C. Lamberg-Karlovsky (Boston, Peabody Museum) "Pyrotechnology Before Ceramics and Metal" was concerned with a serious theme in the study of early origins—the acquisition of fire by man and the first stages of its utilization. The application of accidently obtained fire to the preparation of food appears to be the first biological revolution which gave *Homo sapiens* the ability to flexibly adjust to his natural surroundings. The second experiment was the application of fire, during the Upper Paleolithic, to the firing of some objects related to the cult realm (clay figurines, iron oxides used as dyes). The third stage of the application of fire is marked by its use in the productive process. It represents the first step in the development of pyrotechnology—the preparation of plastic clay and gypsum coatings in pre-ceramic Neolithic sites of the Near East, beginning with the twelfth millennium B.C. The complexity and labor-consuming effort in producing the plaster and the varied application of it offer extensive possibilities for the reconstruction of the socioeconomic pattern. Also the findings of the author that sites with extensive developments in the technology of coatings [on clay objects] yield the earliest evidence of metallurgy and metalworking are quite significant. Thus early pyrotechnology was the indispensable groundwork on which the rise of new production methods was based.

Most of the reports were dedicated to various aspects of the study of early bronzes within the Bronze Age period.

The presentation of T. Stech (Philadelphia, University of Pennsylvania), "Early Copper and Bronze Metallurgy in Mesopotamia and Anatolia," related to the problem of the development of the technology in metalworking and its overall place in the culture, beginning with the first appearance of copper in the eighth–seventh millennium B.C. The author pointed out the characteristic feature of these first experiments in metalworking—the absence of progress in the technology over a very long period of time. It was suggested that the material from which metal was first smelted was lead extracted from galena ore (note the extensive finds of lead vessels in the Jemdet Nasr period of Ur and other sites). Up to the fourth millennium B.C. both regions, Mesopotamia and Anatolia, developed independently in the realm of metallurgy. The situation changed in the third millennium B.C. The analysis of the extensive finds indicates a similarity in the development of metallurgy for both regions—the use of pure copper, arsenical copper (or arsenical bronze in the terminology adopted in the U.S.S.R.), and tin bronze. Artifacts made of the last are principally concentrated in burials (Ur, Kish, Alaça Hüyük, Horoztepe, and later Kültepe). This is the basis for the author's suggestion that tin bronze was a symbolic metal marking the high social status of the interred individual. At the same time the technological quality was ignored by the ancient population. The statement engendered a lively discussion. All told, as related by T. Stech, metallurgy does not appear as an motivating cultural force but only one of its expressions.

In the report by J. Muhly (Philadelphia, University of Pennsylvania) entitled "Copper and Bronze Metallurgy in Cyprus and the Eastern Mediterranean," the problems concerning early metallurgy (principally tin bronze) in the Levant were examined.

The first use of metals, copper and gold, is dated in Greece to the Neolithic (the fifth millennium B.C.), and its widespread distribution to the Early Bronze Age (3100–2200 B.C.). Traces from the working of native copper on Cyprus are known from the Middle Chalcolithic (ca. 3500 B.C.), but the presence of local metallurgy is associated with the cult of Philia dated to the Cypriot Early Bronze Age (2600 B.C.). According to the works of Jak Yakar, the development of metallurgy in Anatolia is analogous to that found in mainland Greece; i.e., the widespread utilization of metal dates to the Early Bronze Age (3100–2200 B.C.).

For all of these regions the transition from arsenical copper to bronze proper (that is, tin alloys) is typical. Although the modes of preparation of arsenical copper are as yet unclear, Muhly holds that it was extracted from arsenic-bearing [copper] ores, such as domeykite, well known in the Near East (Talmessi). Arsenical ores are known on the island of Kythnos and on Cyprus. Information on the sources of tin is more meager. To the author the most convincing is the use of Afghanistan tin ore deposits, particularly in the period Early Minoan III–Middle Minoan I. However, the collection of tin-bronze objects was so large in the second millennium B.C. and particularly in Late Minoan I that one is compelled to think of much closer sources of tin. The exploration [of tin ores] in the Bolkardağ [area] of Cilicia may be mentioned but their utilization during the Bronze Age is quite doubtful; the same holds true for the deposits in Yugoslavia and Czechoslovakia. The probability of shipments of tin from Cornwall to Cyprus still exists.

During the Late Bronze Age, abrupt changes occur in the character of exchanges between Cyprus and territories farther west. As lead-isotope analyses show, during Late Helladic IIIB (thirteenth century B.C.) oxhide ingots from Mycenae were made from Cypriot copper and the same is true for ingots found in Sardinia. Analyses of materials from Cyprus, Crete, Sardinia, Cape Gelidonya, and Ulu Burun confirm the above deduction. During the Late Cypriot I–II periods the center of copper production was the site of Enkomi, identified as the capital of the Alashiya kingdom, the latter mentioned in Egyptian, Hittite, and Mesopotamian texts as a supplier of bronze. During Late Cypriot IIC, however, the scale of production was greatly increased to include a number of sites on the southern coast of the island. This period, 1300–1100 B.C., was characterized by the strengthening of international contacts and the growth of riches. Indeed, at this time the inhabitants of Cyprus reached the level of true urbanization. Trade with territories to the west—the Aegean and Sardinia—was strengthened. The remains of the thirteenth century B.C. shipwrecks at Cape Gelidonya and Ulu Burun reflect this process of expansion of Cypriot copper in a westerly direction.

V. Pigott (Philadelphia, University of Pennsylvania, Museum Applied Science Center for Archaeology) presented in his report "The Development of Metal Use on the Iranian Plateau" the results of 30 years of investigations into the history of applications of copper and its alloys, as well as of iron, beginning with the early stages of metallurgy in the seventh millennium B.C. and [reaching] to the first millennium B.C. As indicated by archaeological and analytical data, the earliest examples of metal artifacts appeared in this region in the sites of the Late Neolithic (Ali Kosh, Zagheh; 6500–6000 B.C.) and were coldworked from native copper.

In the following Eneolithic period (5500–3200 B.C.) copper was of metallurgical origin. Metal artifacts were found in the lower layers of such sites as Susa, Malyan, Tepe Yahya, Tal-i Iblis, Sialk, and Hissar. Arsenical copper ores were used, the sources of which are thought to be the deposits at Talmessi and Meskani, located at the base of the western range of the Dasht-i Kavir.

Traditionally the use of arsenical copper continued here throughout the entire Bronze Age period (3200–1450 B.C.). Tin bronze appears on the plateau only at the beginning of the Iron Age. This fact is particularly noteworthy, as Pigott observes, because tin, enjoying among the Sumerian population of Mesopotamia the high status of an exotic metal, a status shared by lapis lazuli and gold, became the object of trading expeditions over the plateau to the east.

During the Early Iron Age (1450/1350–1100 B.C.) tin bronze became the dominant alloy with a copper base although arsenical copper did not go out of use. Initially iron appeared in the western part of the Iranian Plateau during Early Iron II (1100–800/750 B.C.), within the confines of the tenth and ninth centuries B.C. Its appearance is connected to Assyrian influence. At first the new metal was used as a decorative element, later as a detail [as detailed ornamentation] in bimetallic artifacts. Only at the beginning of Iron III does the functional purpose of iron become quite separate.

Metallographic analyses of eighth century B.C. materials from Hasanlu indicate that in the making of tools and weapons iron was basically used and [also] soft or unevenly carburized raw steel. According to the author there is no proof of purposeful manufacture of [true] steel at this time.

The report of R. Wright (New York University and College of William and Mary), "Technical Investigations of Ceramics from Iran and India: The Transmission of Technology," dealt with an in-depth study of the ways and means of transmitting technological information based on a comparison of the ceramic production of two cultural groups, from Iran and northern India, respectively. Third millennium B.C. vessels of the Emir and Faiz Mohammad types were subjected to macro- and microscopic investigations followed by a statistical analysis. The features of the earthenware, all told rather similar, seemed to be conditioned by the varied successive technical methods of manufacture, although [at a given time] the methods were identical. The author suggested her data should be interpreted according to classical diffusion theory, claiming that it had been unjustifiably rejected in recent times. Wright emphasized that it was indeed the transmission of technological knowledge that stimulated, in her opinion, the initiation of other more profound and intensive contacts between the regions under discussion. The last thesis elicited objections; G. E. Areshyan and V. M. Masson demonstrated that, in their opinion, the transmission of technology dates to a much later period than simple trade exchange and other aspects of cultural ties.

In the report of A. I. Dzhavakhishvili, O. M. Dzhaparidze, and T. V. Kiguradze (Tbilisi), "The Basic Stages of Development of Georgian Cultures in the Neolithic and Bronze Ages," the present-day point of view was offered regarding the cultural development of the region, beginning with the appearance of a productive economy and the distribution of the agricultural complexes of the Shulaveri-Shomutepe culture under the influence of Near Eastern impulses. The first use of copper is associated with the late stage of this culture, which in turn is synchronous with northern Ubaid (Tekhut). In the second half of the fourth millennium B.C. the formation of the Kura-Araxes culture took place in eastern Transcaucasia. With its flourishing is tied the development of Caucasian metallurgy on the basis of local ores from which arsenical bronze [copper] was made, the dominant type of alloy in the given culture. To the authors the most plausible view [of several] is that the Kura-Araxes culture was formed in antiquity by ethnically affiliated Caucasian tribes. The authors associate a crisis in the agricultural economy and its ending with the appearance of early burial mounds [kurgans] of the Martkop and Bedeni types in the second half of the third millennium B.C. At the same time metallurgy and metalworking undergo an upgrading marked specifically by the appearance of tin bronzes. The burial mound tradition is extended to eastern Georgia by the Trialeti culture, which the authors date to the Middle Bronze Age, a culture which is characterized by new forms in the metallic inventory and a high development of the jeweler's art.

In his report "Concerning the Chronology of the Early Metal Epochs in the Caucasus," G. L. Kavtaradze demonstrated the results of the systemization of the series of Georgian archaeological cultures by using calibrated radiocarbon dates and comparison with cultures of the Near East. The known radiocarbon dates for Transcaucasian sites are unequally distributed and principally refer to the Late Bronze Age. The author presented substantial re-datings of practically all of the successive Eneolithic–Bronze Age cultures. Thus he places the Shulaveri-Shomutepe culture in the sixth millennium B.C., and dates the Kura-Araxes to the beginning of the fourth–middle of the third millennium B.C. (not to the third millennium B.C. as it is usually dated). The Martkop and Bedeni cultures are shifted to the first half of the third millennium B.C., the Trialeti to the second half of the third–beginning of the second millennium B.C., with the Trialeti culture being included in the Early Bronze Age.

The pushing of these cultures back in time, in particular the Kura-Araxes culture, is tied to the traditional dating of the early layers of analogous sites in the Near East at the end of the fourth millennium B.C. (Late Uruk–Jemdet Nasr). The overview of the successive cultures of the Transcaucasus offered by G. L. Kavtaradze, given the unquestionable prospects of the C-14 method itself, calls for the accumulation of a new, extensive series of radiocarbon dates, particularly those of the early periods.

In the report of M. R. Abramishvili (Tbilisi), "On the Reconstruction of the History of Metallurgical Development," particular attention was directed to the ties of the Transcaucasus and Anatolia. The author adheres to the point of view of the determinant role of a technological tradition, the accumulation of experience in the development of metallurgy in specific regions. He regards eastern Anatolia as the birthplace of metallurgy, from which, in his opinion, also emerged the methods of manufacturing tin bronze (the early finds from the Eneolithic site of Delisi dating to the end of the fifth–beginning of the fourth millennium B.C.). For the Kura-Araxes culture, both its early and late stages, arsenical bronze is typical, but the tradition of the mass production of tin alloys takes place only with the arrival of kurgan [burial mound] cultures of the Bedeni type. M. R. Abramishvili presumes the origin of the Bedeni and Khirbet-Kerak cultures to reside in hypothetical common ancestral cultures with black-polished ceramics and a highly developed metallurgy. He locates these cultures in northeastern Anatolia. From here the iron industry also reached the Caucasus. However, full-scale proof of these theses will be possible only after extensive investigations in northeastern Anatolia—one of the least studied archaeological regions.

The Late Bronze Age was elucidated in the report of K. N. Pitskhelauri, "The Genesis of Material Culture in the Caucasus during the Late Bronze and Early Iron Ages." As a basic task of his investigation the author singled out the search for the roots of the highly developed metallurgical province of the Caucasus during the Late Bronze Age. He emphasized that the basis for the progressive development of culture in the Caucasus during the third and second millennia B.C. was the achievement of metallurgy. Before the third millennium B.C. the level of development in the Caucasus and the Near East was essentially the same. The situation was profoundly changed in the third millennium B.C. when, with the arrival of new ethnic groups, the early kurgan cultures arose. Two of the factors—the historical-cultural and the productive—resulted in the formation in the middle of the second millennium B.C. of the Caucasian metallurgical province with its specific forms of metal artifacts and composition of alloys determined by the character of the [above-mentioned ethnic] ties. The rapid internal development of the region during the Late Bronze Age seemed to be a precondition for the fast transition to an iron industry. Thus the materials of the Caucasus do not offer a confirmation to the often proposed thesis about the crisis in bronze metallurgy being the reason for the transfer to iron.

A different opinion on the above was held by V. S. Bochkarev (Leningrad). In his report, "The Economic and the Social Role of Metallurgical Production in the Late Bronze Age (based on Data of the Southern Cultures of Eastern Europe)," Bochkarev analyzes the voluminous finds of metal artifacts from the area adjacent to the north coast of the Black Sea. He does so from the point of view of innovations in metalworking. He associates the sharp rise in metallurgy and metalworking in the Late Bronze Age with the widespread use among the tribes of the Timber Grave culture of the technique of thin-walled casting, with the use of tin bronze, and with long-lasting stone casting molds. This gave production a serial character and radically changed all modes of metalworking, both in production and application. Steady importation of metal from the Balkans, Urals, and northwestern Caucasus permitted the establishment of a number of metalworking centers in areas that lacked metal ores. An analysis of the composition of hoards [caches] led the author to the conclusion that metalworking was subordinate to the agricultural cycle. However, the final solution of this problem will be found only after a study of the economic systems of the Late Bronze Age Steppe population. To explain the crisis in Belozerka [eastern Ukraine] metalworking V. S. Bokcharev proposes the invasion of Hallstadt tribes from the Balkans. In his opinion it was just after this [event] that the area along the north coast of the Black Sea outstripped the Caucasus in its transition to the Iron Age. It should be noted that nothing hampered the acquisition of Uralian and Caucasian metal [ore] and thus the proposition that would explain the crisis in the production of bronze by an insufficiency of raw materials cannot be universal. The thesis of the priority of the arrival of the iron industry to the northern shore of the Black Sea compared to the Caucasus also provokes doubts. [The last two sentences are apparently the remarks of the authors of the abstracts.]

R. M. Abramishvili (Tbilisi) in his report, "The Production of Copper-Bronze and Iron Artifacts in the Transcaucasus during the Late Bronze and Early Iron Ages," dwells on the delimitation and characterization of three historically productive regions, encompassing (1) the western and central Caucasus, the western Transcaucasus, and the northwestern part of Georgia; (2) the central Transcaucasus, including eastern Georgia, the western part of the Azerbayzhan SSR, and the entire territory of the Armenian SSR, as well as the eastern part of the northern Caucasus; (3) the southern part of the Murgan steppe and the Talysh.

The basis for the above division was a typological investigation and cartographic presentation of the variety of metal objects and production remains [scrap, et al.] dated to the chronological span of the second half of the second to the first half of the first millennium B.C. The delimited regions are subdivided by the author into large production areas and then into production hearths [centers]. Thus, in the opinion of the author, there functioned on the territory of the Colchis culture several production centers which included southwestern Colchis, the Colchis Lowland, northwestern Colchis, and the northwestern part of eastern Georgia.

In the area of the Samtavrian culture the author singles out two major centers: Shida-Kartli (in the northwestern part of eastern Georgia) and Kvemo-Kartli (in the southern part of eastern Georgia). Within the territory of the Lchashen-Tsitelgorsk culture the investigator finds two centers. One includes the territory of the Armenian SSR, the westernmost part of the

Azerbayzhan SSR, and the regions of eastern Georgia immediately adjacent to it. The other is represented by the Ioro-Alazan Basin.

According to the author, the acquisition of iron in the Transcaucasus begins not later than the middle of the second millennium B.C., at Kvemo-Kartli and adjacent territories, under influences from eastern Anatolia. The earliest iron object is from Kurgan I in the village of Zveli. The kurgan contains materials reflecting contacts with Asia Minor.

The report of T. P. Mudzhiri (Tbilisi) was dedicated to the serious problem of the exploitation of Caucasian ore deposits in the early metal epoch. The author acquainted himself with the results of the investigation of the earliest mines in Georgia, where during the early metal epoch there functioned six mining and metallurgical centers. The largest of these were in Racha, Svanetia, and Abkhazia. The construction of early Georgian mines, as well as techniques and technologies of mining, has parallels in Armenia and the northern Caucasus. This testifies to the generally high technical level and regular, systematic development of the entire mining and metallurgical production of the Caucasus.

The author principally analyzed new methods of classification, periodization, and comparative-historical valuations of the system of early mining and metallurgical production on the basis of an experimentally computed model of production indicators. In particular, this allowed him to establish that a number of mines in Racha, Abkhazia, and the northern Caucasus dated to the second–beginning of the first millennium B.C. and that in their constructional-technological parameters they did not have analogues in the practice of mining in the world at that time.

A. A. Askarov and V. D. Ruzanov (Samarkand) in their report, "Bronze Age Centers of Metallurgy and Metalworking in Central Asia," presented the following scheme of development.

During the Early Bronze Age metalworking was characterized by assemblages from sites of agricultural tribes in three regions: southwestern Turkmenia (the burial mound Parkhay II), the foothill zones of the Kopet Dag (the settlements of Altyn-depe, Namazga-depe, and Anau), and the Zeravshan valley (the settlement of Sarazm in western Tadzhikstan). Genetic ties concerning metal in the preceding epoch of the regions are traced.

During the Middle Bronze Age new centers appeared in the Murgab and the oases along the Amu-Darya—the Murgab and Sapalli centers. There is considerable similarity in the production of these centers but there is a difference in the utilization of tin alloys, with the share of these alloys being four times higher in the Sapalli center. At the end of the third–beginning of the second millennium B.C., mining of local deposits began in the Kyzyl Kum mountains in the north of Central Asia. The mining served as a base for native metallurgy [although the initial influence was from southern centers].

During the Late Bronze Age Central Asia undergoes a division into two zones identified as two large metallurgical provinces: the Irano-Turkmenistan and the Eurasian. A characteristic feature of metalworking among the steppe tribes of Central Asia is the widespread utilization of tin bronzes. The mutual influences of the Eurasian and Irano-Turkmenistan provinces lead, at the end of the second–beginning of the first millennium B.C., to a transformation of metalworking production in the southern regions of Central Asia, where we find a recession in production, a change in the composition of alloys, and also a change in the orientation of ties.

G. F. Korobkova (Leningrad) read a report on the theme "Metallurgy and Stone Tools: The Transformation and Adaptation of a Traditional Technique in the Conditions of Technological Progress." On the basis of trassological investigations [microscopic use-wear studies] and the physical modeling of production processes, the author singled out from the stone inventories of palaeometallic sites in Central Asia and the Caucasus tools that were associated with metallurgical and metalworking productions. Thus, the materials from Altyn-depe show that the functioning of these productions during the Middle and Late Bronze Ages, as well as in preceding periods, was carried out with stone tools. A planigraphic analysis of the concentrations of specific groups of tools pointed out the locations of specialized workshops in the settlement.

The great variety of stone tools used in metalworking is indicated in burial mounds of the Maykop culture. These included hammers, small hammers, anvils, smoothers and straighteners, abraders, etc.

The stone inventory of the Timber Grave and Sabatin cultures of the steppe Bronze Age contains, according to the author, a rich collection of mining, metallurgical, and metalworking instruments which are found within the limits of the settlements. The investigator emphasized that stone tools having undergone a separate transformation and adaptation to the new technical and technological conditions nevertheless retained an important role in the economic and productive systems of the palaeometallic epoch.

The report of V. I. Molodin (Novosibirsk), "Metallurgy of the Krotov Culture," was dedicated to the investigation of the metallurgy of the woodland tribes in Western Siberia during the middle of the second millennium B.C. The high level of development of local bronze casting is characterized both by production, in which we can discern Seyma-[Seima] Turbino, Eurasian, and Andronovo types of artifacts, and by numerous remains of casting instruments and scrap. Of particular interest are the burials of the smiths in the cemeteries Sopka-2 and Rostovka, since they reveal the social status of the buried. The sources of the raw materials for this local metalworking are located in Kazakhstan or the Rudnoy [Ore] Altay [Mountains]. The location of these sources defined the intensive contacts [warfare and eventual absorption] of the Krotov population and the Andronovo tribes.

Laboratories of the Institute of Archaeology of the USSR Academy of Sciences which deal in the natural sciences were represented by three reports covering both non-ferrous and ferrous metallurgy. Two of the reports were based on spectral analysis of early bronzes and were joined [related] by a general method of investigation based on functional-typological, spectral, and mathematical-statistical analyses.

In the first report, "The Circumpontic Metallurgical Province as a System," by E. N. Chernykh, L. I. Avilova, T. B. Bartseva, L. B. Orlovskaya, and T. O. Teneyshvili, an analysis of the Circumpontic Metallurgical Province (CMP) was given, presenting it as a central cultural and production system during the Early and Middle Bronze Ages. This was done on the basis of the morphology of the tools and composition of the alloys, the technology of casting, the depiction of the zones and centers of the smelting of the metal, and the manufacturing of tools from it, the scale of production and its cultural and historical content. The mechanism of the formation of the province is not clear at present. It came into existence in the middle of the fourth millennium B.C. without having a single cultural, ethnic, or productive base. It flourished during the Middle Bronze Age during which metallurgical production experienced an explosive growth, but at the beginning of the Late Bronze Age the CMP dissolved. All of the above was presented as a comparison, in a broad sense, of developmental models of early metallurgy in other regions and periods.

The second report, "The Eurasian Metallurgical Province as a System," by E. N. Chernykh, S. A. Agapov, and S. V. Kuzminykh, was dedicated to the extensive cultural and economic unification of the territory between the Dnieper River and the Altay Mountains during the Late Bronze Age. A single approach to the investigation indicated the presence of particular features of the province in production as well as cultural realms. The Eurasian Metallurgical Province was characterized, first and foremost, by the complexity of its formative mechanism. At its base was the Abashev community and the mobile [nomadic] Seima-Turbino component. The province was also characterized by the extreme dynamism of its development. The Andronovo-Alakul and Timber Grave communities entered during its period of florescence and stabilization. In its late phase, isolation of an extensive community with filleted ceramics [i.e., ceramics with an appliqué "ring" at the rim] took place in the steppe zone and a series of cultures in the forest zone. The province fell apart in the tenth–ninth centuries B.C.

A number of papers were dedicated to the early history of iron manufacturing in various regions of the world.

In the report of Jane Waldbaum (University of Wisconsin, Milwaukee), "The Initial Use of Iron in the Eastern Mediterranean," the results of 20 years of archaeological investigations were presented. Parallel with this various hypotheses relating to the reasons for replacing bronze with iron were reviewed, as well as the sources of its origin and the routes of its distribution and the mutual ties of the new metal with specific cultural and historical conditions.

The author rejects the theory of Hittite monopoly in the production of iron since neither the number of finds nor their quality or conditions of use formed, in the author's opinion, a basis for such a proposition. The technology of smelting iron existed not only in Anatolia among the Hittites but also in Greece during the Late Bronze Age and possibly in Syria-Palestine. The author is guided by data indicating that artifacts of meteoric and metallurgical iron were used in the eastern Mediterranean throughout the Bronze Age and even earlier, though rarely, and then only in ritual and ceremonial functions. Iron became more available after 1200 B.C., at which time it began to be utilized in domestic and military realms. At the beginning of the tenth century B.C. iron became widely distributed but the rate at which it was substituted for bronze varied from region to region.

J. Waldbaum critically views the hypothesis that the dissemination of iron in the eastern Mediterranean was due to the invasions of Dorians or Philistines who were in possession of new, better weapons. The author pointed out that it was exactly in the region regarded as the domicile of the Dorians that iron appeared 300 years later than in Greece and therefore the entry of the Iron Age in the latter was not dependent on northern invaders. The same applies to the Philistines, among whom iron did not become the dominant material any sooner than in other regions of Palestine.

Until recent times the most popular hypothesis about the reason for replacing bronze with iron was an "economic" one which claimed that the access to tin was cut off as a result of political changes around 1200 B.C. Today this hypothesis is being challenged on the basis of documented sources of tin in southeastern Turkey and besides there is no diminution of tin content in artifacts dated later than 1200 B.C.

In the opinion of the author the "ecological" theory also needs a more detailed investigation. This theory asserts that pyrotechnical activities, land clearing, and terraced agriculture led to the deforestation of the Mediterranean. The smelting of iron, demanding less fuel, became more economical. Recently, a "technological" hypothesis appeared claiming priority for Cyprus in the discovery and systematic application of methods for the hardening of iron [into steel] and only then did iron supplant bronze. J.Waldbaum emphasized that at the present there is no conclusive evidence that would support or refute this or that theory.

G. Possehl (Philadelphia, University Museum, University of Pennsylvania) in his report "The Beginning of the Iron Age in India" reviewed such important questions as: (1) What is the probability of an indigenous and independent [of outside influences] transition to the Iron Age in India and Pakistan? (2) What were the technological and economic bases for the establishment of an Iron Age in South Asia? (3) What were the historical and social implications tied in with this process? The reporter [Possehl] pointed out that the level of technical and production development in ancient India (for instance, the high state of development of pyrotechnology and corresponding manipulation of ceramic and faience production in the Harappan civilization) provided objective possibilities for the discovery and development of an iron-working technology. The process may have been enhanced by the presence of large deposits of iron ores in the region. There is convincing archaeological proof that metallurgical iron was known in India at the end of the third–beginning of the second millennium B.C. In the opinion of the author the arrival of the Iron Age was tied not only to technical discoveries and inventions but also to social and historical conditions conducive to the exploitation and widespread introduction of the technology.

The principle advantages of iron, in comparison with copper and particularly bronze, are its large reserves and ease of mining. Copper, and more so bronze, could not be the alloy to be used by all layers of a society. Emphasizing that the appearance of iron and even the discovery of the technology for its smelting do not necessarily mark the beginning of an iron age, G. Possehl drew attention to the need for a set of objective criteria which in cultural and historical contexts will be useful in defining the level or form of technological development that marks the arrival of the Iron Age.

The report of B. Bronson (Chicago, Field Museum of Natural History), "The Transition to Ironmaking in Ancient China," critically examines many debatable questions about the history of development of ferrous metallurgy in China. In particular, these include the possibility of a transition from non-ferrous metallurgy directly to cast iron, thus leaving out the bloom iron method known to the rest of the world. The author, as well as a number of other investigators, assert that in China there existed reasons for this: cast iron was often accidentally produced in Bronze Age shaft furnaces when iron ore was used as a flux; the largest of such furnaces may have operated at temperatures as high as 1350°C, and [bronze] casting technology was more traditional with Chinese smiths than forging. This allowed them to quickly evaluate the properties of cast iron; in resmelting pig iron and cast iron scrap occasional batches of iron [with a lowered carbon content] could be produced.

B. Bronson emphasizes that up to the present there is no reliable archaeological evidence of early bloomeries in all of China; also there is no mention of them in the literature.

Since iron appears in China relatively late (according to the latest data a few finds date to the sixth century B.C. and iron is very rare up to the end of the fifth century B.C.), it cannot be ruled out that the idea of making iron was obtained from abroad. However the development of the technology took a quite original path, one that does not have an analogue in any other region in the world.

The author also showed slides with interesting ethnographic data on ways of obtaining cast iron in the Philippines in the present day, a process reminiscent of early Chinese production.

Soviet scientists interested in the early history of iron presented a number of reports dealing with iron making in the Caucasus.

D. A. Khakhutayshvili (Batumi) in his report, "The Origin and Development of Iron Metallurgy in the Caucasus and Near East," concerned himself with one of the least studied aspects of the history of ferrous metals—the characteristics of iron foundries. He noted that not one of the Old World regions had the requisite number of production centers (dated prior to the seventh century B.C.) that could produce iron. The author offered data on a unique region in the foothill zone of the [coastal area of the] eastern reaches of the Black Sea, where, over the past 30 years, some 400 foundries connected with the production of iron have been discovered. Four basic production processes were grouped into the Colchis mining and metallurgical center. The design of the foundries, raw materials base, and types of fuel were studied. The author offered a chronological sequence for the foundries of various designs and synchronized them with contemporaneous settlements. The earliest complexes were radiocarbon-dated to the eighteenth century B.C.; foundries of similar construction existed into the sixth–fifth centuries B.C.

According to D. A. Khakhutayshvili iron making in the Colchis was based on the tradition of copper and bronze production which, in turn, was based on the production norms of Asia Minor. The absence of traces of iron making in the Near East (among the Hittites, Assyrians, and Mitannians, et al.) before the seventh century B.C., together with the widespread use of iron documented by archaeological finds, gives the author a base to propose that the subtropical foothill zone near the eastern and southeastern coasts of the Black Sea was the principal producer and exporter of iron to the developed countries of the Caucasus and Near East prior to the seventh–sixth centuries B.C.

The report of G. V. Inanishvili (Tbilisi), "The Working of Iron in Ancient Georgia," dealt with the technological analysis of iron artifacts in the Caucasus. The earliest iron artifacts are from Kvemo-Kartli and Shida-Kartli, which date to the fourteenth–twelveth centuries B.C. Metallographic investigations revealed that they were made of low carbon steel; sometimes surface cementation is visible. The structures were fixed by annealing and normalization of the steel. The author regards the period of the ninth to eighteenth centuries B.C. as the epoch of widespread acquisition of iron. Toward the end of the second millennium B.C. a high level of working iron and steel is seen, particularly in the adoption of cementation (carburization) and heat treatment (quenching, tempering, normalization, annealing). During the seventh–sixth centuries B.C. the scale of iron and steel production rises steeply, the forms of tools change, and serial production ensues. As a result, iron finally replaces bronze. Specific quality artifacts are made using complex technological methods (carburization, quenching, tempering) and various types of forging are adopted.

The early history of iron in the northern Caucasus was reviewed in the report of N. N. Terekhova (Institute of Archaeology, USSR Academy of Sciences, Laboratory of Natural Sciences Methods), entitled "Early Iron-making in the Northern Caucasus." On the basis of metallographic investigations of a very large series of a varied iron inventory, principally from the territory of the Kuban culture and the Cis-Kuban [north of the Caucasus range], the author traces the stages of technological developments in working ferrous metals.

In the pre-Scythian period mainly soft raw steel and iron were used. Welding of similar metals was employed, the method of cementation was known. Heat treatment was not consistent.

A sharp qualitative leap took place in metalworking at the end of the seventh–beginning of the sixth century B.C. A characteristic trait was the making of steel of high quality by cementation to blades. Such steel was welded to soft raw iron so that sharp edges could be produced on the implements (axes, knives). Various techniques of cementation of finished products were also used. Differentiated types of thermal application to steel products appear.

The author associates the above innovations in production with a powerful impulse from the Transcaucasus, perhaps an infiltration of groups of master smiths. A further development of iron-working techniques in the northern Caucasus during the sixth–fifteenth centuries B.C. was guided by the increased demand for ferrous materials. The share of complex and time-consuming processes strongly declined and was replaced by more rational processes. In particular single blooms of unevenly carburized raw steel came into widespread use.

Foundry productions from sites in the Cis-Kuban (the Kelermess and Ulskiy kurgans, the Kelermess cemetery) can, by and large, be included in the northern Caucasus sphere although they do have some technological particularities.

Reports dealing with recent archaeological finds were also presented at the symposium. G. E. Areshyan familiarized the participants with the unique assemblage from the Karashamb kurgan on the Razdan River, in which a Trialeti culture human cremation was interred. Among the contents of the inventory were some spectacular pieces, including a silver pole-axe, two coats of mail, a bronze kettle, a quantity of gold ornaments, and ornamented silver cups. On one of the cups five friezes presented mythological, cult, and military scenes. The stated dating of the assemblage—the third dynasty of Ur (twenty-first–twentieth centuries B.C.), based on Mesopotamian and Near Eastern analogues—would push further into antiquity a number of traditional datings for the Trialeti and Kura-Araxes cultures and also for sites of the Martkop and Bedeni type.

M. V. Baramidze (Tbilisi) displayed materials from a bronze foundry discovered in the settlement site of Pichori in Abkhaziya. In a layer dating to the end of the third millennium B.C. a unique cluster of molds was found. It included molds for the making of socketed axes, hoes, a chisel with a [tool] rest, and others, as well as tools made with the molds.

I. Narimanov (Baku) presented data on the new excavations at the settlement site of Leyle-Tepe and metal finds of the Eneolithic, emphasizing the ties of the region with northern Ubaid.

The concluding discussion at the end of the symposium was marked by the deep interest of all of the participants in the materials and hypotheses expressed in the reports. All emphasized the paucity of information about investigations into the early metallurgy in various regions and the long delays in publication. Obvious was the urgent necessity not only for systematic exchange of information, but for practical collaboration which would take place in separate, mutually coordinated programs, with the programs being interdisciplinary. Mutual field investigations of antiquities should be included in the programs, as well as a series of analytical and laboratory results, and the exchange of information of the latest achievements in all facets of early metallurgy.

Translated and annotated by Henry N. Michael
Senior Research Fellow
Museum Applied Science Center for Archaeology
(MASCA)
University of Pennsylvania Museum

List of Contributors

Bennet Bronson
Curator, Asian Archaeology and Ethnology
The Field Museum
Roosevelt Rd. at Lake Shore Dr.
Chicago, IL 60605
USA

Praveena Gullapalli
Department of Anthropology
University of Pennsylvania Museum
33rd and Spruce Sts.
Philadelphia, PA 19104-6324
USA

Henry N. Michael
Museum Applied Science Center for Archaeology
(MASCA)
University of Pennsylvania Museum
33rd and Spruce Sts.
Philadelphia, PA 19104-6324
USA

Jonathan Mark Kenoyer
Department of Anthropology
University of Wisconsin-Madison
1180 Observatory Dr.
Madison, WI 53706
USA

Heather M.-L. Miller
Department of Anthropology and
    Center for South Asia
University of Wisconsin-Madison
1180 Observatory Dr., Rm. 5240
Madison, WI 53706-1393
USA

James D. Muhly
American School of Classical Studies
Souidias 54
Athens 106.76
GREECE

Vincent C. Pigott
Museum Applied Science Center for Archaeology
(MASCA)
University of Pennsylvania Museum
33rd and Spruce Sts.
Philadelphia, PA 19104-6324
USA

Gregory L. Possehl
Asian Section
University of Pennsylvania Museum
33rd and Spruce Sts.
Philadelphia, PA 19104-6324
USA

Tamara Stech
P.O. Box 1451
Stowe, VT 05672
USA

Jane C. Waldbaum
Department of Art History
University of Wisconsin-Milwaukee
P.O. Box 413
Milwaukee, WI 53201
USA